Atlas of Chrysophycean Cysts

Developments in Hydrobiology 99

Series editor
H. J. Dumont

Atlas of Chrysophycean Cysts

by

Katharine E. Duff, Barbara A. Zeeb
and John P. Smol

KLUWER ACADEMIC PUBLISHERS
DORDRECHT / BOSTON / LONDON

Library of Congress Cataloging-in-Publication Data

Duff, Katharine E.
Atlas of chrysophycean cysts / by Katharine E. Duff, Barbara A. Zeeb & John P. Smol ; illustrated by John R. Glew.
p. cm. -- (Developments in hydrobiology ; 99)
Includes bibliographical references (p.) and indexes.
ISBN 0-7923-3039-0 (HB : acid-free paper)
1. Chrysophycean cysts--Atlases. I. Zeeb, Barbara A. II. Smol, J. P. (John P.) III. Title. IV. Series.
QK569.C63D84 1994
589.4'8--dc20 94-30346

ISBN 0-7923-3039-0

Published by Kluwer Academic Publishers,
P.O. Box 17, 3300 AA Dordrecht, The Netherlands.

Kluwer Academic Publishers incorporates
the publishing programmes of
D. Reidel, Martinus Nijhoff, Dr W. Junk and MTP Press.

Sold and distributed in the U.S.A. and Canada
by Kluwer Academic Publishers,
101 Philip Drive, Norwell, MA 02061, U.S.A.

In all other countries, sold and distributed
by Kluwer Academic Publishers Group,
P.O. Box 322, 3300 AH Dordrecht, The Netherlands.

Printed on acid-free paper

Printed in the Netherlands

Table of Contents

Acknowledgements

This book would not have been possible without the encouragement and support of numerous people and organizations. First, we thank Kluwer Academic Publishers, and especially Wil Peters, for their advice and encouragement. Drs. Craig Sandgren, Gertrud Cronberg and Peter Siver provided constructive comments on various parts of the text. Several of the photomicrographs were supplied by Kimberley Brown, Jessica Webb and Alex Wolfe. Sally Taylor and Elisabet Pohjanen provided expert technical assistance. Henry Verstappen assisted with some of the electron microscopy. We are especially grateful to our colleagues at PEARL for their input, support and tolerance throughout the progress of this book.

Financial assistance was provided by a Natural Sciences and Engineering Research Council of Canada (NSERC) grant to JPS and graduate scholarships to KED and BAZ. Further logistical support was provided by Queen's University.

Finally, the three authors would like to acknowledge their tremendous gratitude to their illustrator, John R. Glew, for his patience, perseverance and insight, which greatly improved the quality of this book.

"Chrysophytes are a most intriguing group of freshwater algae..."

Sandgren (1988)

CHAPTER 1

Introduction

Chrysophycean algae (classes Chrysophyceae and Synurophyceae; hereafter referred to as "chrysophytes") are a diverse group of primarily freshwater algae, with well over 1000 known species. They often form a dominant phytoplankton group in temperate (Roijackers, 1986; Siver & Chock, 1986), high latitude (Moore, 1979, 1981; Wallen & Allen, 1982; Eloranta, 1986) and high altitude lakes (Rott, 1988), but are also found in tropical and sub-tropical regions (Cronberg, 1989a). Chrysophytes are typically euplanktonic (Sandgren, 1988), but several epiphytic (Hilliard & Asmund, 1963) and benthic (Dop, 1980) species exist. Although less common in inland saline lakes, some taxa persist in these environments (Pienitz *et al.*, 1992; Cumming *et al.*, 1993). They are considered rare in marine systems (Gayral & Billard, 1986), but new nano- and picoplanktonic forms continued to be described (Peters & Andersen, 1993).

Interest in chrysophytes has been growing steadily, as evidenced by a series of international symposia (Kristiansen & Andersen, 1986; Kristiansen *et al.*, 1989; Sandgren *et al.*, 1994), as well as a number of other books and review articles on their systematics, ecology, biogeography, physiology, biochemistry and paleoecology (Pinter & Provasoli, 1963; Loeblich & Loeblich, 1977; Aaronson, 1980; Adam & Mahood, 1981a; Kristiansen & Takahashi, 1982; Cronberg, 1986a, b; Kristiansen, 1986, 1990; Sandgren, 1988, 1991; Smol, 1988a, 1994). Several monographs have also been completed (e.g. Takahashi, 1978; Wee, 1982; Asmund & Kristiansen, 1986; Siver, 1991a). Kristiansen (1994) provides a historical sketch of chrysophyte researchers.

As several books and reviews have recently been published, we only provide this short overview here. Sandgren (1988) provides a comprehensive review on the distribution, demography, nutrition, ecological interactions and reproductive biology of chrysophytes. Overviews of most aspects of ongoing research are contained in the proceedings of the Third International Chrysophyte Symposium (Sandgren *et al.*, 1994).

Chrysophytes

Chrysophytes are sometimes referred to as the "golden brown algae", which is a result of the predominance of carotenoid pigments, mostly the xanthophyll fucoxanthin, which may represent up to 93% of the total carotenoid (Withers *et al.*, 1981) and 75% of the total pigment content (Bold & Wynne, 1978). Most taxa are flagellate single cells or colonies of flagellate cells, but amoeboid, palmelloid, filamentous and coccoid forms also exist. Chrysophytes possess diverse nutritional strategies, with many taxa exhibiting heterotrophy and phagotrophy; the latter may form an important link between the detritus-based and photosynthetic-based food chains (Bird & Kalff, 1986; Salonen & Jokinen, 1988), particularly in lakes where chrysophytes comprise a preferred food choice for zooplankton (Sandgren, 1988). Chrysophytes are often characterized by marked seasonality (Siver & Chock, 1986; Kristiansen, 1988; Rott, 1988; Sandgren, 1988; Siver & Hamer, 1992), and blooms of these organisms can cause taste and odour problems in drinking water supplies and processed food products (Nicholls, 1994).

Although chrysophytes as a group are found across a broad range of environmental gradients, individual species tend to be restricted in their distributions. Empirical evidence that many chrysophytes have well-defined ecological optima and narrow tolerances for a number of limnological variables, such as pH, temperature, salinity, conductivity and trophic status, has accumulated in recent years (Roijackers & Kessels, 1986; Siver & Chock, 1986; Siver, 1988a, 1989a, b, 1994; Dixit *et al.*, 1989a, 1990a; Cumming *et al.*, 1991, 1992a). This has resulted in heightened interest in the use of chrysophytes in long-term monitoring programs (Siver & Smol, 1993).

Interest in using fossil chrysophytes as indicators of environmental change has mirrored the attention that living assemblages have recently enjoyed. Adam & Mahood (1979a) provide an early annotated bibliography on chrysophyte scales and cysts. Smol (1990) published a working bibliography for most papers published on chrysophyte paleolimnology in the 1980's, and Cronberg (1986a, b) and Smol (1987, 1988a, 1994) have reviewed the use of chrysophytes in paleolimnological studies. Therefore, only a few salient points will be mentioned here.

Chrysophytes are represented in the sedimentary record by two forms of siliceous remains: 1) disarticulated scales, bristles and spines produced by memebers of the Synurophyceae and some Chrysophyceae; and 2) resting cysts (stomatocysts; also called statospores or statocysts), believed to be produced by all chrysophytes. Chrysophyte microfossils are at times extremely abundant in lake sediments (Fig. 1), and in fact a sediment type (chrysopel) has been distinguished, which is characterized by the abundance of cysts.

Chrysophyte scales are useful paleolimnological markers because the shape, size and sculpturing of the individual scales are taxon-specific and, because of their siliceous composition, scales are well-preserved in most sedimentary deposits. The applicability of using chrysophyte scales in paleolimnological research was assessed over a decade ago (Battarbee *et al.*, 1980; Munch, 1980; Smol, 1980), and since that time chrysophyte scale assemblages have been successfully used to infer historical changes in pH, metals, colour, trophic status, conductivity, salinity and other limnological variables (e.g. Smol, 1986; Charles & Smol, 1988; Christie *et al.*, 1988; Dixit *et al.*, 1989a, b, c, 1990a, b; Cumming *et al.*, 1992b; Zeeb & Smol, 1991; Marsicano & Siver, 1993; Siver, 1993). Some of the potential problems of using scales, such as relating scale numbers to cell populations, have been investigated and clarified (Siver, 1991b; Cumming & Smol, 1993).

Chrysophyte scales are clearly useful paleolimnological biomonitors, but scales will not provide a complete record of past chrysophyte populations as only ca. 20% of known

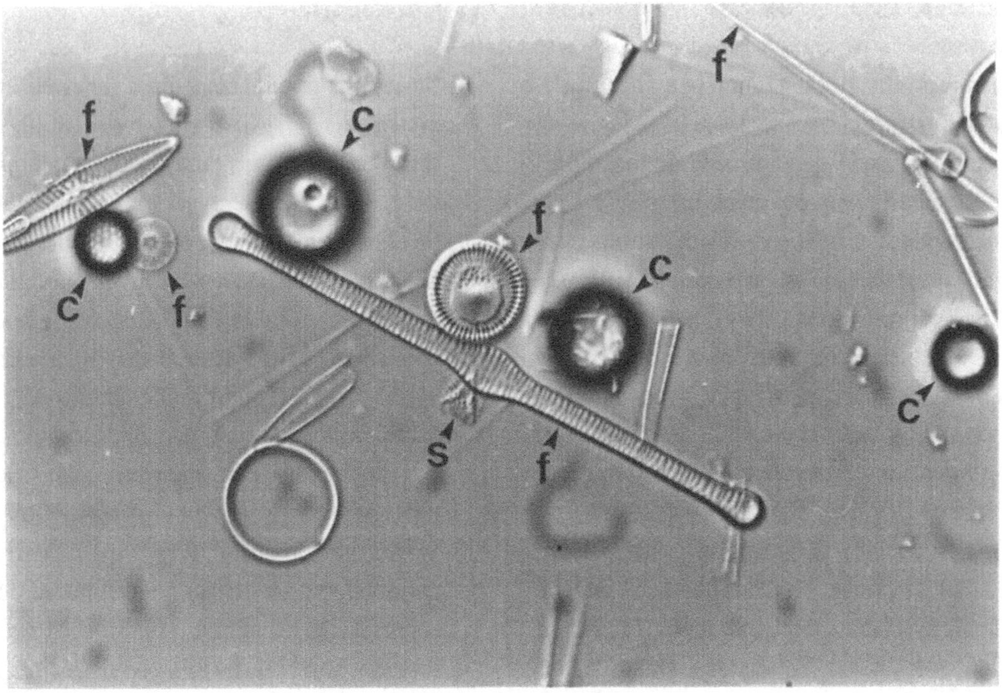

Fig. 1. Light micrograph showing chrysophycean cysts (c), a chrysophycean scale (s) and diatom frustules (f) in a sediment trap sample from Elk Lake, Minnesota. Magnification = x800.

chrysophycean species possess scales (e.g. taxa in the genera *Mallomonas*, *Synura*, *Chrysosphaerella*, *Paraphysomonas*). In many lakes, these scaled taxa do not represent the dominant chrysophyte species (e.g. Siver & Chock, 1986). In some ecosystems, such as high Arctic (Smol, 1983; Douglas & Smol, 1994) and high altitude lakes (Rott, 1988), scaled chrysophytes are very rare, whilst non-scaled chrysophytes may be abundant. Therefore, stomatocysts, which are formed by all chrysophyte taxa, potentially provide a more complete record of the entire chrysophyte community.

Chrysophycean Stomatocysts

The chrysophycean stomatocyst is a hollow siliceous structure, most commonly globose in shape, with a single pore. Cyst morphology is very diverse, but is believed to be species-specific. This presumption has recently elicited some lively debate (Simola, 1991; Sandgren & Smol, 1991; Simola, 1993). Cysts range in diameter from ca. 2 μm to >30 μm, and in shape from spherical to oval to pyramidal. Mature cysts may be smooth-surfaced, or ornamented with a variety of projecting or depressed structural elements. A collar may or may not surround the pore. Details of cyst morphology and descriptive terminology are provided in Chapter 3.

Stomatocysts are produced endogenously within a special membrane system. Interspecific variation in the mechanistic details of encystment may be indicative of diverging evolutionary pathways (Sheath *et al.*, 1975; Hibberd, 1977; Sandgren, 1980a, b, 1981, 1989; Andersen, 1982; Preisig & Hibberd, 1982a, b, 1983). Typically, cyst formation proceeds by the deposition of several successive layers of silica. The primary cyst wall is usually thin and unornamented, and is morphologically similar for many species (Sandgren, 1989). Cysts at primary or intermediate stages of development may be distinct from mature cysts by the same species. For example, certain *Mallomonas* and *Chrysosphaerella* species produce a reticular, honeycomb-like secondary wall [e.g. *Chrysosphaerella longispina* Lauterborn *emend.* Korshikov (Sandgren, 1989); *Mallomonas acaroides* var. *muskokana* Nicholls (Siver, 1991c); *M. caudata* Ivanoff *emend.* Krieger (Cronberg, 1988; Sandgren, 1989); *M. intermedia* Kisselew (Skogstad, 1984)]. However, the mature cysts of these species are either unornamented, or possess a different type of ornamentation (e.g. spines) than the immature cyst. Variability in mature cysts may result from either genetic or environmental factors (Sandgren, 1983a). As an extreme example of the latter, *Dinobryon cylindricum* Imhof cysts may occasionally develop within the lorica, resulting in irregularly oval rather than the usual spherical cysts (Sandgren, 1989). Clearly, developmental variation will limit our ability to accurately distinguish stomatocysts produced by different species; in particular, many species may produce simple, unornamented cysts that may be immature or mature, but which are morphologically identical (for example, see stomatocysts 1, 9 and 120; Chapter 4). However, extremes in developmental variation are likely rare in nature (C.D. Sandgren, pers. commun.), and fully ornamented stomatocysts are generally easily distinguishable (Chapter 4).

Cyst formation, survivorship and germination have been reviewed by Sandgren (1988, 1991). Cysts may be produced by either asexual or sexual (autogamic or hologametic) reproduction; the cysts produced by either mode are morphologically identical (Fig. 2). In contrast to sexual spore production of other algal classes, sexual stomatocyst formation does not appear to be triggered by environmental stressors, at least in the few chrysophycean species studied (Sandgren & Flanagin, 1986; Sandgren, 1988), but is apparently density dependent. Induction of asexual stomatocyst formation is still poorly understood. Asexual formation has not been studied experimentally, and its role in chrysophycean dynamics is uncertain. It seems that cyst formation must be an essential survival adaptation for these seasonally restricted, largely planktonic algae. More work is necessary to establish whether natural populations of chrysophytes employ one or more encystment mechanisms, and whether the type of encystment is species-dependent.

Classification Schemes for Chrysophycean Cysts

Very few stomatocyst morphotypes have as yet been linked to the species that produce them; those that have been are listed in Table 1. Unfortunately, most older literature sources lack adequate, descriptive detail, providing only a sketchy description and possibly a line drawing. To our knowledge, only 32 of more than 150 taxa whose cysts have been mentioned in the literature have been described fully with electron microscopy.

Several taxonomic schemes have been proposed to distinguish cyst morphotypes whose biological affinity is

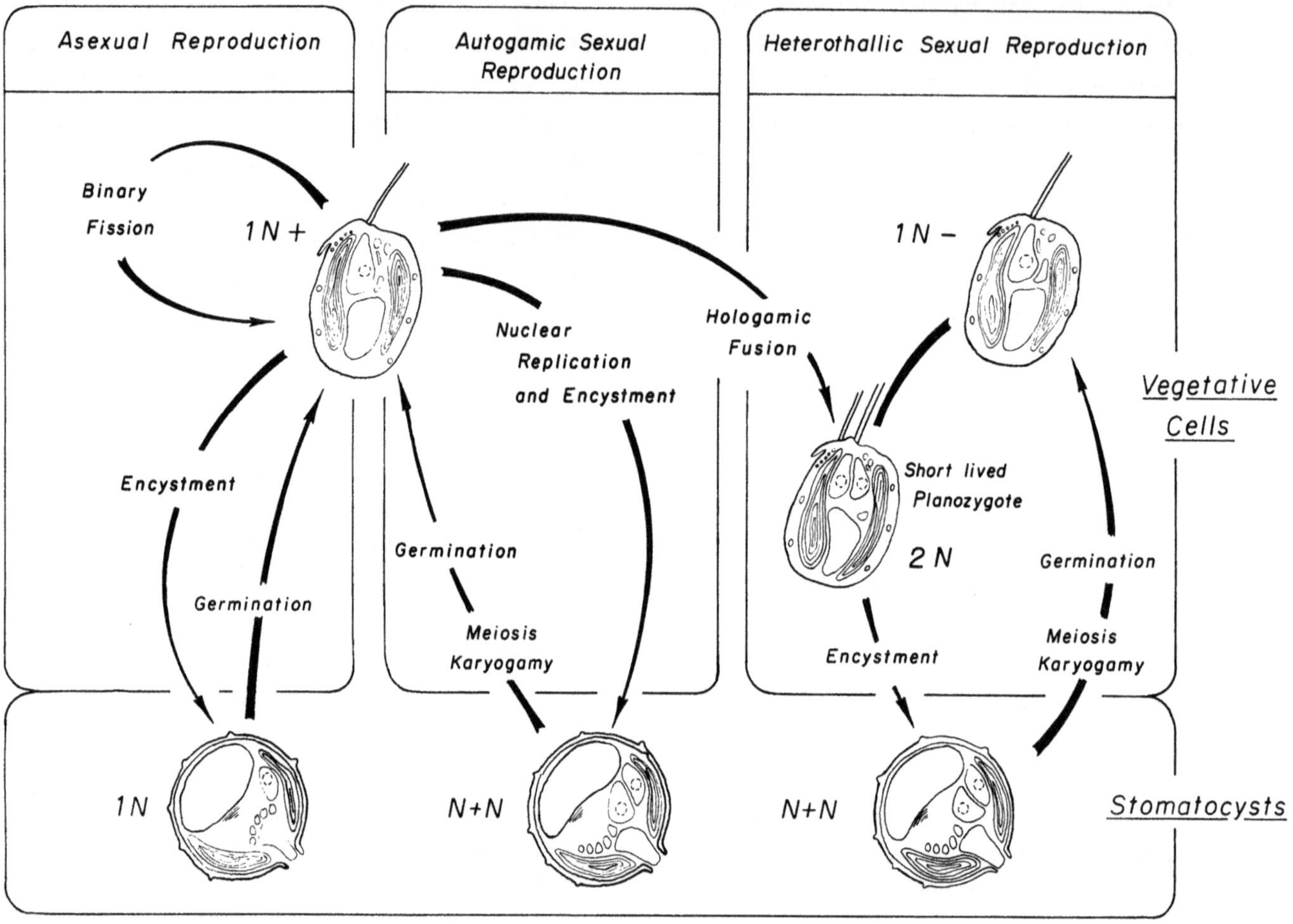

Fig. 2. Generalized life cycle diagram for chrysophycean algae showing both asexual and sexual reproduction. Note that cysts produced by each mode (asexual encystment, autogamic fusion and hologamic fusion) are morphologically identical. Modified from Sandgren (1991).

unknown. Cronberg (1986b) provides a review of these systems, and they are described briefly below.

The earliest schemes were modelled on the Linnaean (binomial) system (e.g. Deflandre, 1932; Nygaard, 1956). Ehrenberg (1854, as cited in Cronberg, 1986b) is believed to be the first person to draw chrysophycean cysts, but he considered them to be diatoms. Pantocsek (1912, 1913, as cited in Cronberg, 1986b) placed the cysts he found in Hungarian lake sediments with the Rhizopoda in two new genera which he described - *Carnegia* and *Echinopyxis*. Frenguelli (1925, 1929, 1931, 1932, 1935a, b, 1936, 1938a, b) originally placed approximately 110 cysts from South American diatomites into the genus *Trachelomonas* (Euglenophyta).

The system proposed by Deflandre (1932, 1936) was the first to be used by other researchers (Frenguelli, 1936). With his system, it was possible to classify the cysts into sixteen different artificial genera according to the form and cell wall structure of the specimen (Deflandre, 1952). His system consisted of the Chrysomonadinea, which are cysts produced by extant identifiable species; the Archaeomonadaceae, an artificial family of cysts produced by unidentifiable fossil, marine forms; and the Chrysostomataceae, an artificial family of unidentified, fossil or recent freshwater cysts. Deflandre's system was used extensively by later paleontological researchers (e.g. Hajós & Radócz, 1969; Hajós, 1970, 1971, 1973, 1974; Gritten, 1977; Srivastata & Binda, 1984; Rull, 1986).

In 1956, Nygaard proposed using the artificial genus *cysta* followed by a descriptive species name as a means to classify unidentified cysts. He used this system to name 77 cyst morphotypes based on light microscopy from the sediments of Lake Gribsø in Denmark. This system was used by several subsequent researchers (e.g. VanLandingham, 1964; Leventhal, 1970; Gritten, 1977; Elner & Happey-Wood, 1978; Carney & Sandgren, 1983; Rybak, 1986). The U.S. Geological Survey Reports by Adam and co-workers (Mahood & Adam, 1979; Adam & Mahood, 1979b, 1980a, b, c, 1981b; Adam & Mehringer, 1980a, b, c; Adam, 1980a, b, c, 1981) provide electron

Table 1: Chrysophycean taxa whose stomatocysts have been mentioned in the literature. The type of information available for each taxon is summarized as follows: WD = a written description is provided; Line = line drawing(s) based on light microscopy; LM = light micrograph(s); SEM = scanning electron micrograph(s); ISWG = the information meets, or at least approaches, the standards outlined by the International Statospore Working Group, i.e. a written description, and a SEM with the pore-collar complex in view, are available. The first reference listed is the earliest ISWG description, if available; otherwise, references are listed chronologically.

Taxon name	WD	Line	LM	TEM	SEM	ISWG	Reference(s)
Antophysa vegetans (F.F. Müller) Stein	X						Pringsheim, 1946
Antophysa vegetans var. *senii* Pringsheim		X					Bourrelly, 1957
Chromulina aerophila Lund	X	X					Lund, 1942
Chromulina placentula Belcher & Swale	X	X					Belcher & Swale, 1967
Chromulina reitereri Wawrik	X	X					Wawrik, 1980
Chromulina sphaeridia Schiller	X	X					Wawrik, 1977
Chromulina sporangifera Lund	X	X					Lund, 1942
Chrysidiastrum catenatum Lauterborn	X				X		Sandgren, 1983b
Chrysococcus furcatus (Dolgoff) Nicholls	X	X	X		X	X	Nicholls, 1981; Hilliard, 1966 (as *Chrysastrella furcata* (Dolgoff) Deflandre)
Chrysolepidomonas anglica Peters & Andersen	X						Peters & Andersen, 1993
Chrysolepidomonas dendrolepidota Peters & Andersen	X	X		X	X	X	Peters & Andersen, 1993
Chrysolykos calceatus Ramberg	X	X					Ramberg, 1978
Chrysolykos planctonicus Mack	X	X					Bourrelly, 1957; Kristiansen, 1965
Chrysosphaera schnelzerii Wawrik	X	X					Wawrik, 1977
Chrysosphaerella brevispina Korshikov *emend.* Nicholls Harris & Bradley	X	X					Bourrelly, 1957 (as *C. conradii* Bourrelly); Harris & Bradley, 1958
Chrysosphaerella longispina Lauterborn	X				X	X	Sandgren, 1989
Chrysostephanosphaera globulifera Scherffel	X	X					Péterfi, 1967
Conradiella iserina Ettl & Permann		X					Bourrelly, 1968
Dinobryon anulatum Hilliard & Asmund	X	X					Hilliard & Asmund, 1963
Dinobryon asmundiae Nygaard	X	X					Nygaard, 1979
Dinobryon bavaricum Imhof	X	X					Krieger, 1932
Dinobryon borgei Lemmermann	X	X					Hilliard, 1966; Nygaard, 1979
Dinobryon crenulatum Wm. & G.S. West	X	X					Asmund, 1955a
Dinobryon crenulatum forma callosum Nygaard	X	X					Nygaard, 1979
Dinobryon cylindricum Imhof	X	X	X	X	X	X	Sandgren, 1980b, 1981, 1983a, b, 1989; Krieger, 1932; Donaldson & Stein, 1984
Dinobryon cylindricum var. *alpinum* (Imhof) Bachmann	X						Krieger, 1932
Dinobryon cylindricum var. *palustris* Lemmermann					X		Smith & White, 1985
Dinobryon divergens Imhof	X	X	X	X	X	X	Sheath *et al.*, 1975; Krieger, 1932; Sandgren, 1980a
Dinobryon faculiferum Willén	X	X					Willén, 1992
Dinobryon hilliardii Nygaard	X	X					Nygaard, 1979
Dinobryon korshikov Matvienko	X	X					Bourrelly, 1957 (as *D. elegantissimum* (Korshikov) Bourrelly); Hilliard, 1966 (as *D. elegantissimum*)

Table 1. continued

Taxon name	WD	Line	LM	TEM	SEM	ISWG	Reference(s)
Dinobryon lindegaardii Nygaard	X	X					Nygaard, 1979
Dinobryon pediforme (Lemmermann) Steinecke	X	X					Krieger, 1932
Dinobryon sertularia Ehrenberg	X		X				Krieger, 1932; Wujek, 1969
Dinobryon sociale Ehrenberg	X	X					Krieger, 1932; Kristiansen, 1964; Hilliard, 1966
Dinobryon sociale var. *stipitatum* (Stein) Lemmermann	X				X		Krieger, 1932; Hickel, 1975
Dinobryon unguentariforme Croome, Ling & Tyler	X		X				Croome *et al.*, 1988
Epipyxis alata Hilliard & Asmund	X	X					Hilliard & Asmund, 1963
Epipyxis alpina Hilliard & Asmund	X	X					Hilliard & Asmund, 1963
Epipyxis borealis Hilliard & Asmund	X	X					Hilliard & Asmund, 1963
Epipyxis condensata (Mack) Hilliard & Asmund	X	X					Hilliard & Asmund, 1963
Epipyxis diplostoma (Jane) Hilliard & Asmund	X	X					Bourrelly, 1957 (as *Hyalobryon diplostoma* Jane *forma astigma* Bourrelly)
Epipyxis gracilis Hilliard & Asmund	X	X					Hilliard & Asmund, 1963
Epipyxis polymorphum Lund	X	X					Lund, 1953 (as *Hyalobryon polymorphum* Lund)
Epipyxis ramosa (Lauterborn) Hilliard & Asmund	X	X					Hilliard & Asmund, 1963
Epipyxis tabellariae (Lemmermann) G.M. Smith	X	X					Hilliard & Asmund, 1963
Epipyxis tubulosa (Mack) Hilliard & Asmund	X	X					Hilliard & Asmund, 1963
Epipyxis utriculus Ehrenberg	X	X					Bourrelly, 1957 (as *Dinobryon utriculus* (Ehrenberg) Klebs); Hilliard & Asmund, 1963
Epipyxis utriculus var. *acuta* (Schiller) Hilliard & Asmund	X	X					Hilliard & Asmund, 1963
Heterochromas vivivara (Ehrenberg) Pascher *forma minor* Bourrelly	X	X					Bourrelly, 1957
Hydurus foetidus (Villars) Trevison	X	X					Bourrelly, 1957
Kephyrion monileferum (Schmid) Bourrelly	X						Hilliard, 1966 (as *Stenokalyx monilifera* Schmid)
Mallomonas acaroides Perty *emend.* Ivanoff	X	X			X	X	Cronberg, 1980 (as *M. acaroides* var. *striatula* Asmund); Bourrelly, 1957; Harris & Bradley, 1960 (also as *M. acaroides* var. *galeata* Harris & Bradley); Asmund & Kristiansen, 1986
Mallomonas acaroides var. *inermis* Fott	X						Asmund & Kristiansen, 1986
Mallomonas acaroides var. *muskokana* Nicholls	X				X	X	Siver, 1991c
Mallomonas actinoloma var. *maramuresensis* Péterfi & Momeu	X						Petérfi & Momeu, 1976; Asmund & Kristiansen, 1986
Mallomonas adamas Harris & Bradley	X	X					Harris & Bradley, 1960; Asmund & Kristiansen, 1986

Taxon name	WD	Line	LM	TEM	SEM	ISWG	Reference(s)
Mallomonas akrokomos Ruttner in Pascher	X	X			X	X	Cronberg, 1980; Krieger, 1932 (as *Mallomonas paucispinosa* Conrad); Conrad, 1933 (also as *M. paucispinosa*); Bourrelly, 1957; Harris, 1958; Hilliard, 1966; Péterfi, 1967; Wawrik, 1980; Asmund & Kristiansen, 1986; Vigna, 1988; Hickel & Maaß, 1989; Sandgren, 1989
Mallomonas allorgei (Deflandre) Conrad	X	X					Conrad, 1933; Harris & Bradley, 1960; Péterfi, 1967; Asmund & Kristiansen, 1986
Mallomonas alpina Pascher & Ruttner *emend.* Asmund & Kristiansen	X	X					Krieger, 1932 (as *Mallomonas tonsurata* var. *alpina*); Conrad, 1933 (as *M. tonsurata* var. *alpina*); Harris & Bradley, 1960 (as *M. monograptus* Harris & Bradley); Asmund & Kristiansen, 1986
Mallomonas anglica (Carter) Huber-Pestalozzi*	X						Harris & Bradley, 1960
Mallomonas annulata (Bradley) Harris	X	X					Harris, 1967; Asmund & Kristiansen,1986
Mallomonas areolata Nygaard	X	X			X	X	Vigna, 1984; Bourrelly, 1957; Momeu & Péterfi, 1983; Asmund & Kristiansen, 1986
Mallomonas bacterium Conrad*	X	X					Krieger, 1932; Conrad, 1933
Mallomonas caudata Ivanoff *emend.* Krieger	X			X	X	X	Cronberg, 1988; Krieger, 1932; Conrad, 1933; Asmund, 1955b; Sandgren, 1980a; Asmund & Kristiansen, 1986; Eloranta, 1989; Sandgren, 1989
Mallomonas clavata Conrad*	X	X					Krieger, 1932; Conrad, 1933
Mallomonas clavus Bradley	X	X					Harris, 1970 (as *M. allantoides* Harris); Momeu & Péterfi, 1983
Mallomonas corcontica (Kalina) Péterfi & Momeu	X						Asmund & Kristiansen, 1986
Mallomonas coronata Bolochonzew*	X	X					Krieger, 1932; Conrad, 1933
Mallomonas crassisquama (Asmund) Fott	X				X		Gretz *et al.*, 1979 (as *M. acaroides* var. *crassisquama* Asmund); Asmund & Kristiansen, 1986
Mallomonas cratis Harris & Bradley	X	X					Harris & Bradley, 1960; Asmund & Kristiansen, 1986
Mallomonas curta (Playfair) Conrad*	X	X					Krieger, 1932; Conrad, 1933
Mallomonas cylindracea Pascher*	X	X					Krieger, 1932; Conrad, 1933
Mallomonas elliptica (Kisselew) Conrad*	X						Conrad, 1933
Mallomonas elongata Reverdin	X						Krieger, 1932; Conrad, 1933; Péterfi, 1965; Asmund & Kristiansen, 1986
Mallomonas eoa Takahashi	X	X	X		X	X	Cronberg, 1973, 1980; Asmund & Kristiansen, 1986
Mallomonas fresenii Kent*	X	X					Krieger, 1932; Conrad, 1933
Mallomonas hamata Asmund	X	X			X	X	Cronberg, 1989b; Asmund & Kristiansen, 1986
Mallomonas heterospina Lund	X	X			X	X	Cronberg, 1989b; Asmund, 1956; Wawrik, 1980; Asmund & Kristiansen, 1986
Mallomonas insignis Penard	X				X	X	Dürrschmidt, 1984; Bourrelly, 1957; Harris, 1958; Péterfi & Momeu, 1976; Asmund & Kristiansen, 1986

Table 1. continued

Taxon name	WD	Line	LM	TEM	SEM	ISWG	Reference(s)
Mallomonas intermedia Kisselew	X	X			X	X	Skogstad, 1984; Lund, 1942; Harris, 1953 (as *M. intermedia* var. *gesticulans* Harris); Harris & Bradley, 1957 (as *M. soleatus* Harris & Bradley); Harris & Bradley, 1960 (as *M. intermedia* var. *gesticulans* and *M. soleatus*); Péterfi, 1967 (as *M. intermedia* var. *gesticulans*); Asmund & Kristiansen, 1986
Mallomonas lata Conrad*	X	X					Krieger, 1932; Conrad, 1933
Mallomonas leboimei Bourrelly	X	X					Harris, 1953
Mallomonas lilloënsis Conrad*	X						Conrad, 1933
Mallomonas litomesa Stokes*	X	X					Krieger, 1932; Conrad, 1933
Mallomonas longiseta Lemmermann*	X	X					Krieger, 1932; Conrad, 1933
Mallomonas lychenensis Conrad	X	X					Harris, 1953; Nygaard, 1956 (as *M. tesselata* Nygaard); Asmund & Kristiansen, 1986
Mallomonas majorensis Skuja*	X	X					Harris & Bradley, 1960; Péterfi, 1967
Mallomonas mangofera Harris & Bradley *emend.* Dürrschmidt	X	X					Harris & Bradley, 1960; Dürrschmidt, 1983; Asmund & Kristiansen, 1986
Mallomonas matvienkoae (Matvienko) Asmund & Kristiansen	X	X					Asmund & Kristiansen, 1986; Harris, 1966 (as *M. elliptica* (Matvienko) Péterfi & Momeu)
Mallomonas mirabilis Conrad*	X						Krieger, 1932; Conrad, 1933
Mallomonas multiunca Asmund	X	X					Harris & Bradley, 1960
Mallomonas oblongispora Lemmermann*	X						Conrad, 1933
Mallomonas ouradion Harris & Bradley	X	X					Kalina, 1969 (as *Mallomonopsis ouradion*); Asmund & Kristiansen, 1986
Mallomonas oviformis Nygaard	X	X					Harris & Bradley, 1960; Harris, 1966 (as *M. elliptica* var. *oviformis* (Nygaard) Harris); Asmund & Kristiansen, 1986
Mallomonas papillosa Harris & Bradley	X	X					Harris & Bradley, 1957; Harris, 1967; Asmund & Kristiansen, 1986 (also as *M. papillosa* var. *ellipsoidea* Harris)
Mallomonas papillosa var. *monilifer* Harris	X	X					Harris, 1967; Asmund & Kristiansen, 1986
Mallomonas phasma Harris & Bradley	X	X					Harris & Bradley, 1960; Asmund & Kristiansen, 1986
Mallomonas producta Ivanoff*	X						Krieger, 1932; Conrad, 1933
Mallomonas pseudocoronata Prescott	X				X	X	Smol, 1984, 1988b; Asmund & Kristiansen, 1986
Mallomonas pulcherrima (Stokes) Lemmermann*	X	X					Krieger, 1932; Conrad, 1933
Mallomonas pumilio Harris & Bradley *emend.* Asmund, Cronberg & Dürrschmidt	X	X			X	X	Asmund *et al.*, 1982; Harris & Bradley, 1957; Asmund & Kristiansen, 1986
Mallomonas pumilio var. *silvicola* Harris & Bradley *emend.* Asmund, Cronberg & Dürrschmidt	X	X					Harris & Bradley, 1960; Asmund *et al.*, 1982
Mallomonas punctifera Korshikov	X						Asmund & Kristiansen, 1986
Mallomonas pyriformis Conrad*	X	X					Krieger, 1932; Conrad, 1933
Mallomonas radiata Conrad*	X				X		Krieger, 1932; Conrad, 1933; Gritten, 1977 (basis for identification uncertain)

Taxon name	WD	Line	LM	TEM	SEM	ISWG	Reference(s)
Mallomonas retifera Dürrschmidt	X				X		Dürrschmidt, 1982a; Asmund & Kristiansen, 1986
Mallomonas schwemmlei Glenk *emend.* Glenk & Fott	X	X					Wawrik, 1980; Asmund & Kristiansen, 1986
Mallomonas spinifera Conrad*	X	X					Krieger, 1932; Conrad, 1933
Mallomonas striata Asmund	X				X	X	Asmund & Kristiansen, 1986
Mallomonas striata var. *serrata* Harris & Bradley	X	X					Harris & Bradley, 1960; Asmund & Kristiansen, 1986
Mallomonas teilingii (Teiling) Conrad	X	X			X	X	Cronberg, 1980; Krieger, 1932; Conrad, 1933; Harris, 1953; Péterfi, 1967; Péterfi & Momeu, 1976; Gritten, 1977; Wawrik, 1980; Asmund & Kristiansen, 1986
Mallomonas tenuis Conrad*	X						Harris & Bradley, 1960
Mallomonas tolerans (Asmund & Hilliard) Asmund & Kristiansen	X	X					Harris, 1966 (as *M. elliptica* var. *salina* Asmund & Hilliard); Asmund & Kristiansen, 1986
Mallomonas tonsurata Teiling *emend.* Krieger	X	X					Krieger, 1932; Harris & Bradley, 1960; Kristiansen, 1964; Asmund & Kristiansen, 1986
Mallomonas torquata Asmund & Cronberg	X		X		X	X	Cronberg, 1980; Asmund & Kristiansen, 1986
Mallomonas transsylvanica Péterfi & Momeu	X	X					Bourrelly, 1957 (as *M. reginae* var. *glabra* Bourrelly); Asmund & Kristiansen, 1986
Mallomonas vannigera Asmund	X	X			X	X	Asmund & Kristiansen, 1986; Asmund, 1977
Ochromonas charkoviensis Matvienko	X						Péterfi, 1967
Ochromonas crenata Klebs	X	X					Péterfi, 1967
Ochromonas globosa Skuja	X	X					Nygaard, 1977
Ochromonas ostreaformis Swale & Belcher	X	X					Swale & Belcher, 1966; Wawrik, 1980
Ochromonas sphaerocystis Matvienko	X				X	X	Andersen, 1982
Ochromonas tuberculata Hibberd	X		X		X	X	Hibberd, 1977
Paraphysomonas antarctica Takahashi	X				X	X	Takahashi, 1987; Takahashi *et al.*, 1986 (as *P. imperforata* Lucas)
Paraphysomonas butcheri Pennick & Clarke					X		Takahashi, 1981
Paraphysomonas caelifrica Preisig & Hibberd	X			X			Preisig & Hibberd, 1983
Paraphysomonas circumvallata ssp. *mediogranulata* Preisig & Hibberd	X			X			Preisig & Hibberd, 1982b
Paraphysomonas corynephora Preisig & Hibberd	X			X			Preisig & Hibberd, 1982a; Preisig & Hibberd, 1983
Paraphysomonas imperforata Lucas					X		Takahashi, 1981
Paraphysomonas spinapunctata Wujek & Gardiner	X			X			Nicholls, 1984 (as *Spiniferomonas septispina* Nicholls)
Paraphysomonas vestita (Stokes) De Saedeler	X				X		Takahashi, 1987
Pseudokephyrion conicum (Schiller) Schmid	X	X					Bourrelly, 1957
Pseudokephyrion gibbosum Ramberg	X	X					Ramberg, 1978
Pseudokephyrion latum (Schiller) Schmid	X	X					Bourrelly, 1957

Table 1. continued

Taxon name	WD	Line	LM	TEM	SEM	ISWG	Reference(s)
Saccochrysis pyriformis Korshikov	X	X					Bourrelly, 1957
Spiniferomonas bourrellyi Takahashi	X		X	X	X	X	Skogstad & Reymond, 1989
Spiniferomonas coronacircumspina (Wujek & Kristiansen) Nicholls	X				X		Preisig & Takahashi, 1978 (as *Chrysosphaerella solitaria* Preisig & Takahashi)
Spiniferomonas crucigera Takahashi					X		Eloranta, 1989
Spiniferomonas trioralis (Takahashi) Preisig & Hibberd	X				X	X	Skogstad, 1986 (as *Chromophysomonas trioralis*); McKenzie & Kling, 1989
Synura curtispina (Petersen & Hansen) Asmund	X				X	X	Sandgren, 1989
Synura mollispina (Petersen & Hansen) Péterfi & Momeu	X				X	X	Sandgren, 1989
Synura petersenii Korshikov	X		X		X	X	Sandgren & Flanagin, 1986; Bourrelly, 1957; Eloranta, 1989; Hickel & Maaß, 1989; Sandgren, 1989
Synura sphagnicola Korshikov	X	X					Harris & Bradley, 1958
Synura splendida Korshikov	X	X					Bourrelly, 1957
Synura uvella Ehrenberg *emend.* Korshikov	X	X					Bourrelly, 1957
Uroglena americana Calkins	X	X		X			Bourrelly, 1957, 1963; Kristiansen, 1964; Sandgren, 1980a
Uroglena botrys (Pascher) Conrad	X	X					Bourrelly, 1957, 1963
Uroglena conimamma Nygaard	X	X					Nygaard, 1977
Uroglena conradii Schiller	X						Bourrelly, 1957
Uroglena conradii var. *gallica* Bourrelly	X	X					Bourrelly, 1957, 1963
Uroglena dendracantha Cronberg	X		X		X	X	Cronberg, 1992
Uroglena europaea (Pascher) Conrad	X	X					Bourrelly, 1957, 1963
Uroglena irregularis Rhodhe & Skuja	X						Bourrelly, 1957
Uroglena lindiae Bourrelly	X	X			X	X	Nygaard, 1977; Bourrelly, 1957, 1963
Uroglena marina Büttner	X	X					Bourrelly, 1957, 1963
Uroglena notabilis Mack	X	X					Bourrelly, 1957, 1963
Uroglena nygaardii Bourrelly	X	X					Bourrelly, 1957, 1963; Nygaard, 1977
Uroglena soniaca Conrad	X	X					Conrad, 1938; Bourrelly, 1957, 1963
Uroglena uplandica (Skuja) Matvienko		X					Bourrelly, 1963 (as *U. volvox* var. *uplandica* Skuja)
Uroglena volvox Ehrenberg	X	X			X	X	Kristiansen, 1980; Bourrelly, 1957, 1963; Cronberg & Kristiansen, 1980; Sandgren, 1980a; Donaldson & Stein, 1984

* not recognized by Asmund & Kristiansen, 1986

micrographs for over 300 cyst morphotypes, but no written descriptions.

Although the above classification schemes were used for some time, the increased use of electron microscopy and the resulting heightened appreciation for the large diversity of morphotypes prompted some recent changes in the way cysts were described. The most significant of these were the guidelines of the International Statospore Working Group (ISWG; Cronberg & Sandgren, 1986). The ISWG, recognizing the confusion likely to result from different names being applied to two life stages (i.e. vegetative cell and cyst) of the same species, developed an alternate system in which cysts are given numbers rather than names. This system is now standard, and has been used throughout this book.

Within the numerical system devised by the ISWG, each researcher, or research group, maintains a personal numerical file of the cysts they have described. A characteristic scanning electron micrograph is required for each cyst type. All available geographical and ecological data are included. The biological affinity is only noted when it is known with reasonable certainty. Publications that follow the ISWG guidelines include Sandgren & Carney (1983), Takahashi *et al.* (1986), Rybak (1987), Takahashi (1987), Rybak *et al.* (1987, 1991), Duff & Smol (1988, 1989, 1991, 1994), Zeeb *et al.* (1990), Carney *et al.* (1992), Duff *et al.* (1992), Pienitz *et al.* (1992), Zeeb & Smol (1993a), Zeeb *et al.* (1994, subm.), and several others describing cysts with known affinities (Table 1).

Chrysophycean Cysts in Paleolimnological Studies

The lack of taxonomic detail and consistency offered by many of the earlier classification systems hindered the use of stomatocysts in paleolimnological studies. It is perhaps not surprising that when chrysophytes were considered in paleolimnological studies, they were simply lumped into one collective category and reported on as a group. The number of cysts was usually reported as a percentage of the number of diatom valves enumerated (e.g. Tippett, 1964; Grönlund *et al.*, 1986; Smol, 1983, 1985; Stoermer *et al.*, 1985; Harwood, 1986). The ratio of chrysophycean cysts to diatom frustules has been proposed as an approximate measure of changes in temperate lake trophic status (Smol, 1985), and in high polar regions, as a possible measure of lake ice cover and hence climate (Smol, 1983, 1988c). Because of the highly seasonal nature of some chrysophyte taxa, layers of cysts in laminated lake sediments (Fig. 3) have been used to differentiate seasons (e.g. Battarbee, 1981; Peglar *et al.*, 1984; Grönlund *et al.* 1986). However, much potential environmental information is inevitably lost with the lumping of such an ecologically diverse group of organisms.

In other studies, past changes in individual cyst types have been linked to changes in paleoenvironmental conditions. Changes in trophic status have been shown to affect chrysophycean assemblage structure (e.g. Nygaard, 1956; Leventhal, 1970; Elner & Happey-Wood, 1978; Carney, 1982; Carney & Sandgren, 1983; Rybak, 1986, 1987; Rybak *et al.*, 1987; Zeeb *et al.*, 1990), as have lake level changes (e.g. Leventhal, 1970; Rybak, 1986) and habitat availability (e.g. Hajós & Radocz, 1969; Hajós, 1973, 1974; Rybak *et al.*, 1987). Other variables that have been correlated with changes in cyst types are: pH (e.g. Nygaard, 1956; Rybak *et al.*, 1987, 1991; Duff & Smol, 1991; Carney *et al.*, 1992); metal concentrations (e.g. Elner & Happey-Wood, 1978); salinity (e.g. Hajós & Radocz, 1969; Hajós 1973, 1974; Pienitz *et al.*, 1992; Cumming *et al.*, 1993); and climatic changes (e.g. Duff & Smol, 1988; Zeeb & Smol, 1993b).

One could argue that there is little value in describing cysts when their taxonomic identities are not known (i.e. which taxa produced them). We disagree. Clearly, an important goal is to eventually determine the taxonomic affinities of the morphotypes. However, even if researchers only refer to numbered morphotypes at this stage, cysts can still be used as powerful markers of environmental change, as well as in other scientific endeavours. For example, if cyst morphotypes can be described in surface sediment calibration or training sets (see Charles & Smol, 1994, for a description of how these calibration sets and transfer functions are developed), and cyst morphotypes can be related to limnological variables of interest, then these data can be used to infer past changes in lake development (e.g. acidification, eutrophication, climatic change, etc.). To date, few attempts have been made to quantify the relationship between cyst distributions and limnological variables (Rull, 1991; Rybak *et al.*, 1991; Carney *et al.*, 1992), but several calibration sets are now being developed in our laboratory (Brown, 1994; Duff, 1994; Zeeb, 1994) and elsewhere. A hopeful outcome of this research will be an intensified interest in the use of stomatocysts as paleolimnological markers.

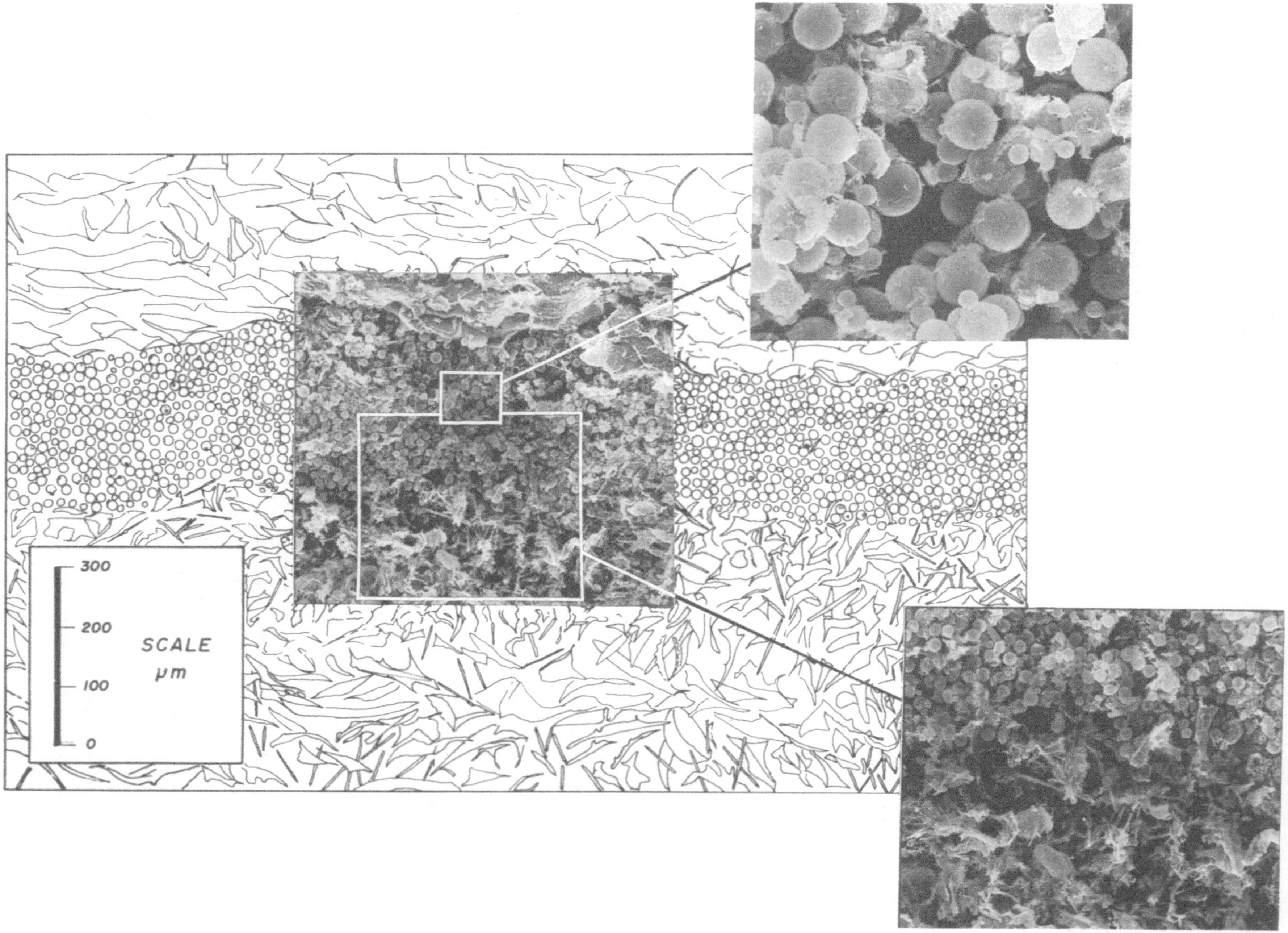

Fig. 3. Schematic diagram of a cyst lamination recovered by tape peel from a frozen core in Long Lake, Michigan. SEM enlargements show diverse morphological features and sizes of stomatocysts.

Rationale for the Atlas

In this Atlas, we have compiled descriptions for almost 250 cyst morphotypes. This is certainly not an exhaustive survey of cyst morphotypes, but it does represent many of the dominant morphotypes and also many rarer forms from a diverse array of environments. We have included cysts from both temperate and arctic locations; from acidic, alkaline and saline lakes and ponds; from sites spanning a large trophic gradient; from relatively pristine to highly disturbed lakes; from preserved algal material, modern sediments, and Holocene and Pliocene sedimentary deposits. Our specimens are largely from North America, but many of these morphotypes are common elsewhere (e.g. Nygaard, 1956; Leventhal, 1970; Rull, 1986; Rybak, 1986, 1987).

By providing full documentation of each cyst type, including scanning electron and light microscopy and line illustrations, we hope to facilitate the accurate identification of individual cyst morphotypes. We hope this Atlas will be useful to paleoecologists and micropaleontologists who wish to include stomatocysts in their studies. In particular, stomatocysts can easily be enumerated along with other siliceous microfossils, such as diatom frustules and chrysophycean scales (Fig. 1). However, we also expect that this book will be used by researchers working with living chrysophycean algae, and those interested in the morphology and ultrastructure of cyst morphotypes. Hopefully, these descriptions will also accelerate the continued effort to link morphotypes to the algae that produce them.

The Atlas is divided into five chapters. Following this brief introduction, Chapter 2 outlines our methods. A description of the morphological terms we use to describe the cysts is given in Chapter 3. Chapter 4 forms the bulk of the Atlas, with ISWG descriptions of 155 cyst morphotypes. Only those morphotypes for which we had a sufficient number of SEM specimens (i.e. at least two, and usually more) are included here; they are described using SEM, LM and line drawings, as well as a written description following ISWG guidelines. Ten cyst morphotypes are described for the first time and most descriptions of previously described cysts are amended. The Atlas concludes with Chapter 5, which contains brief notes concerning 94 cyst morphotypes that were described in our previous papers, but for which we had insufficient specimens to warrant a full description in this volume. Stomatocyst morphotypes that have been discontinued because they were invalidly described, or because the specimens have been reassigned, are also discussed in Chapter 5.

As we show in this Atlas (often by correcting our own previous publications), the description of stomatocysts has been somewhat in a state-of-flux. We do feel, however, that now is the time to consolidate our observations. We hope that this Atlas will serve as a point-of-departure for further studies. Our overall goal is to hopefully remove some of the mystery surrounding cysts, and to encourage other workers to begin using these important indicators.

CHAPTER 2

Methods

Sample Collection and Preparation

The reader is referred to the original papers cited in Chapters 4 and 5 for full sampling and laboratory procedures. In general, the methodology was as follows.

In most cases, cysts were described from lake sediments, which were collected using a sediment corer (e.g. Livingstone, 1955; Glew, 1991; most of the commonly used equipment is illustrated in Smol & Glew, 1992). Some of the surface sediments, such as those from British Columbia, were collected using an Ekman dredge. In addition, some of the Elk Lake samples were collected from sediment traps (Bradbury, 1988), whilst some of the arctic specimens were collected from habitats such as mosses, epilithon, and other periphytic habitats, in addition to surface sediments.

The raw samples were prepared following standard protocols for cleaning and examining siliceous microfossils (e.g. Battarbee, 1986); Smol (1983) provides a summary of the methodology normally followed in our lab. If the samples contained carbonates, the material was pretreated in a 5% hydrochloric acid solution. The samples were then digested in strong acids (e.g. sulphuric acid). To complete the digestion, the samples were heated in a boiling water bath (usually ca. 1 hour), cooled, and then centrifuged several times and washed with distilled water. The overall goal of the above treatments is to remove as much extraneous material (e.g. organic matter, carbonates) from the sediment matrix as possible, but leave a siliceous slurry that contains the cleaned chrysophyte cysts.

For light microscopy (LM), subsamples of the siliceous slurries were either pipetted onto square glass coverslips or known aliquots were allowed to settle in Battarbee (1973) trays. Slurries were allowed to evaporate on a slide warmer set at low heat. Once dry, the coverslips were mounted onto glass slides using Hyrax® mounting medium, a permanent medium with a high refractive index (R.I. = 1.65-1.71). Other media, such as Naphrax®, could also be used. Canada Balsam has also been recommended for stomatocysts (Cronberg, 1986b) because it appears to minimize the "halo effect" caused by light refraction in the cyst wall.

For scanning electron microscopy (SEM), slurries were air-dried onto small (12 mm) circular glass coverslips or directly onto aluminum foil. The dried slurries were affixed to aluminum stubs using double-sided adhesive tape, and the stubs were sputter-coated with approximately 20 nm of gold or platinum using, among others, a Denton Vacuum Desk II cold sputter unit.

Microscopy and Illustrations

A variety of light microscopy units were used to study and describe the cysts. These included: 1) a Leitz 20 Dialux microscope equipped with Smith interference optics, a 100x oil immersion lens (N.A. = 1.32) and a Wild Photoautomat MPS 45 camera; and 2) a Nikon Optiphot microscope equipped with Nomarski interference optics, a 100x oil immersion lens (N.A. = 1.25), and a Nikon FX-35 camera. We recommend differential interference contrast as the best available LM optics for observing the surface features used to distinguish stomatocyst morphotypes. A low-contrast LM film such as Pan-X or TechPan (50 to 100 ASA) will also help to minimize the "halo effect".

Most of the scanning electron microscopy was performed using an Hitachi S-2500, equipped with both a 35 mm camera and a Polaroid camera, and operated at 20 kV and a 15 mm working distance. For many specimens, multiple pictures were taken from different angles (i.e. tilted and rotated).

Line illustrations were drawn free-hand based on several SEM and LM specimens. They are our interpretations, meant to complement but not copy the photographs. We have attempted to provide alternate views for each of our cyst types and to provide a visual representation of as many variations as possible.

CHAPTER 3

Terminology

The terminology used in this Atlas relies heavily upon that developed by Cronberg & Sandgren (1986) and further modified by Dr. C.D. Sandgren (pers. commun.). Much of this terminology will be familiar to palynologists and other taxonomists, although some size categories are different. We have modified the terminology and introduced some new terms (defined below) to identify features that appear to be unique to chrysophyte cysts. Descriptive terms are further illustrated in Figs. 4A-G.

Cyst Body Morphology (Fig. 4A, B)

The **cyst body** can be divided into an anterior and posterior hemisphere. The **pore** is always located in the anterior hemisphere, but not always at the anterior pole. The **length** is defined as the distance between the anterior and the posterior pole. The **width** is the maximum dimension perpendicular to the length, and may or may not equal the **equator**, which is the dimension perpendicular to the length located half-way along the length. If more than one layer of silica is present, the innermost layer is called the **primary wall**, with overlaying layers being the **secondary wall**, **tertiary wall**, etc. (Fig. 4A). The outermost wall is used to define the length and width.

Cyst shape is most commonly spherical, but may be **oval** to **oblate** or **obovate** to **ovate**. These categories are defined according to the length:width ratio of the cyst, and the position of the width with respect to the equator. A few cysts are **pyramidal** in shape (Fig. 4B). Minor variations in cyst shape from those described above are possible; for example, **hemispherical** cysts that are flattened on one side have been observed by C.D. Sandgren (pers. commun.).

Pore/Collar Morphology (Fig. 4C-F)

Pores are circular in outline. A **regular pore** extends perpendicularly through the cyst wall; i.e. it has outer and inner **pore margins** of equal diameter. The pore margins may be sharply delineated or rounded. A pore with an outer pore margin larger than the inner pore margin is either **conical** (with straight sides) or **concave** (with curved sides). A **pseudoannulus** is a thin shelf of silica projecting from the inner pore margin; it may be **planar** (flat) or **swollen** (Fig. 4C). We recommend tilting SEM specimens to determine if the pseudoannulus is planar or swollen, because a swollen pseudoannulus will appear planar in apical view; for example, see stomatocyst 49 (Chapter 4). Pore features never project beyond the curved profile of the outer cyst wall.

A **collar** is a siliceous thickening surrounding the pore. Some stomatocysts lack a collar entirely, and on others, the pore may be surrounded by **anterior projections** that simulate a collar, but do not constitute a **true collar** because they do not form a continuous ring of silica. When a true collar is present, it is defined according to its cross-sectional outline, as **cylindrical**, **conical**, or **obconical** (Fig. 4D). The junction between the collar and cyst body is the **collar base**; the terminal end of the collar is the **apex**. The collar shape may be modified somewhat at either the base or the apex; for example, a cylindrical collar may have a conical base or an obconical apex. Short ridges, known as **struts**, may project from the outer collar margin onto the cyst wall. The apex may be **acute, rounded** or **ornamented**. We no longer use the terms **marginal rim** or **marginal ridge**, which have been used in previous papers to describe a low, wide, conical collar with a rounded or acute apex. The inner and sometimes the outer collar margins may be ornamented with very thin spiralling or vertical ridges called **striations**. The inner collar margin may be separated from the outer pore margin by a shelf of silica called the **annulus**; this should not be confused with the pseudoannulus described above. The annulus may be swollen, flat planar, or sloping planar in cross-sectional view (Fig. 4D).

A **false complex collar** is an unusual type of simple collar. It consists of a single collar with an apical groove that may appear as two separate collars on mature SEM specimens (Fig. 4D). The structure of the collar is best shown on broken, tilted or less developed SEM specimens, or on LM specimens.

A **true complex collar** consists of more than one separate collar surrounding the pore. Each collar must project from the outermost cyst wall. The collar nearest to the pore is referred to as the **primary collar**, with successive collars being secondary, tertiary, and so on. Two collars may be separated by a section of the cyst wall referred to as the **interannulus**, which may be planar or swollen. Thus, the interannulus between the primary and secondary collars is the **primary interannulus**, that between the secondary and tertiary collars is the **secondary interannulus** (regardless of whether a primary interannulus exists), and so on (Fig. 4E). For example, stomatocysts 203 and 204 (Chapter 4) are very similar cysts that have false and true complex collars, respectively.

Occasionally, a siliceous **plug** may obscure the pore, or a lenticular **cap** may be found covering the collar or the pore (Fig. 4F).

Ornamentation (Fig. 4G)

Projecting elements

The cyst body may be ornamented with a variety of projecting elements. Elements with a roughly circular base include scabrae, verrucae, conula, and spines (see below). Elements with an elongated base are called ridges, circuli, or reticula (see below).

Scabrae (papillae, granules) are defined as projections with a base diameter ≤ 0.2 μm. These features are observable with a high resolution LM (Chapter 2). **Verrucae** (nodules) and **conula** have a basal diameter that is >0.2 μm and is greater than or equal to the projection height; verrucae are apically rounded whereas conula are pointed. **Spines** are projections with a length greater than or equal to the basal diameter. They may be **baculate** (cylindrical) or **echinate** (pointed). Spines may **bifurcate**; primary bifurcations are most common, but secondary and even tertiary bifurcations may be found. Additional projecting elements may ornament the spine or the spine apex. Spines are sometimes twisted, or rope-like, in appearance, and are often hollow. The spine base is sometimes striated; the striations may continue along part or all of the spine length, and may be straight, wavy or spiralling.

In this Atlas, we have separated cysts with **ridges** based on their maximum ridge length. Short ridges are less than one-half of the cyst circumference; this characteristic is based on careful observation of many SEM specimens, but may not be taxonomically significant. Ridges may be straight or curved, and may be branched. They are often **lunate** (semi-circular) in lateral view. They may be ornamented with vertical striations or struts. A **circulus** is a ridge that forms a closed ring. Circuli are further defined according to their orientation, as **latitudinal**, **longitudinal** or **tangential**. We no longer use the term **flange**, which would refer only to circuli in a longitudinal, pole-to-pole or latitudinal, equatorial position. A **reticulum** is a network of ridges; the spaces between the ridge walls are referred to as **lacunae**, and may be circular or polygonal (e.g. triangular, hexagonal, etc). A **regular reticulum** possesses lacunae of exactly the same size and shape. A **variable reticulum** has lacunae that are only approximately the same size and shape. An **irregular reticulum** has lacunae that exhibit substantial variation in size and shape. It is often difficult to distinguish between a cyst that is ornamented with a reticulum, and a cyst that is ornamented with many depressions. For simplicity, we consider the ornamentation pattern to be a reticulum if the maximum width of the lacunae is greater than or equal to the reticular ridge width.

Indentations

Indentations consist of either **psilae (punctae)**, **depressions (craters)**, or **fossae (grooves)**. Psilae and depressions are circular in outline; psilae are ≤ 0.2 μm in diameter, whereas depressions are >0.2 μm in diameter. In our experience, a psilate surface pattern (often called **microtextured**) usually indicates degradation or developmental immaturity. Fossae are elongated indentations; they may be straight or curved and may be branched. They are a relatively rare form of ornamentation.

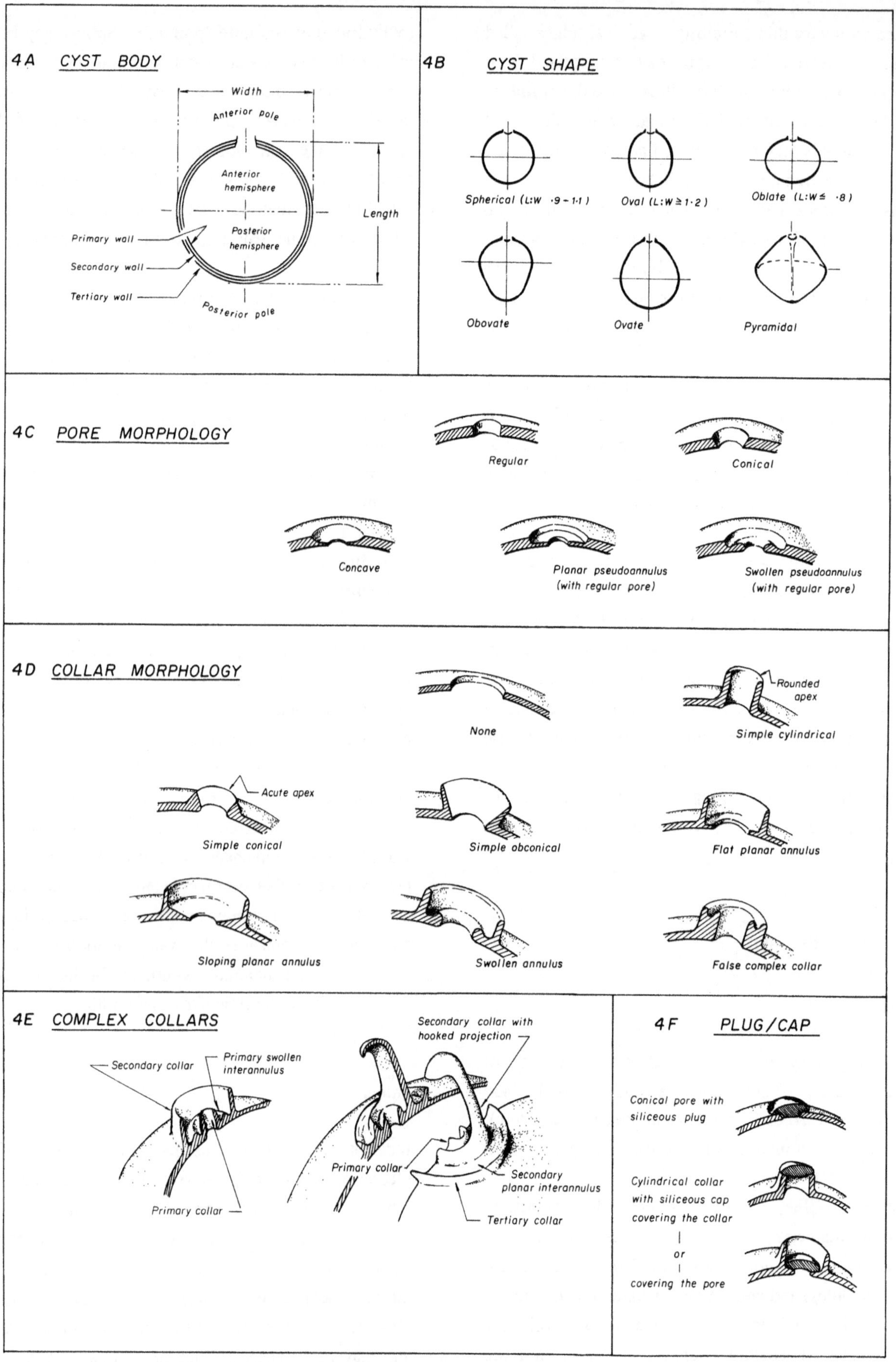

Fig. 4. Terminology used to describe the morphological features of chrysophycean cysts specifically relating to: A. cyst body; B. cyst shape; C. pore morphology; D. collar morphology; E. complex collars; F. plug/cap; G. ornamentation (facing page).

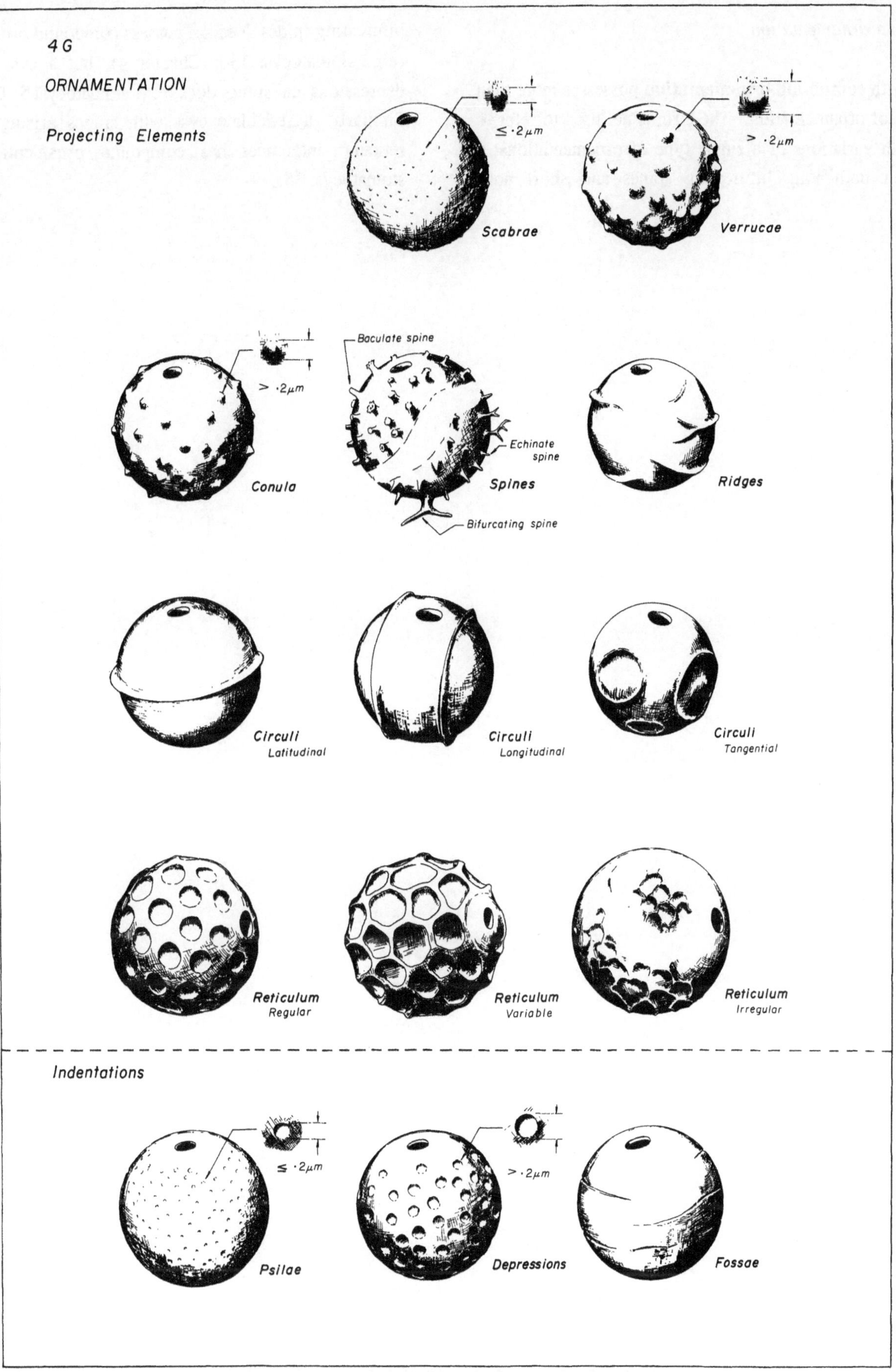
4 G
ORNAMENTATION
Projecting Elements
≤ ·2μm
Scabrae
> ·2μm
Verrucae
> ·2μm
Conula
Baculate spine
Echinate spine
Spines
Bifurcating spine
Ridges
Circuli
Latitudinal
Circuli
Longitudinal
Circuli
Tangential
Reticulum
Regular
Reticulum
Variable
Reticulum
Irregular
Indentations
≤ ·2μm
Psilae
> ·2μm
Depressions
Fossae

Compound ornamentation

A cyst with **compound ornamentation** possesses more than one type of ornamentation. The term is not used to refer to cysts with variations of a single type of ornamentations; a cyst with both long bifurcating spines and short non-bifurcating spines does not possess compound ornamentation (e.g. stomatocyst 130; Chapter 4), but a cyst with both depressions and spines does (e.g. stomatocyst 5; Chapter 4). Similarly, a reticulate cyst with spines arising from the reticular interstices has compound ornamentation (e.g. stomatocyst 35).

CHAPTER 4

Cyst Descriptions

In this chapter, we present full descriptions of those cysts for which we had sufficient SEM specimens (see Chapter 1), using the following format:

1) Each morphotype has been assigned a unique number according to the ISWG guidelines (Cronberg & Sandgren, 1986). Numbers have been assigned consecutively in each of our previous publications, and we have continued this practice for cysts that are described here for the first time.

2) The authors' names and the year of the original publication follow the morphotype number. The only exception is Smol's (1984) stomatocyst 1, produced by *Mallomonas pseudocoronata* Prescott, as the description of this cyst predated ISWG guidelines.

3) The number of SEM specimens that were used for the description is then given in parentheses.

4) The biological affinity is discussed.

5) For newly described morphotypes, we have included the negative number of the specimen we chose as most characteristic.

6) The locality from which the characteristic specimen was retrieved is given. Fig. 5 shows the locations from which all characteristic specimens referred to in this Atlas were obtained.

7) The morphotype is described based on SEM specimens. Several morphotypes have been divided into two or more *formae*. We define *formae* as distinct groups of specimens that possess varying degrees of ornamentation, such that *forma* A is the least ornamented, *forma* B is relatively more ornamented, and so on. We believe that *formae* may represent stages in cyst development, and therefore they do not warrant assignment to separate morphotypes.

8) The criteria by which LM identification can be made are discussed. In some cases, two or more cysts are indistinguishable with LM, and are referred to in the text as a collective category (e.g. stomatocysts 114/115/218).

9) Any literature references to the morphotype are listed geographically. North American references are presented first, followed by citations from other continents.

10) Any pertinent ecological data are summarized.

The cysts are presented in groups, based on their common morphological characteristics. Unornamented cysts are described first, followed by cysts with projections only, then by cysts with indentations only, and finally by cysts with compound ornamentation, according to the element order above. Table 2 describes the groups in more detail with corresponding page numbers; it can be used as a quick guide for cyst identification. At the end of Chapter 4, line drawings of each cyst type or collective category are presented according to their relative sizes (Fig. 154). We anticipate that this figure will facilitate rapid LM identification.

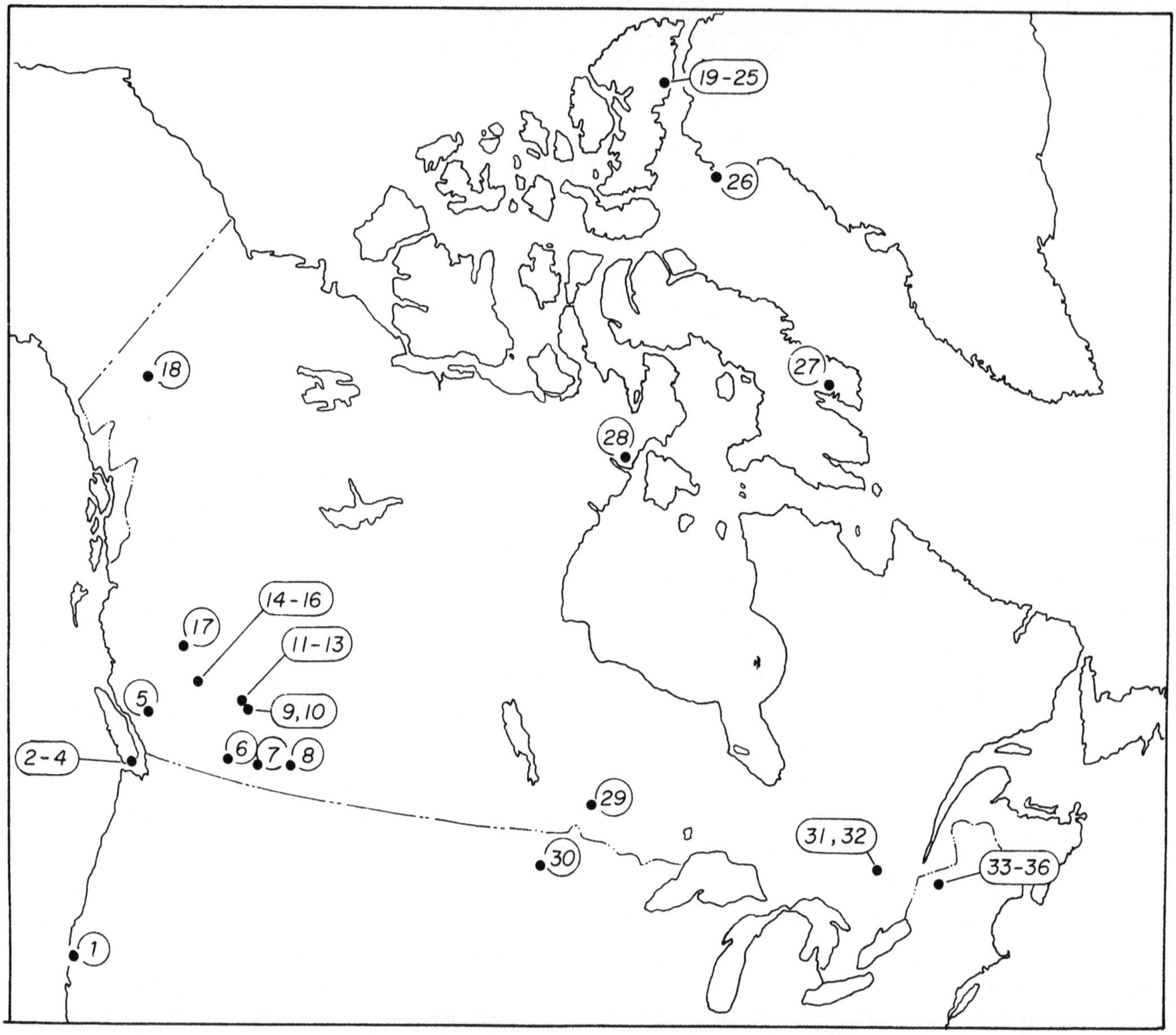

Fig. 5. Locations of sites from which the characteristic specimens of all chrysophycean stomatocyst morphotypes discussed in this Atlas were collected.

1 Sonoma Volcanics, Napa County (38°N; 122°W), California, U.S.A.
2 Florence Lake (48°30'N; 123°30'W), British Columbia, Canada
3 Durrance Lake (48°30'N; 123°25'W), British Columbia, Canada
4 Glen Lake (48°30'N; 123°30'W), British Columbia, Canada
5 Elkin Lake (51°40'N; 123°45'W), British Columbia, Canada
6 Davis Lake (49°50'N; 120°45'W), British Columbia, Canada
7 Wilgress Lake (49°10'N; 118°40'W), British Columbia, Canada
8 Champion Lake #3 (49°15'N; 117°30'W), British Columbia, Canada
9 King Lake West (51°45'N; 120°50'W), British Columbia, Canada
10 Gardom Lake (50°40'N; 119°10'W), British Columbia, Canada
11 Hawkins Lake (51°55'N; 120°55'W), British Columbia, Canada
12 Lac de Roche (51°45'N; 120°45'W), British Columbia, Canada
13 Spout Lake (52°00'N; 121°20'W), British Columbia, Canada
14 Lang Lake (52°00'N; 121°00'W), British Columbia, Canada
15 Green Lake (52°30'N; 121°00'W), British Columbia, Canada
16 Spanish Lake (52°35'N; 121°20'W), British Columbia, Canada
17 Dahl Lake (53°50'N; 123°15'W), British Columbia, Canada
18 Unnamed saline lake (U60) (62°45'N; 136°38'W), near Pelly Crossing, Yukon Territory, Canada
19 "Rock Basin Lake" (unofficial name) (78°29.5'N; 76°44.8'W), Ellesmere Island, Northwest Territories, Canada
20 "Camp Pond" (unofficial name), Cape Herschel (78°37'N; 74°42'W), Ellesmere Island, Northwest Territories, Canada
21 "Elison Pond" (unofficial name), Cape Herschel (78°37'N; 74°42'W), Ellesmere Island, Northwest Territories, Canada
22 "Horseshoe Pond" (unofficial name), Cape Herschel (78°37'N; 74°42'W), Ellesmere Island, Northwest Territories, Canada
23 "Col Pond" (unofficial name), Cape Herschel (78°37'N; 74°42'W), Ellesmere Island, Northwest Territories, Canada
24 "Beach Ridge Pond" (unofficial name), Cape Herschel (78°37'N; 74°42'W), Ellesmere Island, Northwest Territories, Canada
25 "Willow Pond" (unofficial name), Cape Herschel (78°37'N; 74°42'W), Ellesmere Island, Northwest Territories, Canada
26 Nordvestø (76°44'N; 73°13'W), Carey Islands, Greenland
27 Tasikutaaq Lake (66°23'N; 65°57'W), Baffin Island, Northwest Territories, Canada
28 Hawk Lake (63°N; 90°W), Northwest Territories, Canada
29 Lake 227 (49°42'N; 93°42'W), Experimental Lakes Area, Ontario, Canada
30 Elk Lake (47°11'N; 95°13'W), Clearwater County, Minnesota, U.S.A.
31 Little Round Lake (44°47.5'N; 76°41.5'W), Ontario, Canada
32 Lake Opinicon (44°33'N; 76°26'W), Ontario, Canada
33 Upper Wallface Pond (44°08.8'N; 74°03.3'W), Adirondack Park, New York, U.S.A.
34 "Unnamed Pond" (unofficial name) (43°11'N; 74°37'W), Adirondack Park, New York, U.S.A.
35 Dry Channel Pond (42°22'N; 74°28'W), Adirondack Park, New York, U.S.A.
36 Lake Frances (44°45'N; 74°22'W), Adirondack Park, New York, U.S.A.

Table 2: Stomatocyst groups presented in this chapter.

Group	Page numbers
Unornamented stomatocysts	
Spherical stomatocysts **lacking a collar** or ornamentation	24-30
Oval stomatocysts **lacking a collar** or ornamentation	30-32
Spherical or oval stomatocysts with a **simple conical collar** but no ornamentation	32-41
Spherical or oval stomatocysts with a **simple cylindrical collar** but no ornamentation	41-55
Spherical stomatocysts with a **simple obconical collar** but no ornamentation	55-57
Spherical or oval stomatocysts with a **true or false complex collar** but no ornamentation	57-61
Stomatocysts ornamented with **projections**	
Stomatocysts with or without collar; ornamentation consists of **anterior projections simulating simple or complex collar**	61-64
Stomatocysts ornamented with scabrae, verrucae or conula	
Stomatocysts ornamented with **scabrae**	64-67
Stomatocysts ornamented with **verrucae only**	67-73
Stomatocysts ornamented with **either verrucae or conula**	73-75
Stomatocysts ornamented with **conula only**	76-81
Stomatocysts ornamented with spines	
Stomatocysts ornamented with **short, thin spines** that are roughly **equidistant** and cover the entire cyst surface	81-88
Stomatocysts with **short spines** that are usually concentrated in the **posterior hemisphere**	88-93
Stomatocysts with **long, non-bifurcating spines**	93-97
Stomatocysts with **bifurcating spines**	97-105
Stomatocysts ornamented with ridges, circuli or a reticulum	
Stomatocysts with **short ridges**; length of ridges is less than half of the cyst circumference	105-114
Stomatocysts with **long ridges**; ridge length is equal to or greater than half of the cyst circumference	114-118
Stomatocysts with one or more **circuli**	118-120
Stomatocysts with a **reticulum**	121-131
Stomatocysts ornamented with **indentations**	131-134
Stomatocysts with **compound ornamentation**	
Stomatocysts ornamented with **spines and ridges**	134-135
Stomatocysts ornamented with **spines and a reticulum**	135-139
Stomatocysts ornamented with **spines and indentations**	140-144
Stomatocysts ornamented with **ridges and a reticulum**	144-145
Stomatocysts ornamented with **ridges and indentations**	145-148
Stomatocysts ornamented with **more than two types** of ornamentation	148-149

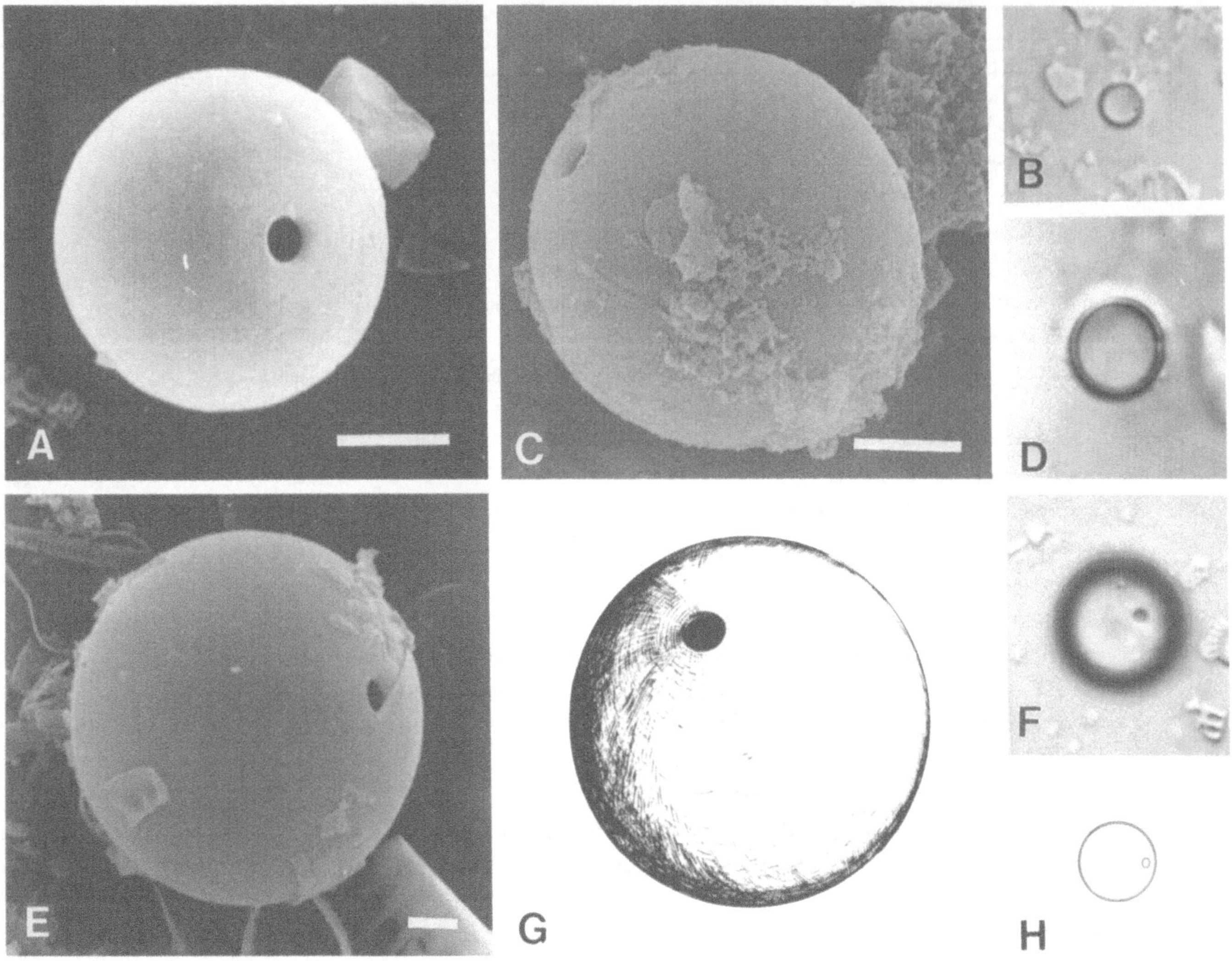

Fig. 6. Stomatocysts 1, 9, and 15. A: Stomatocyst 1, SEM. B: Stomatocyst 1, LM. C: Stomatocyst 9, SEM. D: Stomatocyst 9, LM. E: Stomatocyst 15, SEM. F: Stomatocyst 15, LM. G: Stomatocysts 1, 9 and 15, line drawing of SEM. H: Stomatocyst 9, line drawing of LM. Scale bars = 2 μm. LM x2000.

Spherical stomatocysts lacking a collar or ornamentation

Stomatocyst 1, Duff & Smol, 1988 *emend.* Zeeb & Smol, 1993a (29) Figs. 6A, B

BIOLOGICAL AFFINITY: This cyst is probably the mature or immature form produced by a number of chrysophytes. At least two *Paraphysomonas* species have cysts of this type (Preisig & Hibberd, 1982a, b, 1983).

LOCALITY: Elk Lake (47°11'N; 95°13'W), Clearwater County, Minnesota, U.S.A. Piston core, varve year 3437 B.P.

SEM DESCRIPTION: Zeeb & Smol (1993a) updated the original description by Duff & Smol (1988) and provided a new characteristic specimen. They restricted the size range of this cyst to ≤5.9 μm. It belongs to a group of cysts (including stomatocysts 9 and 15) that differ only on the basis of size. Stomatocysts 43 and 109 (Chapter 5) are now included within stomatocyst 1.

It is a spherical stomatocyst with a smooth or psilate (if immature or eroded) surface. There is no collar development. The regular pore (diameter 0.3-0.7 μm) is flush with the cyst body, or very slightly depressed with no sharp delineation between the upper pore margin and the cyst surface (pore diameter : cyst diameter ratio 0.10-0.13).

LM DESCRIPTION: Stomatocyst 1 (Fig. 6B) is distinguished from stomatocysts 29 and 46 on the basis of pore morphology, and from stomatocyst 9 on the basis of size. It cannot be distinguished from stomatocyst 148 with LM.

REFERENCES: This cyst has been found in the Northwest Territories, Canada (Duff & Smol, 1988); Ontario, Canada (Zeeb *et al.*, 1990 - stomatocyst 109; Zeeb *et al.*, 1994); British Columbia, Canada (Duff & Smol, 1994; Zeeb, 1994); Minnesota, U.S.A. (Zeeb & Smol, 1993a); and New York, U.S.A. (Duff & Smol, 1991 - stomatocyst 43; Duff & Smol, unpubl.).

ECOLOGY: In most paleolimnological studies, this cyst was counted as a collective category, and so the ecological information that could be gleaned from its stratigraphic distribution is limited. In Elk Lake, Minnesota, it flourished during the warm, dry climate of the mid-Holocene (Zeeb & Smol, 1993b). It also increased in relative abundance after nutrient (P and N) additions to a manipulated lake (Zeeb *et al.*, 1994). In a survey of freshwater lakes from British Columbia, Canada, it was more common in large, clear, circumneutral lakes (Duff, 1994). More studies are likely to demonstrate that this cyst is widely distributed across broad ecological gradients, consistent with our belief that it is produced by several different species.

Stomatocyst 9, Duff & Smol, 1988 *emend.* Zeeb & Smol, 1993a (19) Figs. 6C, D, and H

BIOLOGICAL AFFINITY: Like stomatocyst 1, this cyst is probably produced by several species, including *Chrysolepidomonas dendrolepidota* Peters & Andersen (1993).

LOCALITY: "Rock Basin Lake" (unofficial name) (78°29.5'N; 76°44.8'W), Ellesmere Island, Northwest Territories, Canada. Piston core, core depth 15-16 cm.

SEM DESCRIPTION: This cyst is identical to stomatocyst 1, except that it is larger (diameter 6.0-8.9 μm; pore diameter 0.5-0.8 μm; pore diameter : cyst diameter ratio 0.08-0.11). Specimens with spines, as described in Duff & Smol (1988), are no longer included within stomatocyst 9.

LM DESCRIPTION: Stomatocyst 9 (Figs. 6D, H) is distinguished from stomatocysts 120 and 189 on the basis of pore morphology and from stomatocysts 1 and 15 on the basis of size.

REFERENCES: This cyst has been observed in the Northwest Territories, Canada (Duff & Smol, 1988; Duff *et al.*, 1992); the Yukon Territory, Canada (Pienitz *et al.*, 1992); British Columbia, Canada (Duff & Smol, 1994; Zeeb, 1994); Minnesota, U.S.A. (Zeeb & Smol, 1993a); Michigan, U.S.A. (Sandgren & Carney, 1983 - cyst 7); California, U.S.A. (Zeeb *et al.*, subm.); and Greenland (Brown *et al.*, 1994). Some specimens of Carney *et al.* (1992 - California, U.S.A.) cyst 14 are probably stomatocyst 9.

ECOLOGY: This cyst appears to be tolerant of a wide range of environmental variables including pH (Duff & Smol, 1988; Duff *et al.*, 1992; Carney *et al.*, 1992; Duff, 1994), temperature (Duff & Smol, 1988; Duff *et al.*, 1992; Zeeb & Smol, 1993a), and salinity (Pienitz *et al.*, 1992). It displayed a negative correlation with Secchi disk depth in a survey of freshwater lakes (Duff, 1994). Stomatocyst 9 was among the most common cysts in a three million year old sediment deposit (Zeeb *et al.*, subm.).

Stomatocyst 15, Duff & Smol, 1988 *emend.* Zeeb & Smol, 1993a (4) Figs. 6E, F

BIOLOGICAL AFFINITY: This cyst may be produced by several species. *Chrysosphaerella brevispina* Korshikov *emend.* Harris *et* Bradley is thought to produce a cyst of this type (Harris & Bradley, 1958; Bourrelly, 1957 - *C. conradii* Bourrelly; H. Kling, pers. commun.).

LOCALITY: "Rock Basin Lake" (unofficial name) (78°29.5'N; 76°44.8'W), Ellesmere Island, Northwest Territories, Canada). Piston core, core depth 15-16 cm.

SEM DESCRIPTION: This stomatocyst is identical in appearance to stomatocysts 1 and 9, except that it is larger (diameter $\geq$9.0 μm). The pore is ca. 1.0 μm in diameter.

LM DESCRIPTION: Stomatocyst 15 (Fig. 6F) is distinguished from stomatocysts 42 and 150 on the basis of pore morphology, and from stomatocyst 9 on the basis of size.

REFERENCES: This cyst was observed in the Northwest Territories, Canada (Duff & Smol, 1988); British Columbia, Canada (Duff & Smol, 1994; Zeeb, 1994); Ontario, Canada (Zeeb *et al.*, 1994); Minnesota, U.S.A. (Zeeb & Smol, 1993a); and Greenland (Brown *et al.*, 1994). It includes some specimens of Carney *et al.* (1992) cyst 14 (California, U.S.A.).

ECOLOGY: This cyst responded positively to nutrient additions to a manipulated lake (Zeeb *et al.*, 1994). Carney *et al.* (1992) cyst 14 is described as pH-indifferent.

Stomatocyst 29, Duff & Smol, 1989 *emend.* Zeeb & Smol, 1993a (23) Figs. 7A, B

BIOLOGICAL AFFINITY: This cyst may be produced by more than one species.

LOCALITY: Tasikutaaq Lake (66°23'N; 65°57'W), Baffin Island, Northwest Territories, Canada. Percussion core, core depth 26.5 cm.

SEM DESCRIPTION: This is a small, spherical stomatocyst (diameter 3.0-5.9 μm) with a shallow, concave pore (inner diameter 0.3-0.7 μm) and a smooth to psilate surface. It is the smallest member of a group of identical cysts (stomatocysts 29, 120 and 42) that differ only on the basis of size. The size limits for these cysts are: ≤5.9 μm, 6.0-8.9 μm and ≥9.0 μm, respectively. Stomatocyst 29 now includes stomatocyst 44 (Chapter 5), some specimens of stomatocyst 45 (Chapter 5), and some of the specimens originally assigned to stomatocyst 120 by Duff *et al.* (1992).

LM DESCRIPTION: Stomatocyst 29 (Fig. 7B) can be distinguished from stomatocysts 1 and 46 by the pore morphology, and because the wall thickness is usually less than that of stomatocyst 46. It is distinguished from stomatocyst 120 on the basis of size only.

REFERENCES: Stomatocyst 29 has been found in Ontario, Canada (Zeeb *et al.*, 1994); the Northwest Territories, Canada (Duff & Smol, 1989); British Columbia, Canada (Zeeb, 1994); New York, U.S.A. (Duff & Smol, 1991 - misidentified as stomatocyst 44 or 45; Duff & Smol, unpubl.); Minnesota, U.S.A. (Zeeb & Smol, 1993a); and Greenland (Brown *et al.*, 1994).

ECOLOGY: This cyst may be produced by a cold-tolerant taxon (or taxa); it has been found in arctic lakes and ponds (Duff & Smol, 1988; Duff *et al.*, 1992), alpine lakes (Duff & Smol, 1991, unpubl.) and during periods of inferred cold climate in Elk Lake, Minnesota (Zeeb & Smol, 1993b). A cyst of this type may be produced by a eutrophic species (Zeeb *et al.*, 1994).

Stomatocyst 120, Duff & Smol in Duff *et al.*, 1992 *emend.* Zeeb & Smol, 1993a (30) Figs. 7C, D, and H

BIOLOGICAL AFFINITY: More than one species may produce a cyst of this type. Immature cysts (<9.0 μm) of *Chrysosphaerella longispina* Lauterborn *emend.* Nicholls (Sandgren, 1989) are identical to stomatocyst 120.

LOCALITY: "Camp Pond" (unofficial name), Cape Herschel (78°37'N; 74°42'W), Ellesmere Island, Northwest Territories, Canada. Net tow, collected July 28, 1986.

SEM DESCRIPTION: This stomatocyst is identical in appearance to stomatocyst 29, except that it is larger (cyst diameter 6.0-8.9 μm; inner pore diameter 0.5-0.9 μm). Stomatocyst 120 now includes some specimens of stomatocyst 45 (Chapter 5).

LM DESCRIPTION: Stomatocyst 120 (Figs. 7D, H) is distinguished from stomatocysts 29 and 42 on the basis of size, and from stomatocysts 9 and 189 on the basis of pore morphology. Small (<9.0 μm) specimens of stomatocyst 49 are indistinguishable from stomatocyst 120. See the description of stomatocyst 149 for the features used to distinguish it from stomatocyst 120 with LM.

REFERENCES: This stomatocyst has been observed in the Northwest Territories, Canada (Duff *et al.*, 1992); British Columbia, Canada (Duff & Smol, 1994; Zeeb, 1994); Ontario, Canada (Rybak *et al.*, 1987 - cyst 9; Zeeb *et al.*, 1994); New York, U.S.A. (Duff & Smol, 1991 - misidentified as stomatocyst 45; unpubl.); California, U.S.A. (Adam & Mahood, 1979b - type 51; Zeeb *et al.*, subm.); Minnesota, U.S.A. (Zeeb & Smol, 1993a); and Greenland (Brown *et al.*, 1994).

ECOLOGY: This cyst is found primarily in circumneutral to alkaline water (Duff *et al.*, 1992; Zeeb & Smol, 1993b; Rybak *et al.*, 1987). In one study, it increased in relative abundance after nutrient additions to the lake (Zeeb *et al.*, 1994), and it was positively correlated with total phosphorus in freshwater British Columbia lakes (Duff, 1994). It may also be common in shallow lakes (Duff *et al.*, 1994; Duff, 1994).

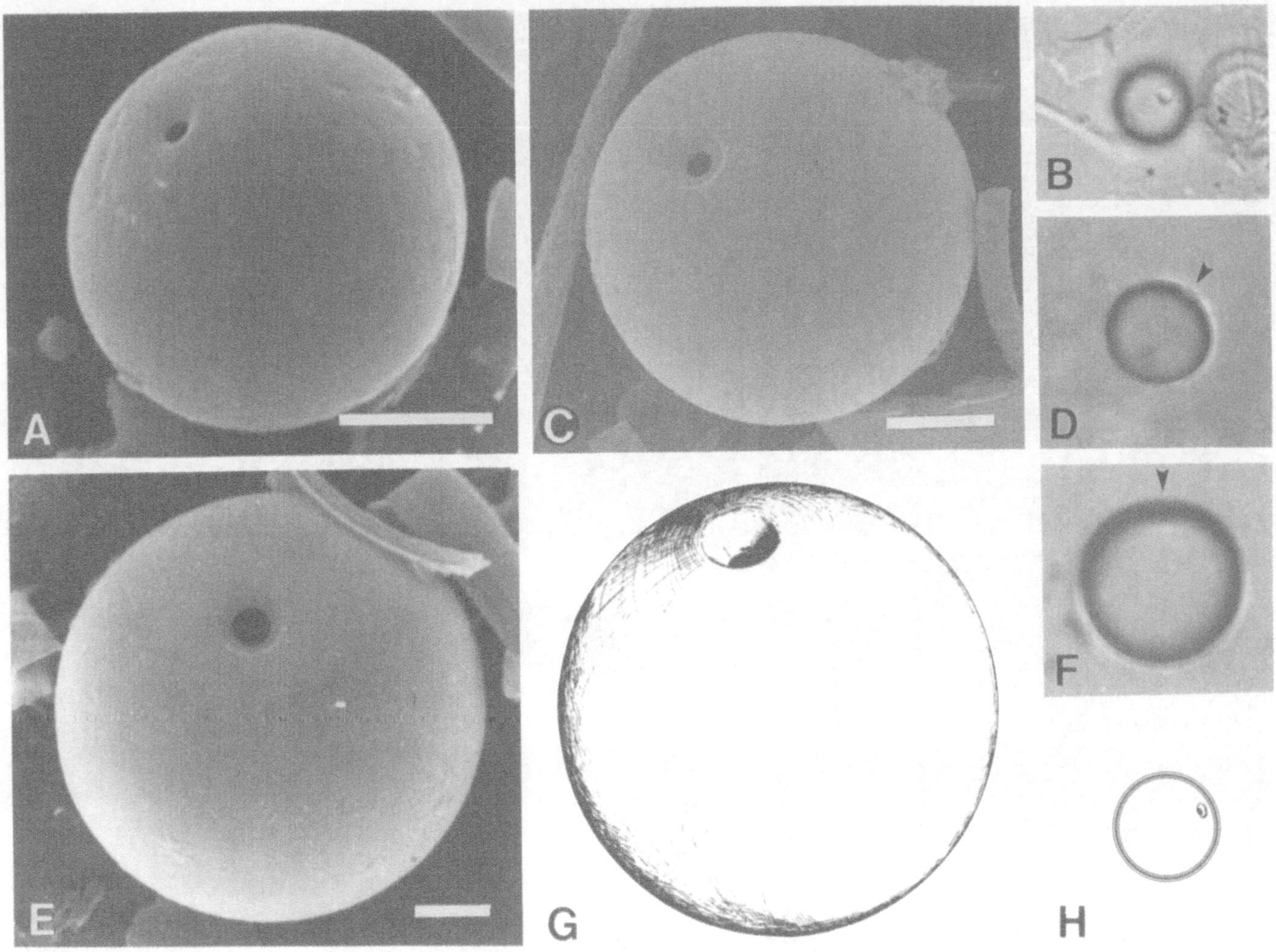

Fig. 7. Stomatocysts 29, 120, and 42. A: Stomatocyst 29, SEM. B: Stomatocyst 29, LM. C: Stomatocyst 120, SEM. D: Stomatocyst 120, LM. E: Stomatocyst 42, SEM. F: Stomatocyst 42, LM. G: Stomatocysts 29, 120 and 42, line drawing of SEM. H: Stomatocyst 120, line drawing of LM. Pores indicated by arrows. Scale bars = 2 μm. LM x2000.

Stomatocyst 42, Duff & Smol, 1989 (5) Figs. 7E, F

BIOLOGICAL AFFINITY: This cyst is probably produced by more than one species. *Synura petersenii* Korshikov produces an identical stomatocyst within this size range (Sandgren & Flanagin, 1986). Large ($\geq$9.0 μm), immature cysts of *Chrysosphaerella longispina* Lauterborn *emend.* Korshikov (Sandgren, 1989) are also identical. However, scales of these species were not found in the arctic locations where we observed stomatocyst 42.

LOCALITY: Tasikutaaq Lake (66°23'N; 65°57'W), Baffin Island, Northwest Territories, Canada. Percussion core, core depth 75 cm.

SEM DESCRIPTION: This stomatocyst is identical to stomatocyst 120, except that it is larger (diameter 9.0-19.0 μm; inner pore diameter 0.3-2.1 μm). Given the large size range, it is possible that this group will be split further when more SEM specimens are available.

LM DESCRIPTION: Stomatocyst 42 (Fig. 7F) is differentiated from stomatocysts 15 and 150 on the basis of pore morphology, and from stomatocyst 120 on the basis of size. Large ($\geq$9.0 μm) specimens of stomatocyst 49 are indistinguishable from stomatocyst 42.

REFERENCES: This stomatocyst has been observed in the Northwest Territories, Canada (Duff & Smol, 1989; Duff *et al.*, 1992); British Columbia, Canada (Duff, 1994; Zeeb, 1994); Michigan, U.S.A. (Sandgren & Carney, 1983 - cyst 6); California, U.S.A. (Mahood & Adam, 1979 - type 9); and Poland (Rybak, 1987 - cyst 4).

ECOLOGY: Stomatocyst 42 has been observed in slightly acidic to alkaline lakes or ponds (Mahood & Adam, 1979;

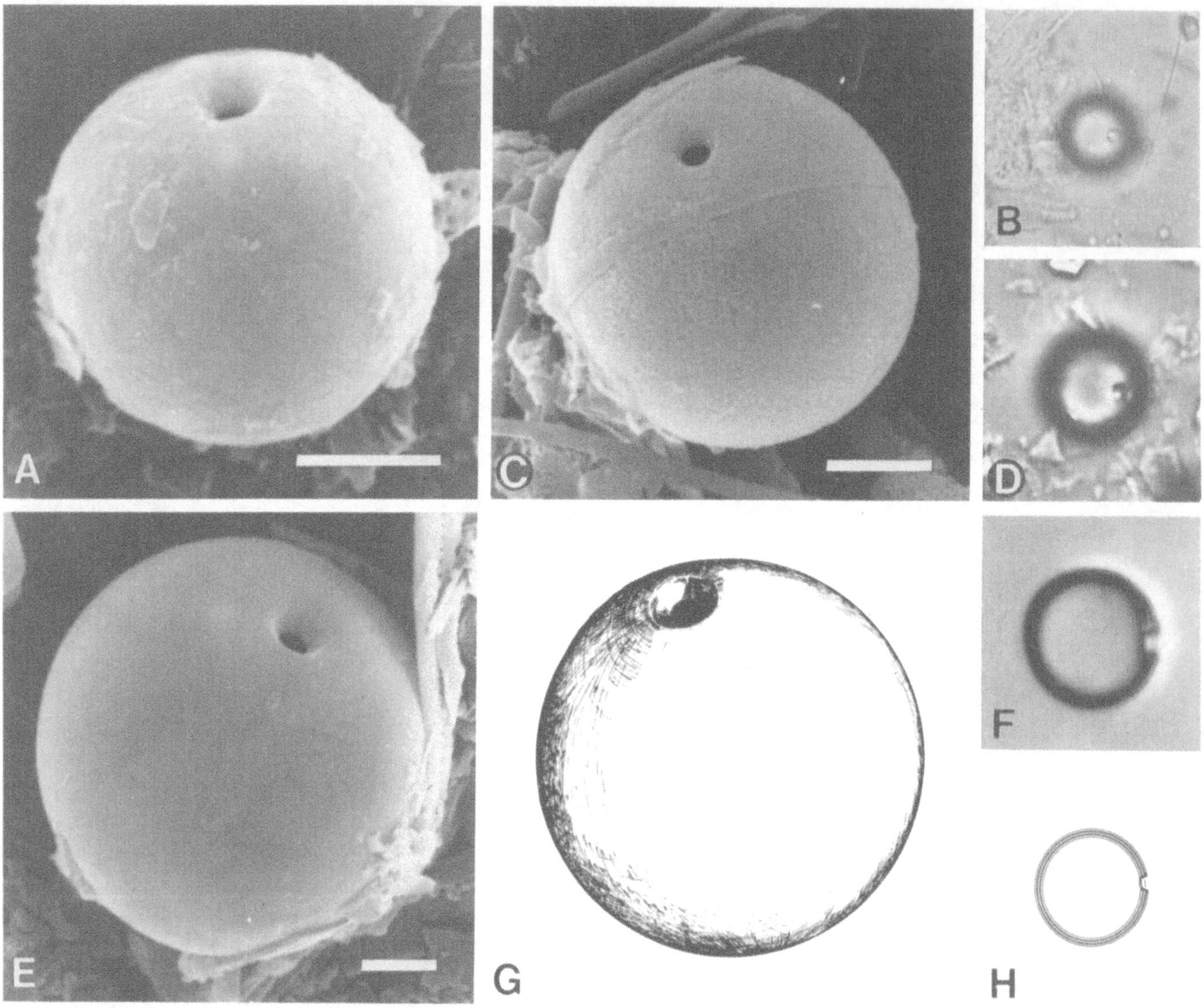

Fig. 8. Stomatocysts 46, 189, and 150. A: Stomatocyst 46, SEM. B: Stomatocyst 46, LM. C: Stomatocyst 189, SEM. D: Stomatocyst 189, LM. E: Stomatocyst 150, SEM. F: Stomatocyst 150, LM. G: Stomatocysts 46, 189, and 150, line drawing of SEM. H: Stomatocyst 189, line drawing of LM. Scale bars = 2 μm. LM x2000.

Sandgren & Carney, 1983; Rybak, 1987; Duff *et al.*, 1992; Duff, 1994). Its distribution was positively correlated with conductivity in British Columbia, Canada (Duff, 1994).

Stomatocyst 46, Duff & Smol, 1991 (3) Figs. 8A, B

BIOLOGICAL AFFINITY: unknown

LOCALITY: Upper Wallface Pond (44°08.8'N; 74°03.3'W), Adirondack Park, New York, U.S.A. Gravity core, core depth 0.0-0.5 cm.

SEM DESCRIPTION: This is a spherical, unornamented cyst (diameter 3.0-5.9 μm) with a deep, conical pore (diameter 0.6-0.7 μm) and multiple (up to four) cyst walls. Stomatocyst 46 is the smallest of three identical cysts, including stomatocyst 189 (6.0-8.9 μm) and stomatocyst 150 ($\geq$9.0 μm); the size classes have been based on natural groupings of our SEM specimens.

The cysts identified as stomatocyst 46 by Duff *et al.* (1992) have been reclassified as either stomatocyst 189 or stomatocyst 150.

LM DESCRIPTION: Stomatocyst 46 (Fig. 8B) is differentiated from stomatocysts 1 and 29 on the basis of pore morphology, and from stomatocyst 189 on the basis of size.

REFERENCES: This cyst has been observed in the Yukon Territory, Canada (Pienitz *et al.*, 1992), British Columbia,

Canada (Duff & Smol, 1994; Zeeb, 1994), New York, U.S.A. (Duff & Smol, 1991), and Greenland (Brown *et al.*, 1994).

ECOLOGY: This cyst is likely produced by a species tolerant of high (>10 mg/L) salinities (Pienitz *et al.*, 1992).

Stomatocyst 189, Zeeb & Smol in Zeeb *et al.*, subm. (13) Figs. 8C, D, and H

BIOLOGICAL AFFINITY: unknown

LOCALITY: Sonoma Volcanics, Napa County (38°N; 122°W), California, U.S.A. Pliocene (2-3 m.y.) sample

SEM DESCRIPTION: This stomatocyst is identical in appearance to stomatocyst 46, except that it is larger (diameter 6.0-8.9 μm; outer pore diameter 0.8-1.4 μm; inner pore diameter 0.5-0.8 μm; pore depth 0.2-0.5 μm).

LM DESCRIPTION: Stomatocyst 189 (Figs. 8D, H) is distinguished from stomatocysts 9 and 120 on the basis of pore morphology, and from stomatocysts 46 and 150 on the basis of size.

REFERENCES: This cyst has been found in British Columbia, Canada (Duff & Smol, 1994; Zeeb, 1994); the Northwest Territories, Canada (Duff *et al.*, 1992 - misidentified as stomatocyst 46); California, U.S.A. (Zeeb *et al.*, subm.); New York, U.S.A. (Duff & Smol, unpubl.); and Greenland (Brown *et al.*, 1994).

ECOLOGY: Stomatocyst 189 is strongly associated with oligotrophic, alkaline lakes; it may be produced by a cold-tolerant species (Duff *et al.*, 1992; Duff, 1994).

Stomatocyst 150, Zeeb & Smol, 1993a (6) Figs. 8E, F

BIOLOGICAL AFFINITY: unknown

LOCALITY: Elk Lake (47°11'N; 95°13'W), Clearwater County, Minnesota, U.S.A. Piston core, varve year A.D. 1617.

SEM DESCRIPTION: This stomatocyst is identical to stomatocysts 46 and 189, except that it is larger (diameter ≥9.0 μm; inner pore diameter 0.7-1.2 μm; outer pore diameter 1.5-1.8 μm; pore depth 0.5-0.8 μm).

LM DESCRIPTION: Stomatocyst 150 (Fig. 8E) is distinguished from stomatocysts 15 and 42 on the basis of pore morphology, and from stomatocyst 189 on the basis of size. Stomatocyst 205 is nearly identical, except that it is ornamented around the pore.

REFERENCES: This cyst has been observed in British Columbia, Canada (Duff & Smol, 1994; Zeeb, 1994); the Northwest Territories, Canada (Duff *et al.*, 1992 - misidentified as stomatocyst 46); Minnesota, U.S.A. (Zeeb & Smol, 1993a); Michigan, U.S.A. (Sandgren & Carney, 1983 - cyst 5); and New York, U.S.A. (Duff & Smol, unpubl.).

ECOLOGY: Stomatocyst 150 has been reported most often in alkaline lakes and ponds (Sandgren & Carney, 1983; Duff *et al.*, 1992; Zeeb & Smol, 1993a; Duff, 1994). In their paleolimnological study, Zeeb & Smol (1993b) suggested that stomatocyst 150 was associated with relatively cold and wet winters. It was common in small, oligotrophic lakes in British Columbia, Canada (Duff, 1994).

Stomatocyst 49, Duff & Smol, 1991 *emend.* Zeeb & Smol, 1993a (10) Fig. 9

BIOLOGICAL AFFINITY: Stomatocyst 49 closely resembles mature cysts of *Chrysosphaerella longispina* Lauterborn *emend.* Nicholls; less mature cysts of *C. longispina* appear to be identical to stomatocysts 120 or 42, depending on size (Sandgren, 1989).

LOCALITY: Upper Wallface Pond (44°08.8'N; 74°03.3'W), Adirondack Park, New York, U.S.A. Gravity core, core depth 0.0-0.5 cm.

SEM DESCRIPTION: This is a smooth, spherical to slightly oblate stomatocyst (diameter 6.8-10.2 μm). It has a concave pore (outer diameter 1.1-2.0 μm; inner diameter 0.4-0.8 μm) with a swollen pseudoannulus (Fig. 9A) that appears planar in apical view (Fig. 9B).

Stomatocyst 49 now includes most of the SEM specimens previously assigned to stomatocyst 48 (Chapter 5). The remainder of the stomatocyst 48 specimens have been reassigned to either stomatocyst 42 or stomatocyst 120, and the number (i.e. 48) has been discontinued.

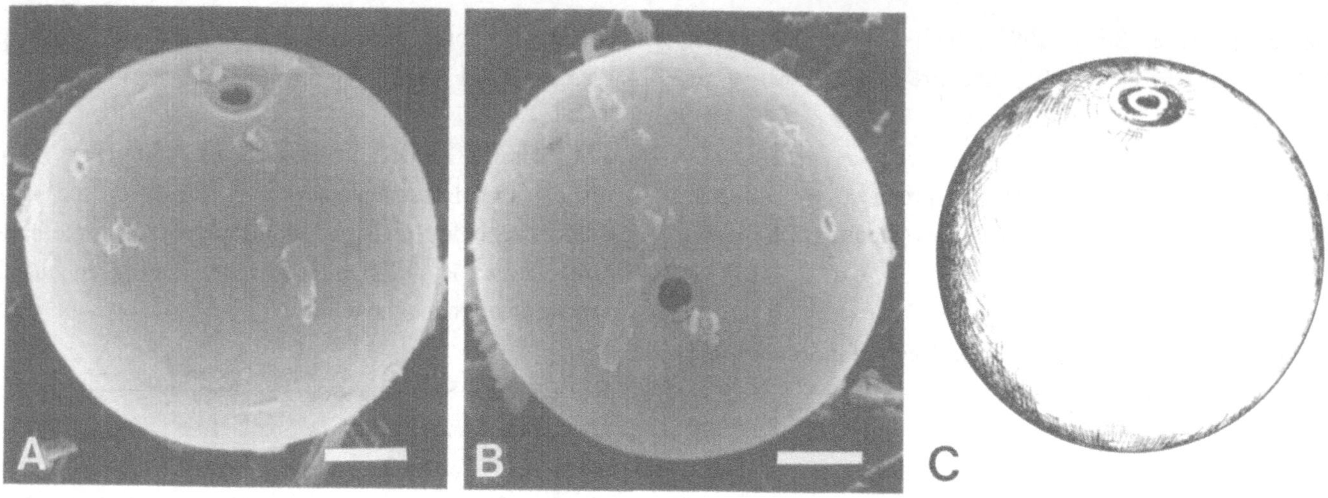

Fig. 9. Stomatocyst 49. A: SEM. B: SEM, apical view. C: Line drawing of SEM. Scale bars = 2 μm. See Figs. 7D, F, and H for LM.

LM DESCRIPTION: This stomatocyst cannot be distinguished from stomatocysts 42 or 120 with LM (Figs. 7 D, F, and H).

REFERENCES: Stomatocyst 49 has been observed in Ontario, Canada (Rybak *et al.*, 1991 - cyst 4), Minnesota, U.S.A. (Zeeb & Smol, 1993a), California, U.S.A. (Zeeb *et al.*, subm.), and New York, U.S.A. (Duff & Smol, 1991, unpubl.). The cyst identified by Rybak (1986 - Poland) as *cysta globata* Nygaard may include both stomatocysts 49 and 42; it is difficult to determine whether Nygaard's (1956) specimens are equivalent to either stomatocyst 49 or 42 without SEM confirmation.

ECOLOGY: *Chrysosphaerella longispina* is a widely distributed taxon, found primarily in acidic to circumneutral waters, with an estimated optimum pH of 5.4-6.3 (Dixit *et al.*, 1988, 1989a; Eloranta, 1989; Siver, 1989a; Siver & Hamer, 1990, 1992). It is most abundant during summer to early winter (Davis, 1972; Siver & Hamer, 1992). Stomatocyst 49 has been found in both acidic (Duff & Smol, 1991, unpubl.) and alkaline (Zeeb & Smol, 1993a) locations, and was classified as pH-indifferent by Rybak *et al.* (1991).

Oval stomatocysts lacking a collar or ornamentation

Stomatocyst 19, Duff & Smol, 1988 (8) Fig. 10

BIOLOGICAL AFFINITY: This stomatocyst appears to be similar to the cysts produced by several *Epipyxis* species, notably *Epipyxis tubulosa* (Mack) Hilliard & Asmund (Hilliard & Asmund, 1963).

LOCALITY: "Rock Basin Lake" (unofficial name) (78°29.5'N; 76°44.8'W), Ellesmere Island, Northwest Territories, Canada. Piston core, core depth 15-16 cm.

SEM DESCRIPTION: This is an oval stomatocyst (5.3-7.6 x 6.7-8.6 μm; width:length ratio 0.64-0.88) with a regular or shallow concave pore (diameter 0.6-0.8 μm) and a smooth to psilate surface.

LM DESCRIPTION: This stomatocyst is distinguished from similar cysts (e.g. stomatocyst 118) by its lack of collar. However, it is possible that immature specimens of stomatocyst 118 may be misidentified as stomatocyst 19.

REFERENCES: This stomatocyst has been found in the Northwest Territories, Canada (Duff & Smol, 1988; Duff *et al.*, 1992); the Yukon Territory, Canada (Brown & Smol, unpubl.); British Columbia, Canada (Duff & Smol, 1994; Zeeb, 1994); New York, U.S.A. (Duff & Smol, unpubl.); Connecticut, U.S.A. (Gritten, 1977 - *cysta decollata* Playfair); and Washington, U.S.A. (Smith & White, 1985 - Fig. 22).

ECOLOGY: This cyst has been found in several arctic (Duff & Smol, 1988; Duff *et al.*, 1992; Brown & Smol, unpubl.) and alpine (Duff & Smol, 1994, unpubl.; Smith & White, 1985) sites, suggesting that it is produced by a cold-tolerant taxon. It has been observed across a wide range of pH, and in both lakes and ponds.

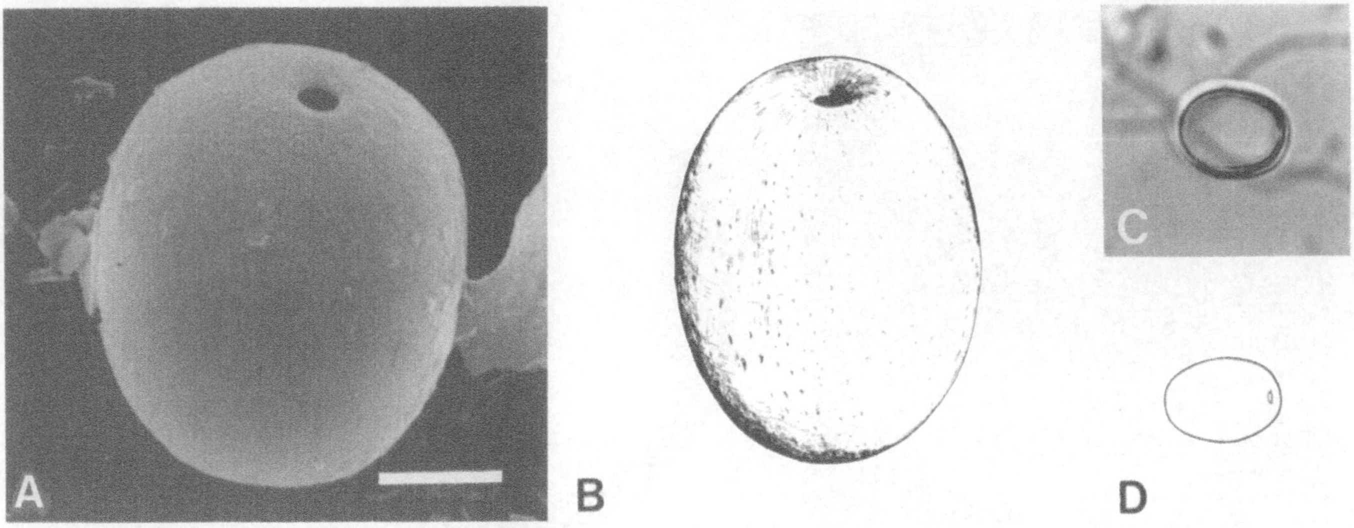

Fig. 10. Stomatocyst 19. A: SEM. B: Line drawing of SEM. C: LM. D: Line drawing of LM. Scale bars = 2 μm. LM x2000.

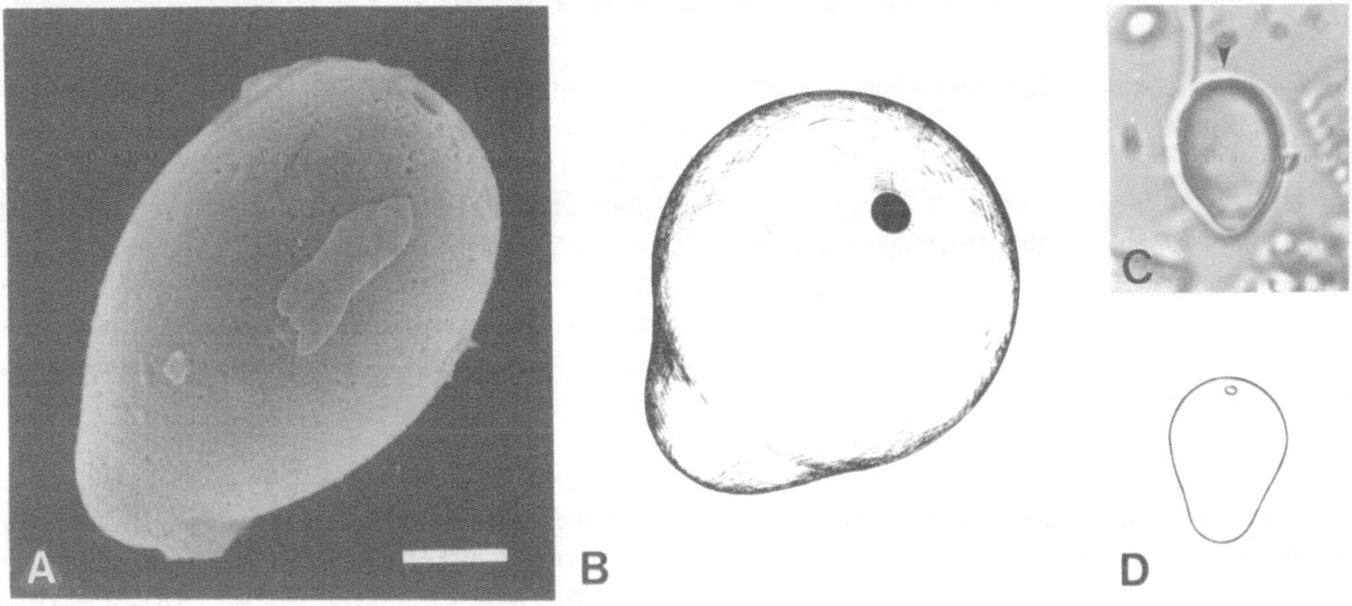

Fig. 11. Stomatocyst 121. A: SEM. B: Line drawing of SEM. C: LM. Arrow indicates location of pore. D: Line drawing of LM. Scale bars = 2 μm. LM x2000.

Stomatocyst 121, Duff & Smol in Duff *et al.*, 1992 (5) Fig. 11

BIOLOGICAL AFFINITY: unknown

LOCALITY: "Elison Pond" (unofficial name), Cape Herschel (78°37'N; 74°42'W), Ellesmere Island, Northwest Territories, Canada. Epilithon, collected August 3, 1986.

SEM DESCRIPTION: This stomatocyst is oval to obovate with a rounded extension of the posterior pole (5.7-7.6 x 6.9-10.8 μm) and a smooth surface. The pore is regular (diameter 0.6-0.8 μm).

LM DESCRIPTION: Stomatocyst 121 can be distinguished from similar cysts (e.g. stomatocyst 19) by the posterior extension; careful focussing is required to identify the cyst in apical view.

REFERENCES: We have observed this stomatocyst in the Northwest Territories, Canada (Duff *et al.*, 1992) and British Columbia, Canada (Zeeb, 1994). Rull (1986 - Spain) *forma* 61 appears to be identical.

ECOLOGY: This cyst has been observed in small, shallow, alkaline, alpine, or arctic ponds (Duff *et al.*, 1992; Rull, 1986), and may be produced by a littoral (possibly periphytic) chrysophyte. It was particularly common in an arctic pond

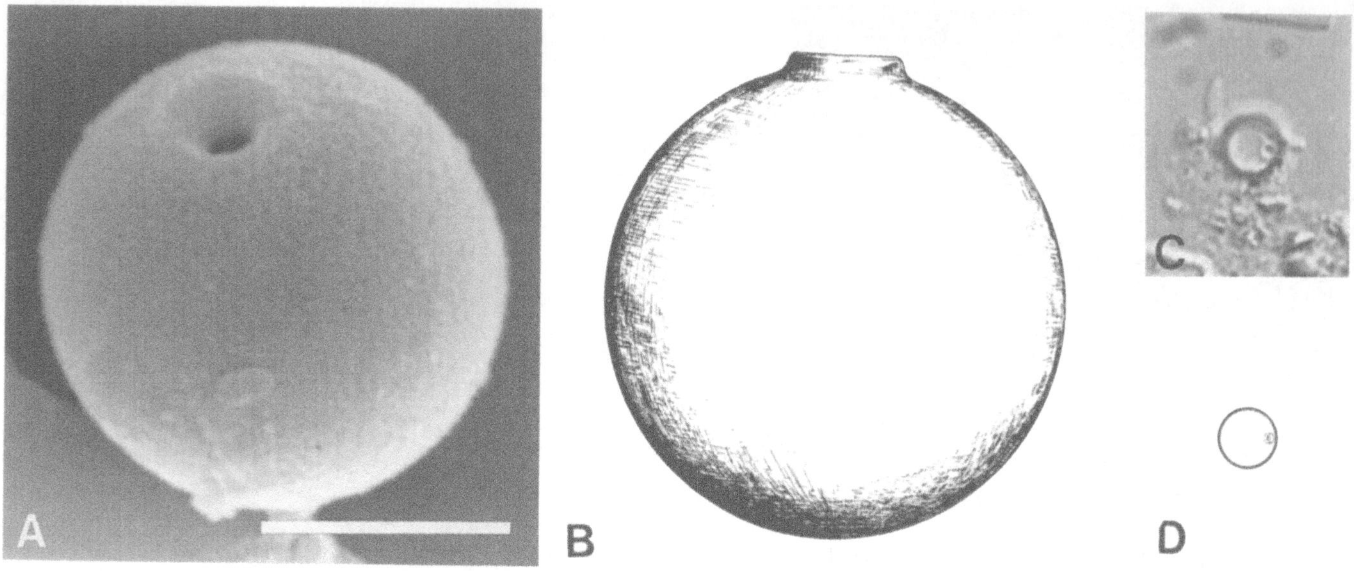

Fig. 12. Stomatocyst 110. A: SEM. B: Line drawing of SEM. C: LM. D: Line drawing of LM. Scale bar = 2 μm. LM x2000.

that received modest nutrient enrichment from nesting birds (Duff *et al.*, 1992).

Spherical or oval stomatocysts with a simple conical collar but no ornamentation

Stomatocyst 110, Zeeb *et al.* 1990 (18) Fig. 12

BIOLOGICAL AFFINITY: The range of collar morphology indicates that stomatocyst 110 may be produced by more than one chrysophycean species.

LOCALITY: Little Round Lake (44°47.5'N; 76°41.5'W), Ontario, Canada. One metre piston core, core depth 20 cm.

SEM DESCRIPTION: This stomatocyst is very small and spherical (diameter 2.6-4.8 μm, always <5.0 μm). The surface is smooth to psilate. The collar is conical to obconical with a gradual basal margin (collar diameter 0.7-1.4 μm; height 0.2-0.4 μm; collar diameter : cyst diameter ratio 0.21-0.34). The pore is regular (diameter 0.3-0.5 μm). The inner collar margin is abrupt to sloping and there is no annulus, although some immature specimens may appear to possess one.

LM DESCRIPTION: With LM, stomatocyst 110 cannot be distinguished reliably from stomatocysts 50, 52, and small (<5.0 μm) specimens of stomatocyst 51.

REFERENCES: This stomatocyst type is widespread. It has been reported from Ontario, Canada (Zeeb *et al.*, 1990), British Columbia, Canada (Duff & Smol, 1994; Zeeb, 1994), the Northwest Territories, Canada (Duff *et al.*, 1992; Webb & Smol, unpubl.), and New York, U.S.A. (Duff & Smol, unpubl.). Rybak (1987) cyst 1 (Poland) is probably identical. Small (<5.0 μm) specimens of Sandgren & Carney (1983) cyst 3 (Michigan, U.S.A.) would be classified as stomatocyst 110.

ECOLOGY: Limited paleolimnological evidence suggests that this stomatocyst is an indicator of oligotrophy (Rybak, 1987; Zeeb *et al.*, 1990), although it was more common in productive lakes in Duff's (1994) survey. It displayed a clear affinity for acidic, low conductivity lakes (Duff, 1994). It is produced by a cold-tolerant taxon (Duff *et al.*, 1992; Webb & Smol, unpubl.).

Stomatocyst 51, Duff & Smol, 1991 *emend.* Duff *et al.* (this volume) (2) Fig. 13

BIOLOGICAL AFFINITY: unknown

NEGATIVE NUMBER: J.P. Smol 3001

LOCALITY: "Horseshoe Pond" (unofficial name), Cape Herschel (78°37'N; 74°42'W), Ellesmere Island, Northwest Territories, Canada. Moss sample, collected July 19, 1986.

SEM DESCRIPTION: Stomatocyst 51 was originally

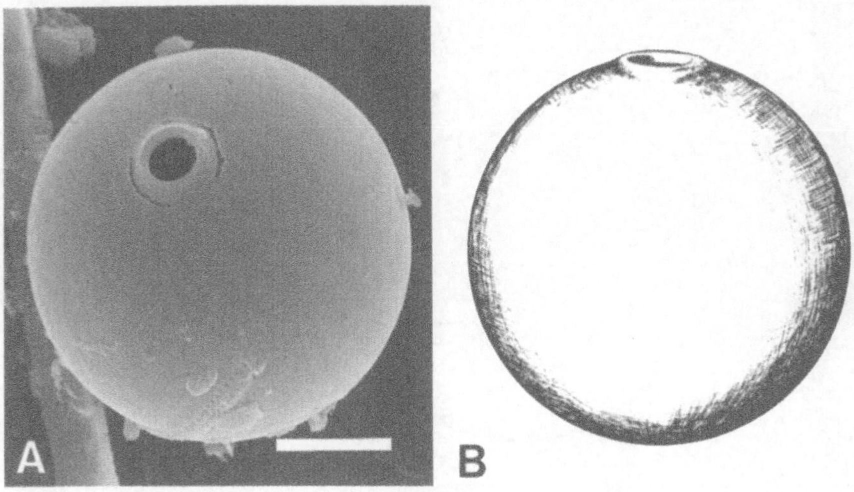

Fig. 13. Stomatocyst 51. A: SEM. B: Line drawing of SEM. Scale bar = 2 μm. See Figs. 12C, D and 14C, D for LM.

described by Duff & Smol (1991) based on one SEM specimen only; here, we present an emended description based on further specimens. A new characteristic specimen is included because the original characteristic specimen appeared to have low ridges; these were not observed on our other SEM specimens, and may have been a developmental artefact of that particular specimen.

This is a spherical stomatocyst (diameter 4.3-6.6 μm) with a smooth surface. The relatively large, regular pore (diameter 0.6-0.8 μm) is surrounded by a low conical collar with a rounded outer margin and a more acute inner margin (diameter 1.0-1.5 μm; collar diameter : cyst diameter ratio 0.14-0.19); there is no annulus.

This stomatocyst is distinguished from stomatocyst 188 with SEM because the width of the collar is less than the diameter of the pore.

LM DESCRIPTION: With LM, small (<5.0 μm) specimens cannot be distinguished from stomatocysts 50, 52, and 110 (Figs. 12C, D); large (>5.0 μm) specimens cannot be distinguished from stomatocysts 123 (Figs. 14C, D) and 188.

REFERENCES: Stomatocyst 51 has been found in the Northwest Territories, Canada (Duff *et al.*, 1992 - misidentified as stomatocyst 124), Ontario, Canada (Zeeb & Smol, unpubl.), and New York, U.S.A. (Duff & Smol, 1991).

ECOLOGY: Autecological details are not yet available.

Stomatocyst 123, Duff & Smol in Duff *et al.*, 1992 (3) Fig. 14

BIOLOGICAL AFFINITY: Unknown, although *Synura curtispina* (Petersen *et* Hansen) Asmund produces a similar stomatocyst (Sandgren, 1989).

LOCALITY: "Col Pond" (unofficial name), Cape Herschel (78°37'N; 74°42'W), Ellesmere Island, Northwest Territories, Canada. Moss sample, collected July 29, 1986.

SEM DESCRIPTION: This stomatocyst is spherical (diameter 6.4-8.5 μm) with a smooth and apparently thin single wall. The regular pore (diameter 0.6-0.8 μm) has a small conical collar with an acute apex and a sloping inner margin (diameter 0.9-1.7 μm; collar diameter : cyst diameter ratio 0.14-0.20).

LM DESCRIPTION: With LM, this stomatocyst cannot be distinguished from stomatocyst 188 or large (>5.0 μm) specimens of stomatocyst 51.

REFERENCES: We have observed this cyst in the Northwest Territories, Canada (Duff *et al.*, 1992), British Columbia, Canada (Duff & Smol, 1994; Zeeb, 1994), and Greenland (Brown *et al.*, 1994).

ECOLOGY: This stomatocyst was particularly common in an arctic pond with extremely high conductivity values due to periodic marine influxes (Duff *et al.*, 1992). In a survey of temperate lakes, however, its distribution was uncorrelated with conductivity (Duff, 1994).

Stomatocyst 188, Brown & Smol in Brown *et al.*, 1994 (6) Fig. 15

BIOLOGICAL AFFINITY: unknown

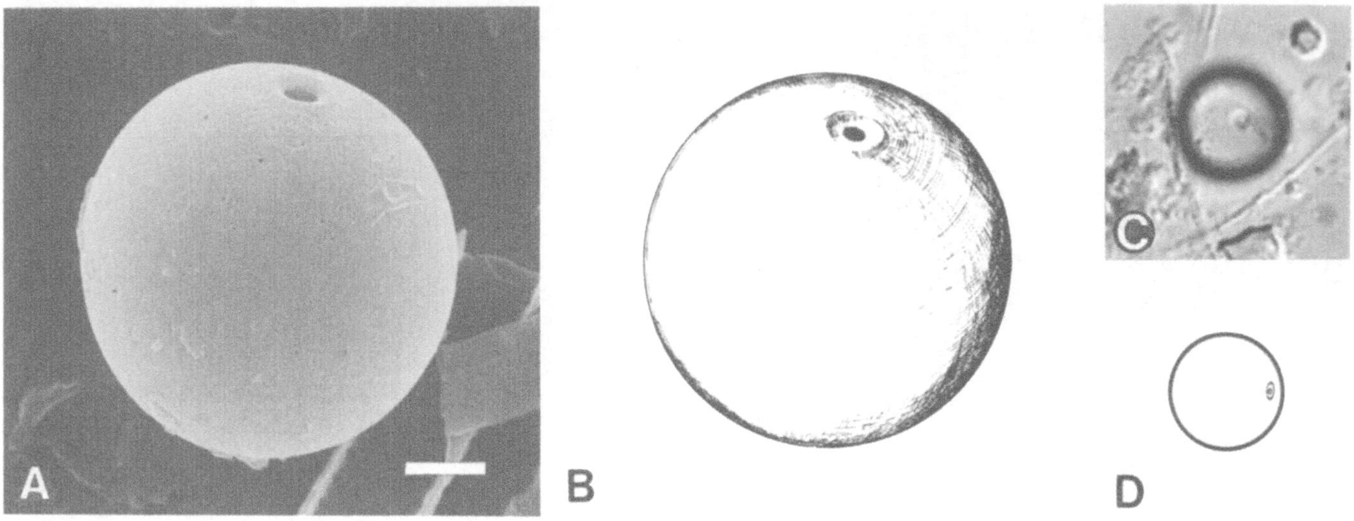

Fig. 14. Stomatocyst 123. A: SEM. B: Line drawing of SEM. C: LM. D: Line drawing of LM. Scale bar = 2 μm. LM x2000.

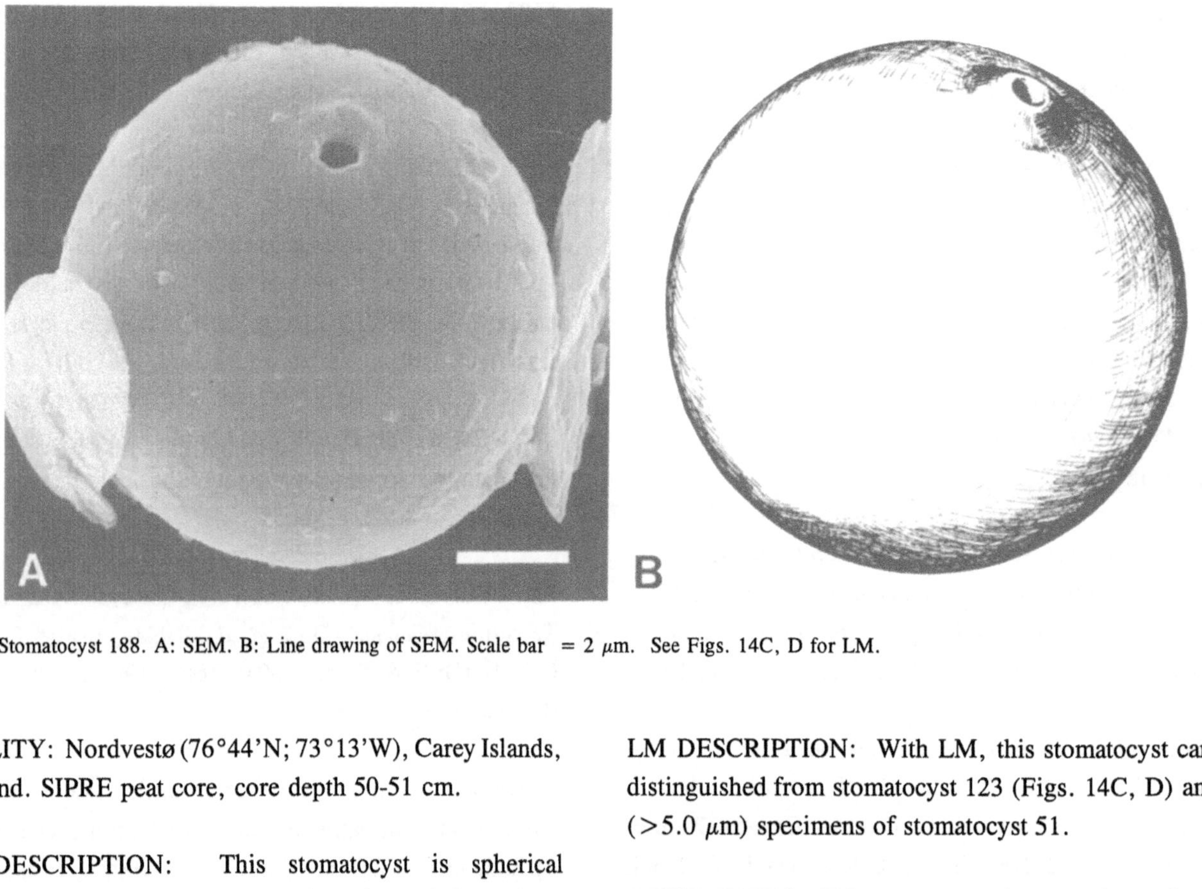

Fig. 15. Stomatocyst 188. A: SEM. B: Line drawing of SEM. Scale bar = 2 μm. See Figs. 14C, D for LM.

LOCALITY: Nordvestø (76°44'N; 73°13'W), Carey Islands, Greenland. SIPRE peat core, core depth 50-51 cm.

SEM DESCRIPTION: This stomatocyst is spherical (diameter 6.2-9.2 μm) with a smooth surface. It has a low, broadly rounded conical collar (basal diameter 1.5-2.4 μm; collar diameter : cyst diameter ratio 0.18-0.29) and a regular pore (diameter 0.6-0.9 μm). With SEM, this stomatocyst is distinguished from stomatocyst 51 by its collar width, which is greater than the pore diameter.

Several specimens identified by Duff *et al.* (1992) as stomatocyst 123 have been reclassified as stomatocyst 188.

LM DESCRIPTION: With LM, this stomatocyst cannot be distinguished from stomatocyst 123 (Figs. 14C, D) and large (>5.0 μm) specimens of stomatocyst 51.

REFERENCES: This stomatocyst has only been observed in Greenland (Brown *et al.*, 1994), and the Northwest Territories, Canada (Duff *et al.*, 1992 - misidentified as stomatocyst 123).

ECOLOGY: Ecological details for this cyst are as yet unavailable.

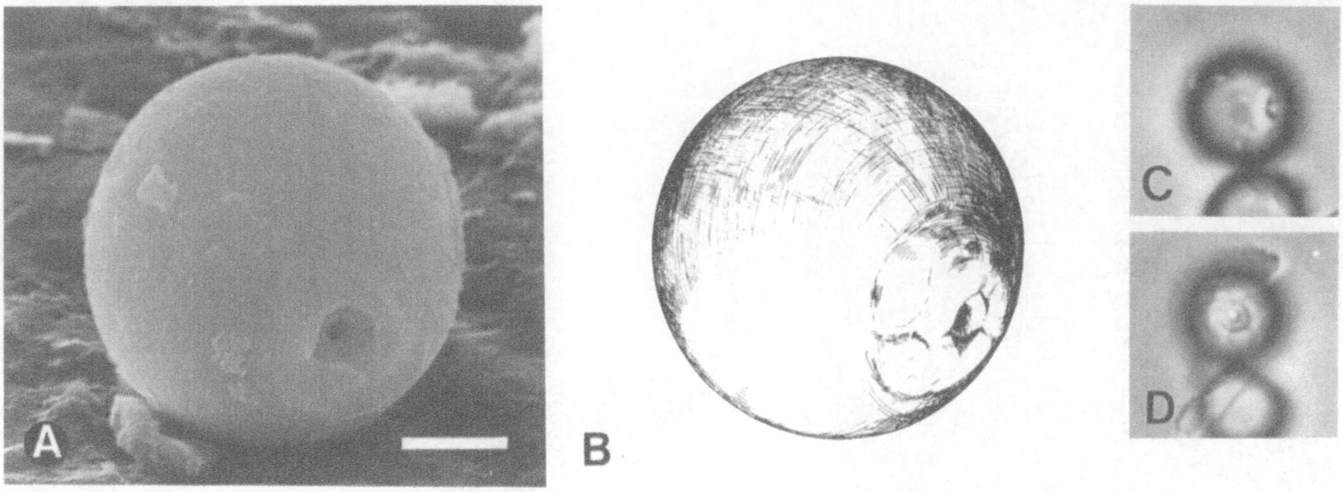

Fig. 16. Stomatocyst 181. A: SEM. B: Line drawing of SEM. C: LM. D: LM, apical view. E: Line drawing of LM. Scale bar = 2 μm. LM x2000.

Stomatocyst 181, Brown & Smol in Brown *et al.*, 1994 (5) Fig. 16

BIOLOGICAL AFFINITY: unknown

LOCALITY: Nordvestø (76°44'N; 73°13'W), Carey Islands, Greenland. SIPRE peat core, core depth 121-122 cm.

SEM DESCRIPTION: This stomatocyst is spherical and smooth surfaced (cyst diameter 6.1-7.5 μm). The collar is conical, with a broadly rounded apex (collar base diameter 2.1-3.0 μm; apical diameter 0.7-1.7 μm; collar height 0.7-0.8 μm). The outer collar margin is gradual, but the inner collar margin is steeply sloping. The outer collar base may be irregular in outline. The pore is regular (diameter 0.4-0.5 μm).

LM DESCRIPTION: This cyst can be distinguished from stomatocysts 146/156 (Figs. 17C, D) by the irregular collar outline.

REFERENCES: This stomatocyst has been observed from British Columbia, Canada (Zeeb, 1994), California, U.S.A. (Zeeb *et al.*, subm.), and Greenland (Brown *et al.*, 1994). Duff & Smol (unpubl.) have observed one SEM specimen of a similar, but much larger, cyst (diameter 13.5 μm) from the Northwest Territories, Canada.

ECOLOGY: The ecological distribution of this cyst is undetermined.

Stomatocyst 156, Zeeb & Smol, 1993a (6) Fig. 17

BIOLOGICAL AFFINITY: unknown

LOCALITY: Elk Lake (47°11'N; 95°13'W), Clearwater County, Minnesota, U.S.A. Piston core, varve year A.D. 1851.

SEM DESCRIPTION: This stomatocyst is spherical (diameter 7.0-10.0 μm) and smooth surfaced. There is a short, conical collar with a rounded outer margin, an abrupt vertical inner margin, and a swollen annulus surrounding the pore (collar diameter 3.0-4.1 μm; height 0.2-0.7 μm; annulus diameter 2.1-2.5 μm; pore diameter 0.5-0.8 μm).

LM DESCRIPTION: Stomatocyst 156 cannot be distinguished from stomatocyst 146 with LM.

REFERENCES: Stomatocyst 156 has been observed in Ontario, Canada (Zeeb & Smol, unpubl.); British Columbia, Canada (Duff & Smol, 1994; Zeeb, 1994); Minnesota, U.S.A. (Zeeb & Smol, 1993a); New York, U.S.A. (Duff & Smol, unpubl.); and Michigan, U.S.A. (Sandgren & Carney, 1983 - cyst 10).

ECOLOGY: Autecological data are not yet available.

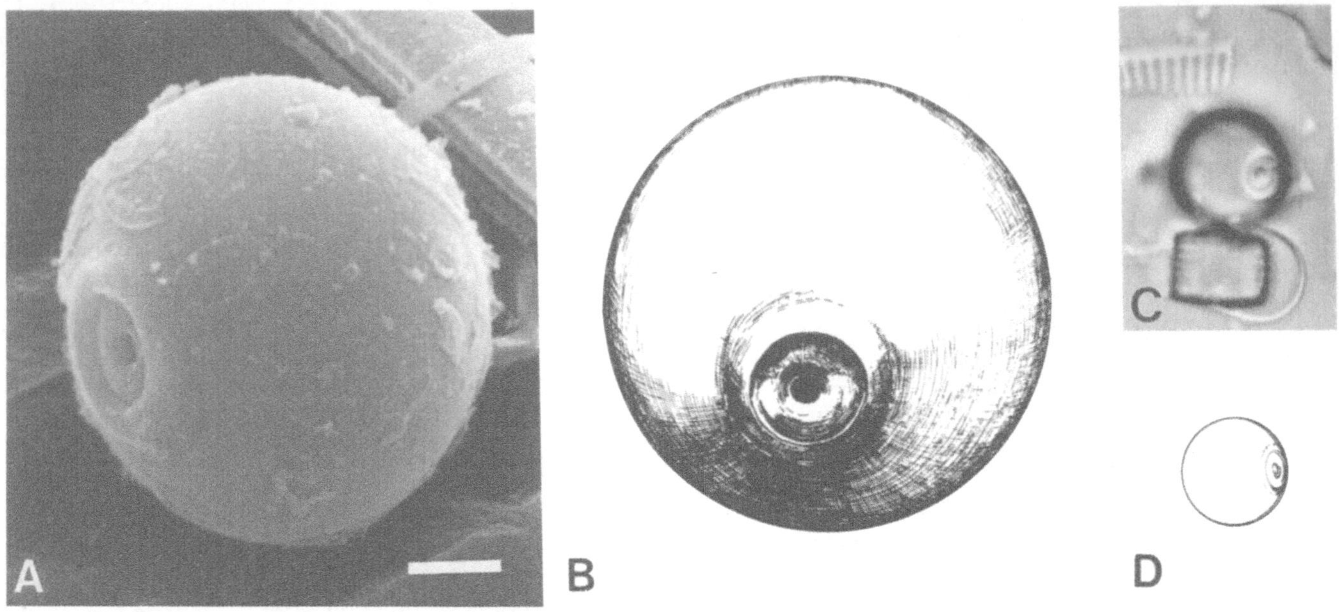

Fig. 17. Stomatocyst 156. A: SEM. B: Line drawing of SEM. C: LM. D: Line drawing of LM. Scale bar = 2 μm. LM x2000.

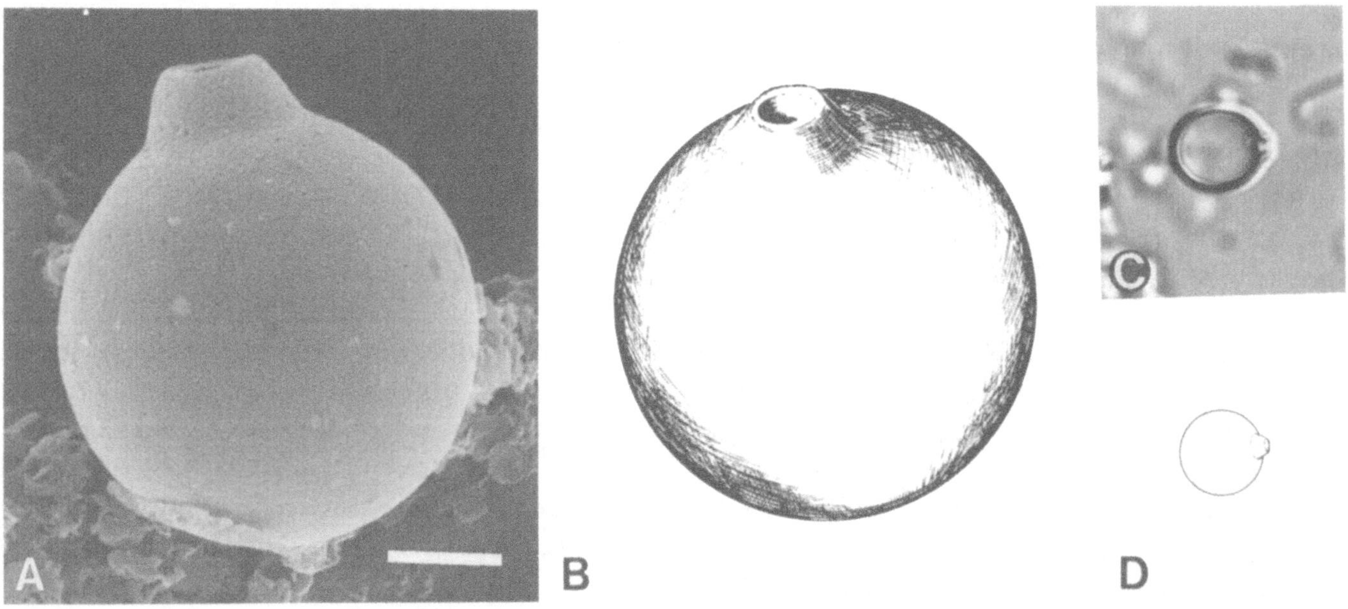

Fig. 18. Stomatocyst 126. A: SEM. B: Line drawing of SEM. C: LM. D: Line drawing of LM. Scale bar = 2 μm. LM x2000.

Stomatocyst 126, Duff & Smol in Duff *et al.*, 1992 (3) Fig. 18

BIOLOGICAL AFFINITY: unknown

LOCALITY: "Beach Ridge Pond" (unofficial name), Cape Herschel (78°37'N; 74°42'W), Ellesmere Island, Northwest Territories, Canada. Surface sediment sample, collected July 19, 1986.

SEM DESCRIPTION: This stomatocyst is spherical to widely oval (diameter 5.4-6.9 μm) with a smooth surface. There is a wide, conical collar with an abrupt delineation between the outer collar base and the cyst body (basal diameter 1.7-2.9 μm; apical diameter 1.5-1.7 μm; height 0.9-1.1 μm). The apex is fairly thick, and the inner margin is straight. A very narrow, sloping to flat planar annulus is located at a variable height within the collar; the annulus was never observed with LM.

Most of the original specimens of stomatocyst 126 have now been reclassified as stomatocyst 199.

LM DESCRIPTION: This cyst can be distinguished by the collar shape. Identification is more difficult in apical view.

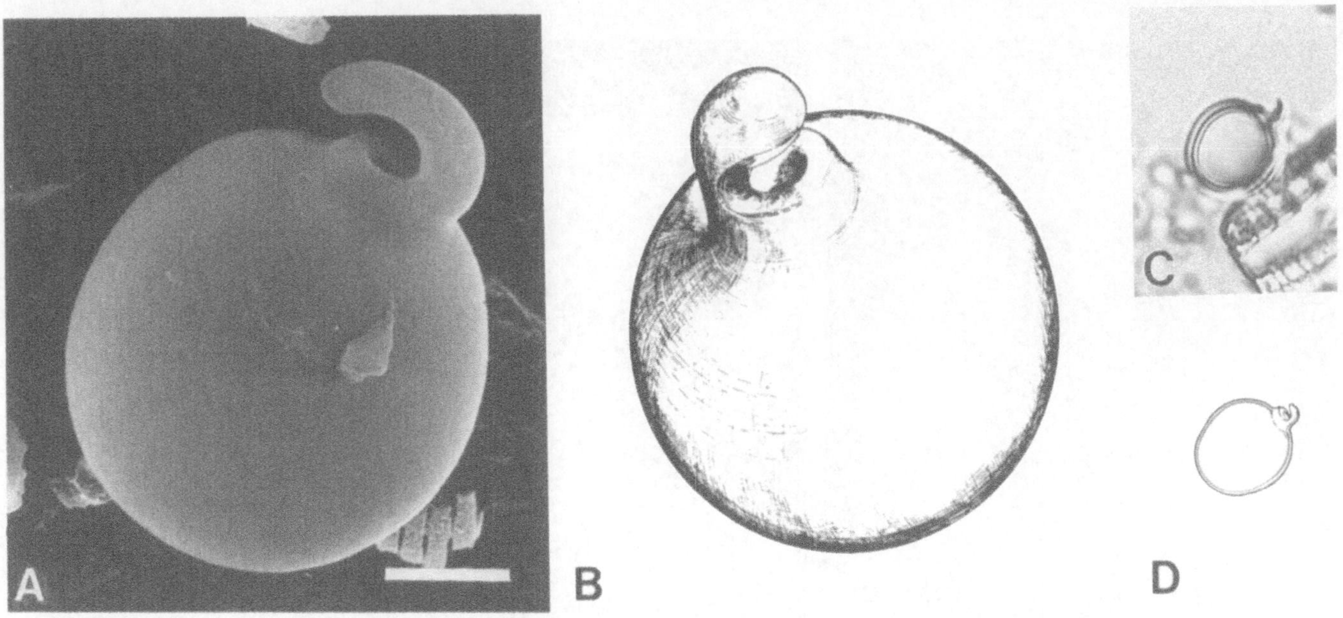

Fig. 19. Stomatocyst 134. A: SEM. B: Line drawing of SEM. C: LM. D: Line drawing of LM. Scale bar = 2 μm. LM x2000.

REFERENCES: We have observed stomatocyst 126 in the Northwest Territories, Canada (Duff *et al.*, 1992), and British Columbia, Canada (Zeeb, 1994).

ECOLOGY: Autecological data are not yet available.

Stomatocyst 134, Duff & Smol in Duff *et al.*, 1992 (7) Fig. 19

BIOLOGICAL AFFINITY: unknown

LOCALITY: "Camp Pond" (unofficial name), Cape Herschel (78°37'N; 74°42'W), Ellesmere Island, Northwest Territories, Canada. Net tow, collected July 28, 1986.

SEM DESCRIPTION: This is a spherical to oval or sometimes obovate stomatocyst (diameter 5.0-8.0 μm). The collar has a low, conical to cylindrical base with a thin, curved projection that hooks over the pore (collar diameter 1.2-2.5 μm; height of collar base 0.5-0.9 μm; projection height to 3.8 μm).

LM DESCRIPTION: This cyst is identified by the collar shape and projection. It may be confused with immature specimens of stomatocysts 135 or 187, if the outer collars of these cysts are undeveloped.

REFERENCES: This stomatocyst has been found in the Northwest Territories, Canada (Duff *et al.*, 1992), British Columbia, Canada (Zeeb, 1994), Spain (Rull, 1986 - *forma* 38), and the Kerguelen Islands, Southern Indian Ocean (Andrieu, 1936 - *Carnegia kriegeri* Andrieu).

ECOLOGY: Stomatocyst 134 may be produced by a littoral species that is tolerant of cold water (Rull, 1986; Duff *et al.*, 1992). It was recovered from a moss sample by Andrieu (1936).

Stomatocyst 41, Duff & Smol, 1989 (53) Fig. 20

BIOLOGICAL AFFINITY: *Dinobryon cylindricum* Imhof (Donaldson & Stein, 1984). We have not observed ornamented forms such as those reported by Sandgren (1980b, 1981, 1983a, b, 1989).

LOCALITY: Tasikutaaq Lake (66°23'N; 65°57'W), Baffin Island, Northwest Territories, Canada. Percussion core, core depth 75 cm.

SEM DESCRIPTION: This is a smooth, spherical to obovate stomatocyst (diameter 9.5-12.6 μm; average dimensions 10.8 x 11.5 μm). The collar is widely conical with a hooked apex, which may be radially striated when mature (collar diameter 2.4-4.5 μm; length to 9.6 μm; height to flexure 2.0-3.5 μm). The collar appears to develop by the accretion of siliceous globules at the apex, and immature specimens may appear rough in texture (Fig. 20B).

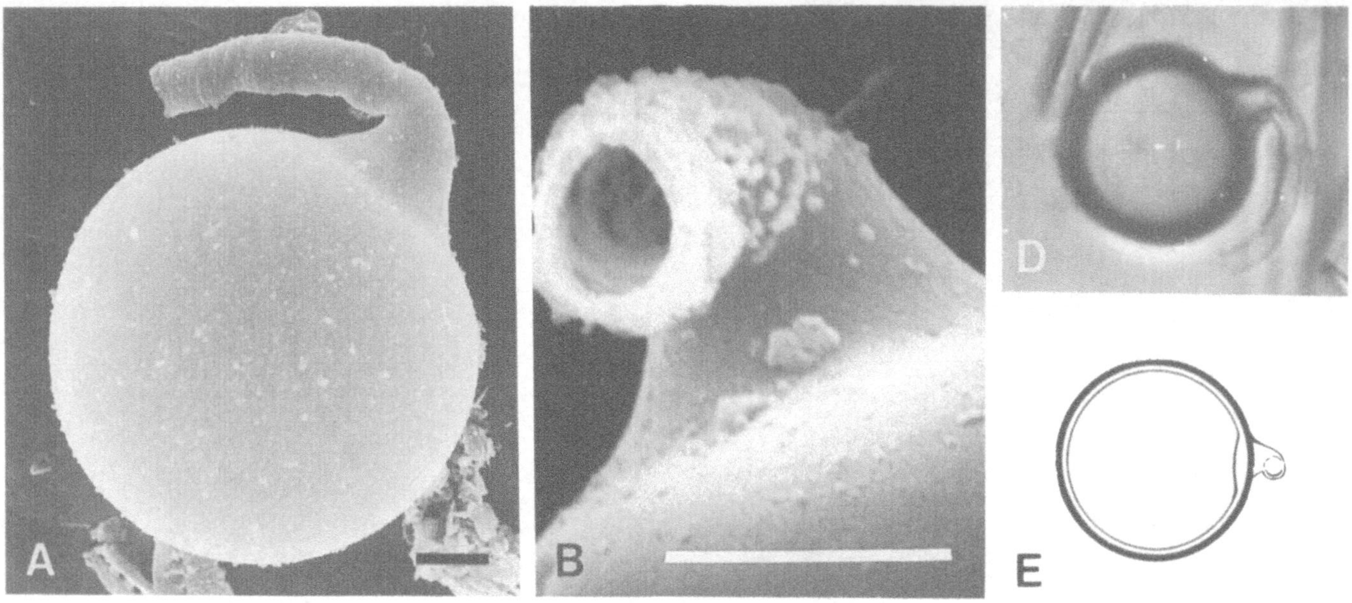

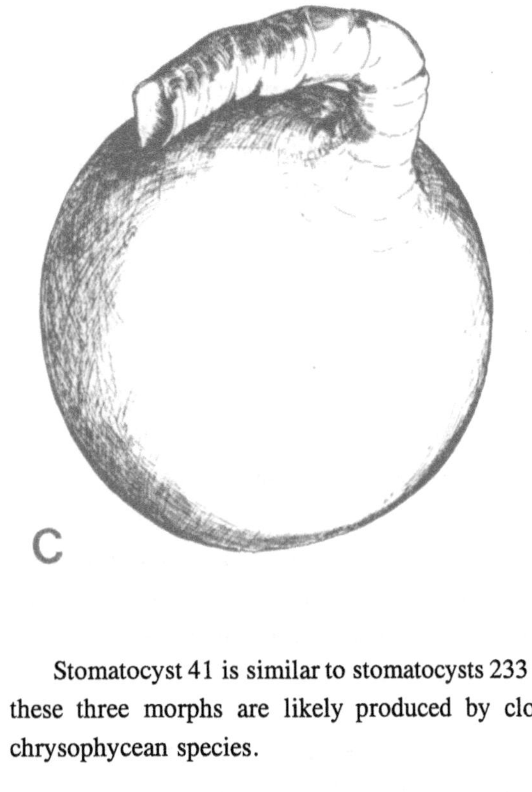

Fig. 20. Stomatocyst 41. A: SEM. B: SEM, close-up of collar. C: Line drawing of SEM. D: LM. E: Line drawing of LM. Scale bars = 2 μm. LM x2000.

Stomatocyst 41 is similar to stomatocysts 233 and 78, and these three morphs are likely produced by closely related chrysophycean species.

LM DESCRIPTION: This stomatocyst is easily recognized when mature by its hooked collar. If the collar is undeveloped or broken, the cyst can be recognized by its shape (often obovate), multiple cyst walls, smooth surface and wide collar base. It is distinguished from stomatocyst 233 (Figs. 54C, D) by its smooth surface, and from stomatocyst 78 (Figs. 62D-F) by its smooth surface and absence of a bipartite collar.

REFERENCES: *Dinobryon cylindricum* is an extremely common, cosmopolitan species, and not surprisingly the cyst has been observed in many studies. It has been found in Ontario, Canada (Rybak *et al.*, 1987 - cyst 3 as depicted in Fig. 5 but not Fig. 6; Zeeb *et al.*, 1990 - unidentified stomatocyst 2 (Chapter 5); Rybak *et al.*, 1991 - cyst 27); British Columbia, Canada (Donaldson & Stein, 1984; Duff & Smol, 1994; Zeeb, 1994); the Northwest Territories, Canada (Duff & Smol, 1989); Minnesota, U.S.A. (Zeeb & Smol, 1993a); Connecticut, U.S.A. (Gritten, 1977 - *cysta curvicollis* Nygaard); Michigan, U.S.A. (Carney & Sandgren, 1983 - *cysta curvicollis*; Sandgren & Carney, 1983 - cyst 9); California, U.S.A. (Adam & Mahood, 1979b, 1980b - type 56; Carney *et al.*, 1992 - cyst 15); New York, U.S.A. (Smol, 1988a - Plate 1C); and Denmark (Nygaard, 1956 - *cysta curvicollis*).

ECOLOGY: *Dinobryon cylindricum* apparently prefers cool oligotrophic or mildly eutrophic waters (Rybak, 1986; Rybak *et al.*, 1987; Eloranta, 1989). The cyst has been found in arctic (Duff & Smol, 1989), alpine (Carney *et al.*, 1992; Duff & Smol, 1994), and temperate sites (Zeeb *et al.*, 1990; Carney & Sandgren, 1983; Rybak *et al.*, 1987, 1991; Duff & Smol, 1994), and has been classified as alkaliphilic (Rybak *et al.*, 1991; Carney *et al.*, 1992; Duff, 1994). Carney & Sandgren (1983) found the cyst almost exclusively in pre-European settlement sediments in a Michigan lake. However, Zeeb *et al.* (1990) observed it during a post-settlement and more eutrophic time period in Little Round Lake, Ontario.

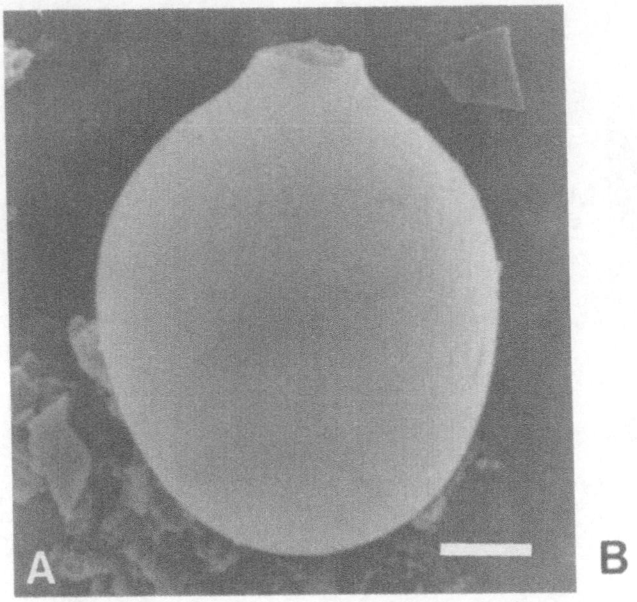

Fig. 21. Stomatocyst 199. A: SEM. B: Line drawing of SEM. Scale bar = 2 μm.

Stomatocyst 199, Duff & Smol, 1994 (7) Fig. 21

BIOLOGICAL AFFINITY: unknown

LOCALITY: Wilgress Lake (49°10'N; 118°40'W), British Columbia, Canada. Ekman dredge, surface (1 cm) sample.

SEM DESCRIPTION: This stomatocyst is oval to almost spherical (6.8-10.3 x 7.7-12.2 μm; width:length ratio <0.90). The surface is smooth. The collar is conical (apical diameter 1.5-2.6 μm; height ca. 0.7-1.1 μm) with a very gradual outer margin and no clear delineation between the cyst body and the outer basal collar margin. The inner collar details are unknown.

With SEM, this stomatocyst can be distinguished from similar stomatocysts (i.e. stomatocysts 126 and 154) by the gradual basal collar margin, and by the conical, not cylindrical, collar shape.

LM DESCRIPTION: Stomatocyst 199 has not yet been identified with LM. It may be indistinguishable from stomatocysts 126 (Figs. 18C, D) and 154 (Figs. 41C, D).

REFERENCES: We have observed stomatocyst 199 in British Columbia, Canada (Duff & Smol, 1994), the Northwest Territories, Canada (Duff *et al.*, 1992 - misclassified as stomatocyst 126), and New York, U.S.A. (Duff & Smol, unpubl.).

ECOLOGY: No autecological details are yet available.

Stomatocyst 155, Zeeb & Smol, 1993a (4) Fig. 22

BIOLOGICAL AFFINITY: unknown

LOCALITY: Elk Lake (47°11'N; 95°13'W), Clearwater County, Minnesota, U.S.A. Piston core, varve year 4284 B.P.

SEM DESCRIPTION: This is a large, unornamented oval stomatocyst (16.4-18.6 x 20.7-23.2 μm) with a low conical collar (collar diameter 1.8-2.4 μm; height <1.0 μm; pore diameter ca. 1.3 μm).

LM DESCRIPTION: With LM, this stomatocyst is easily distinguished by its large size and oval shape.

REFERENCES: This stomatocyst has only been observed in British Columbia, Canada (Duff & Smol, 1994; Zeeb, 1994), and Minnesota, U.S.A. (Zeeb & Smol, 1993a).

ECOLOGY: The only reports of this rare stomatocyst are from alkaline lakes (Zeeb & Smol, 1993a; Duff & Smol, 1994).

Stomatocyst 202, Duff & Smol, 1994 (7) Fig. 23

BIOLOGICAL AFFINITY: Both *Mallomonas hamata* Asmund and *M. heterospina* Lund produce cysts of this type (Cronberg, 1989b).

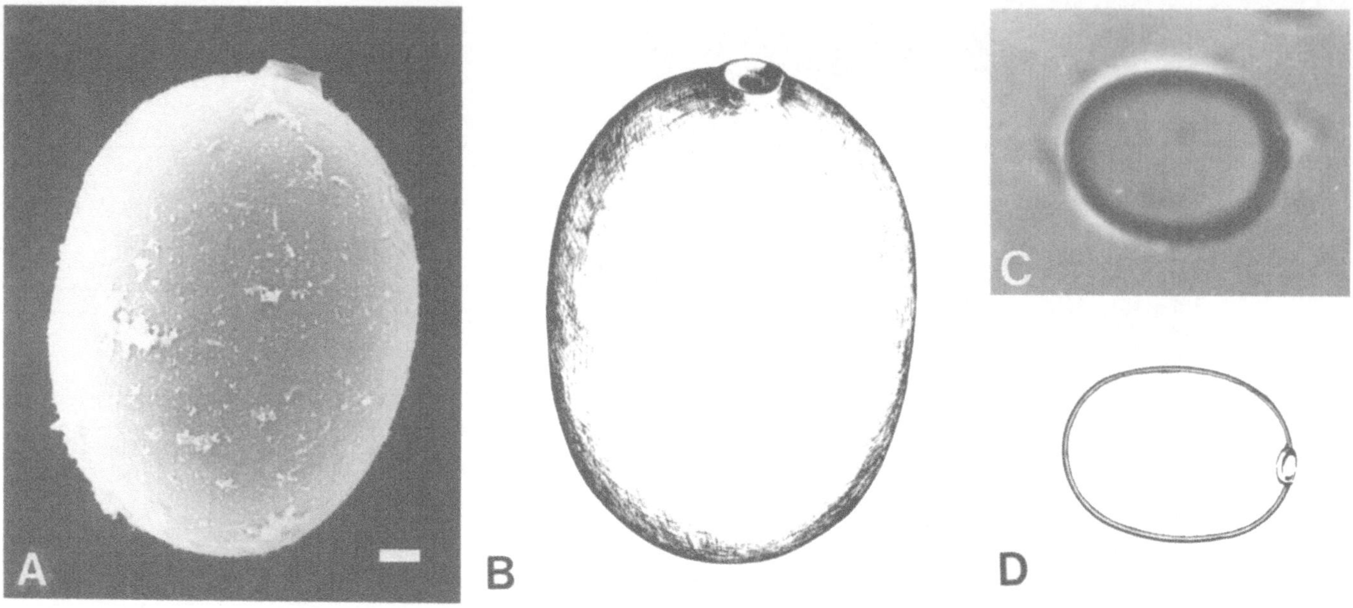

Fig. 22. Stomatocyst 155. A: SEM. B: Line drawing of SEM. C: LM. D: Line drawing of LM. Scale bar = 2 μm. LM x1500.

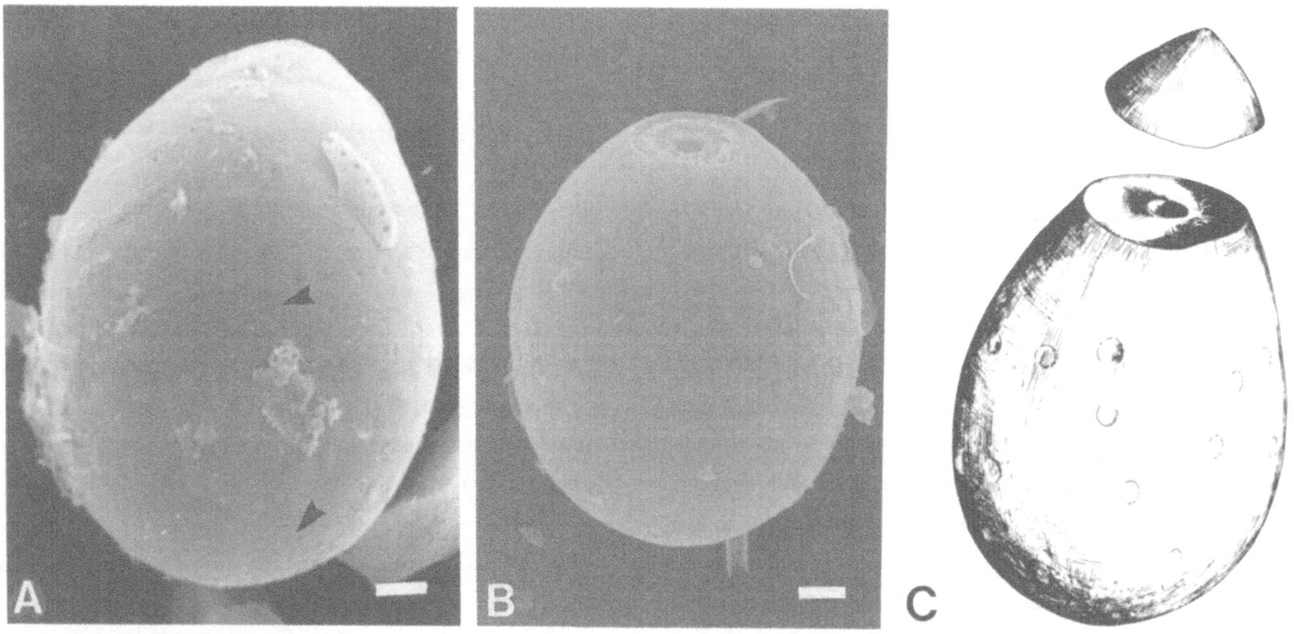

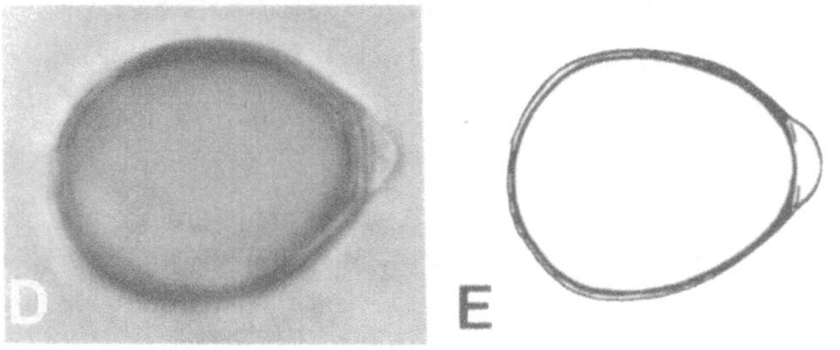

Fig. 23. Stomatocyst 202. A: SEM; pore covered by cap. Depressions indicated by arrows. B: SEM; pore visible. C: Line drawing of SEM. D: LM. E: Line drawing of LM. Scale bars = 2 μm. LM x2000.

LOCALITY: Florence Lake (48°30'N; 123°30'W), British Columbia, Canada. Ekman dredge, surface (1 cm) sediment sample.

SEM DESCRIPTION: This is an oval to obovate stomatocyst (13.5-16.7 x 15.4-21.9 μm). The pore is regular (diameter ca. 0.9 μm), and is surrounded by a flat planar annulus and a slight conical collar (diameter 4.5-7.7 μm); the outer collar margin is indistinguishable from the cyst body. The anterior pole is usually covered with a siliceous conical cap (base diameter 5.3-7.9 μm; height 1.8-2.3 μm; Fig. 23A). The cyst is generally smooth, but may have small circular depressions (diameter 0.6-0.9 μm) scattered on the surface (Figs. 23A, C).

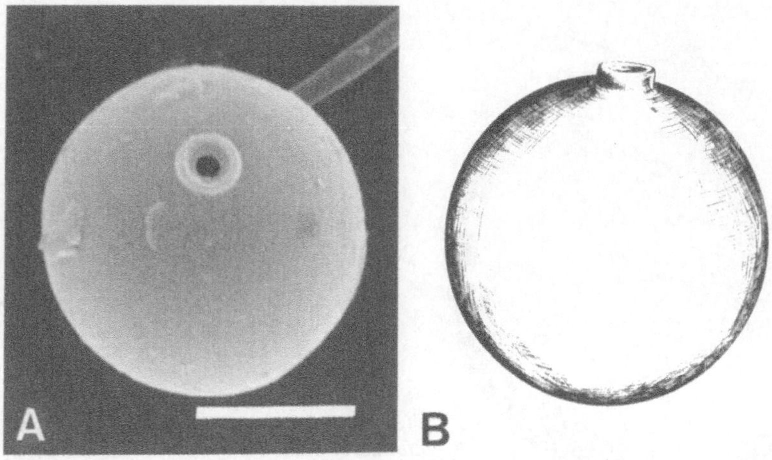

Fig. 24. Stomatocyst 50. A: SEM. B: Line drawing of SEM. Scale bar = 2 μm. See Figs. 12C, D for LM.

LM DESCRIPTION: This stomatocyst can be recognized by its shape, and by the anterior cap, if present.

REFERENCES: This cyst has been observed in British Columbia, Canada (Duff & Smol, 1994), Ontario, Canada (Zeeb & Smol, unpubl.), New York, U.S.A. (Duff & Smol, 1991 - Fig. 153; Duff & Smol, unpubl.), and Sweden (Cronberg, 1989b).

ECOLOGY: *Mallomonas hamata* is an oligotrophic, eurythermal, acidophilic species, which may be absent from organically acidified sites (Cronberg, 1989b; Siver, 1991a). *M. heterospina* is a meso-eutrophic taxon with an acidophilic or circumneutral distribution and a tolerance for humic acids (Cronberg, 1989b; Siver, 1991a). It is a cold-water species that encysts in spring (Asmund, 1956; Siver, 1991a). In a paleolimnological study, stomatocyst 202 was associated with very low pH (<5.0), although it was very rare (Duff & Smol, 1991); in this lake, it was most likely produced by *M. hamata*. Stomatocyst 202 was most common in shallow, alkaline, productive freshwater lakes in British Columbia, Canada (Duff, 1994); in this survey, it was most likely produced by *M. heterospina*.

Spherical or oval stomatocysts with a simple cylindrical collar but no ornamentation

Stomatocyst 50, Duff & Smol, 1991 (29) Fig. 24

BIOLOGICAL AFFINITY: unknown

LOCALITY: Upper Wallface Pond (44°08.8'N; 74°03.3'W), Adirondack Park, New York, U.S.A. Gravity core, core depth 10.0-10.5 cm.

SEM DESCRIPTION: This is a small, spherical stomatocyst (diameter 2.8-5.2 μm). The collar is abruptly cylindrical and quite narrow (diameter 0.5-1.0 μm; height 0.1-0.5 μm). A planar sloping annulus may or may not surround the regular pore (diameter 0.2-0.4 μm).

Stomatocyst 50 is now separated from stomatocyst 52 (Fig. 25) on the basis of the collar diameter : cyst diameter ratio, which is always ≤0.21. This distinction is somewhat arbitrary but is based on natural groupings within our SEM specimens. Hence, some of the original specimens of stomatocyst 52 (Duff & Smol, 1991), including the original characteristic specimen, have been reassigned to stomatocyst 50. Most of the specimens of stomatocyst 124 (Chapter 5) have been reclassified as stomatocyst 50, and the later number (i.e. 124) has been discontinued.

LM DESCRIPTION: Stomatocyst 50 cannot be distinguished with LM from stomatocysts 110 (Figs. 12C, D), 52 and small (<5.0 μm) specimens of stomatocyst 51.

REFERENCES: This cyst has been observed in the Northwest Territories, Canada (Duff *et al.*, 1992 - misidentified as stomatocyst 124), and New York, U.S.A. (Duff & Smol, 1991, unpubl.). Some specimens of *cysta minima* Nygaard (1956 - Denmark) may be stomatocyst 50 (see discussion under stomatocyst 52).

ECOLOGY: Stomatocyst 50 has been found in both acidic (Duff & Smol, 1991, unpubl.) and alkaline (Duff *et al.*, 1992) sites.

Stomatocyst 52, Duff & Smol, 1991 *emend.* Duff *et al.* (this volume) (12) Fig. 25

BIOLOGICAL AFFINITY: This stomatocyst is likely

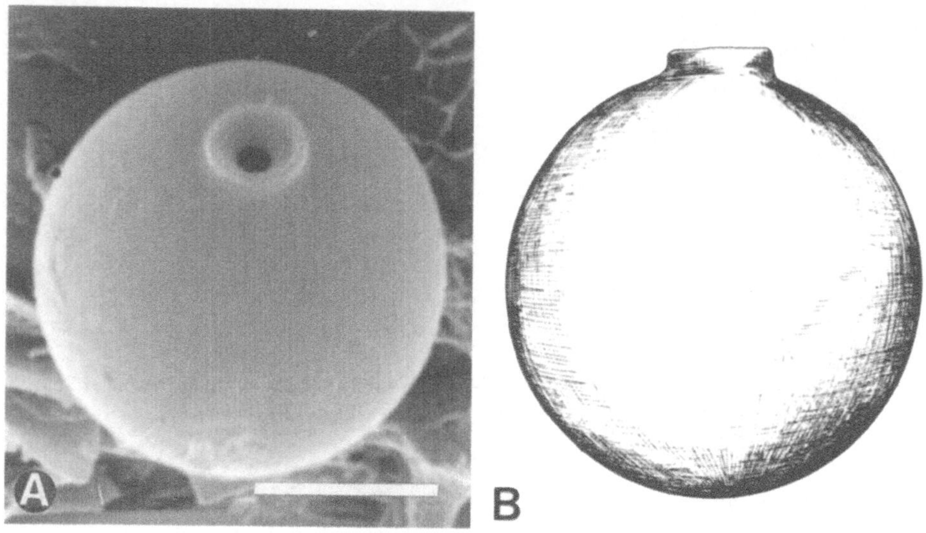

Fig. 25. Stomatocyst 52. A: SEM. B: Line drawing of SEM. Scale bar = 2 μm. See Figs. 12C, D for LM.

produced by more than one species. Nygaard (1956) suggested that it was produced by a *Chromulina* species, based on the descriptions of Huber-Pestalozzi (1941).

NEGATIVE NUMBER: J.P. Smol 592

LOCALITY: "Unnamed Pond" (unofficial name) (43°11'N; 74°37'W), Adirondack Park, New York, U.S.A. Gravity core, surface (0-0.25 cm) sediment sample.

SEM DESCRIPTION: This is an emended description of stomatocyst 52, based on new observations. The size range and diagnostic criteria have been redefined. It is identical to stomatocysts 234, 127, and 197 (Chapter 5), except that is smaller. The new size ranges for these cysts are <5.0 μm, 5.0-9.9 μm, 10.0-15.0 μm, and >15.0 μm, respectively.

Stomatocyst 52 is a small, spherical stomatocyst (diameter 4.0-4.8 μm; always <5.0 μm) with a smooth surface. The collar is abruptly cylindrical (diameter 0.7-1.2 μm; height 0.2-0.6 μm), with a thickened apex and a sloping planar annulus surrounding a regular pore (diameter 0.3-0.4 μm). This stomatocyst is distinguished from stomatocyst 50 (Fig. 24) by its larger collar diameter : cyst diameter ratio (0.22-0.27). Using this criterion, the characteristic specimen in Duff & Smol (1991) is now classified as stomatocyst 50, and a new characteristic specimen is included here.

LM DESCRIPTION: With LM, this stomatocyst is indistinguishable from stomatocysts 110 (Figs. 12C, D), 50, and small (<5.0 μm) specimens of 51.

REFERENCES: It is difficult to link this cyst with morphotypes described in previous studies because of uncertainties in the taxonomy; it has frequently been lumped with cysts we now describe as separate (e.g. stomatocysts 50, 53, and 234). Of the original specimens described by Duff & Smol (1991), one has been reassigned to stomatocyst 50, and two have been reassigned to stomatocyst 234. The specimens identified as stomatocyst 52 by Duff et al. (1992) and by Duff & Smol (1994) have all been reclassified as stomatocyst 234, or as some other morphotype (e.g. stomatocyst 146). We have observed stomatocyst 52, as defined here, from several sites in New York, U.S.A. (Duff & Smol, 1991, unpubl.), and from Greenland (Brown *et al.*, 1994).

We believe that *cysta minima* Nygaard (1956 - Denmark), with a size range of 3.9-5.0 μm, is the same as our stomatocyst 52, although it may also include stomatocyst 50. However, the cyst identified by Carney & Sandgren (1983 - Michigan, U.S.A.) as *cysta minima* probably also includes specimens of both stomatocysts 234 and 53. Sandgren & Carney (1983 - Michigan, U.S.A.) cyst 3 includes both stomatocysts 52 and 234. Adam & Mehringer (1980c - Oregon, U.S.A.) type 319 is similar, but they did not publish the size range used to differentiate their morph.

ECOLOGY: Based on limited information, this cyst appears to be present over a wide range of pH and trophic status, but may be somewhat more common in acidified sites (Duff & Smol, 1991, unpubl.). However, if this cyst is produced by several species, as we believe, we would expect it to be associated with a large ecological tolerance (i.e. a "generalist" distribution).

Stomatocyst 234, Duff *et al.* (this volume) (13) Fig. 26

BIOLOGICAL AFFINITY: More than one species likely produces cysts of this type. Several species of

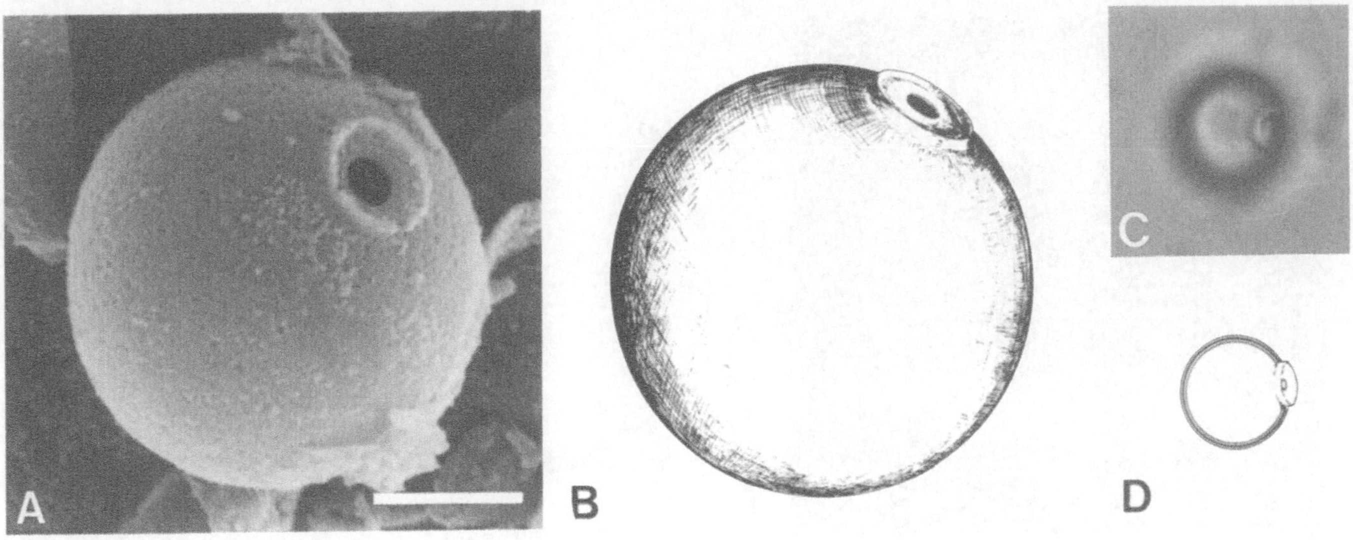

Fig. 26. Stomatocyst 234. A: SEM. B: Line drawing of SEM. C: LM. D: Line drawing of LM. Scale bar = 2 μm. LM x2000.

Paraphysomonas are described as producing stomatocysts that roughly conform to this type (e.g. *P. antarctica* Takahashi (Takahashi, 1987); *P. corynephora* Preisig & Hibberd (Preisig & Hibberd, 1982a, 1983); *P. vestita* (Stokes) De Saedeler (Takahashi, 1987)).

NEGATIVE NUMBER: J.P. Smol 2999

LOCALITY: Hawk Lake (63°N; 90°W), Northwest Territories, Canada. Box core, 2.7-3.9 cm sediment depth.

SEM DESCRIPTION: This stomatocyst is identical to stomatocysts 52, 127 and 197 (Chapter 5), except that the size is different; the size ranges for these cysts are given under stomatocyst 52. Stomatocyst 234 is spherical (observed diameter 5.3-7.9 μm) with a smooth surface. The collar is low and cylindrical (diameter 1.2-2.1 μm; height 0.1-0.3 μm; collar diameter : cyst diameter ratio 0.22-0.35). The outer margin is abrupt but the very base may be gradual; the inner margin is generally steep, and the apex is usually acute. There is a gently sloping to flat planar annulus surrounding a regular pore (diameter 0.4-0.9 μm).

LM DESCRIPTION: This stomatocyst cannot be distinguished from stomatocysts 53 and 152 with LM.

REFERENCES: This stomatocyst is widespread. It has been observed in the Northwest Territories, Canada (Duff *et al.*, 1992 - misclassified as stomatocyst 52; Brown *et al.*, 1994; Webb & Smol, unpubl.), British Columbia, Canada (Duff & Smol, 1994 - misclassified as stomatocyst 52; Zeeb, 1994), New York, U.S.A. (Duff & Smol, 1991 - misclassified as stomatocyst 52; Duff & Smol, unpubl.), and California, U.S.A. (Zeeb *et al.*, subm.). We believe *cysta modica* Nygaard (1956 - Denmark) is identical, but SEM confirmation is needed. The cyst identified by Carney & Sandgren (1983 - Michigan, U.S.A.) as *cysta modica* probably also includes some specimens of stomatocyst 53, whereas their *cysta minima* Nygaard probably includes both stomatocysts 52 and 234. Similarly, some specimens of Sandgren & Carney (1983) cyst 3 could be attributed to stomatocyst 234.

ECOLOGY: This cyst has been found in both acidic (Duff & Smol, 1991) and alkaline (Carney & Sandgren, 1983; Duff *et al.*, 1992; Duff & Smol, 1994) sites. It is apparently tolerant of a wide range of environmental gradients, probably because it is produced by more than one species.

Stomatocyst 127, Duff & Smol in Duff *et al.*, 1992 (9) Fig. 27

BIOLOGICAL AFFINITY: unknown

LOCALITY: "Col Pond" (unofficial name), Cape Herschel (78°37'N; 74°42'W), Ellesmere Island, Northwest Territories, Canada. Net tow, collected July 24, 1986.

SEM DESCRIPTION: This stomatocyst is identical to stomatocysts 52, 234, and 197 (Chapter 5), except that the size is different; the size ranges for these cysts are given under stomatocyst 52. Stomatocyst 127 is a spherical stomatocyst (observed diameter 10.2-15.0 μm). Its surface is generally smooth, although most arctic specimens appeared

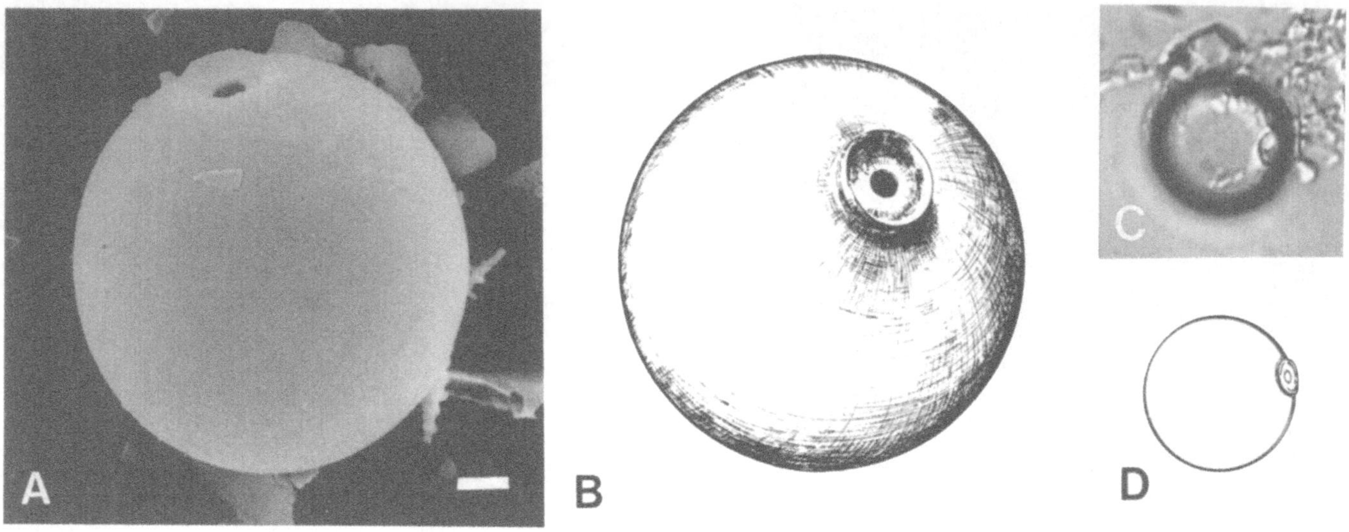

Fig. 27. Stomatocyst 127. A: SEM. B: Line drawing of SEM. C: LM. D: Line drawing of LM. Scale bar = 2 μm. LM x2000.

degraded (Duff *et al.*, 1992). The collar is abruptly cylindrical to conical, with a straight outer margin and a generally acute apex (diameter 2.2-3.2 μm; height 0.3-0.9 μm; collar diameter : cyst diameter ratio 0.19-0.25). A gently sloping to flat planar annulus surrounds a regular pore (diameter 0.6-1.3 μm).

LM DESCRIPTION: This stomatocyst is distinguished from stomatocysts 234 and 197 on the basis of size. It is differentiated from stomatocyst 116 *forma* A by the presence of an annulus.

REFERENCES: Stomatocyst 127 has been observed in the Northwest Territories, Canada (Duff *et al.*, 1992; Webb & Smol, unpubl.); Ontario, Canada (Zeeb & Smol, unpubl.); British Columbia, Canada (Duff & Smol, 1994; Zeeb, 1994); New York, U.S.A. (Duff & Smol, unpubl.); and Michigan, U.S.A. (Sandgren & Carney, 1983 - cyst 4). Rybak *et al.* (1991 - Ontario, Canada) cysts 7 and 9 are probably identical.

ECOLOGY: This cyst has been observed in a wide range of environments; its ecological optima and tolerances are unclear, but it was associated with shallow, alkaline, highly conductive lakes in British Columbia, Canada (Duff, 1994).

Stomatocyst 152, Zeeb & Smol, 1993a (7) Fig. 28

BIOLOGICAL AFFINITY: unknown

LOCALITY: Elk Lake (47°11'N; 95°13'W), Clearwater County, Minnesota, U.S.A. Piston core, varve year A.D. 1708.

SEM DESCRIPTION: This stomatocyst is spherical to slightly oblate (diameter 6.5-9.6 μm), and smooth surfaced. There is a cylindrical to slightly conical collar, with a straight outer collar margin and an acute apex (diameter 2.0-3.3 μm; height ≤1.0 μm; collar diameter : cyst diameter ratio 0.30-0.48); the collar width is always greater than the collar height. Zeeb & Smol (1993a, b) divided stomatocyst 152 into two size classes for counting (A: 6.5-7.4 μm; B: >9.0 μm); however, we do not have enough SEM specimens at this time to justify splitting stomatocyst 152 into two separate morphotypes.

LM DESCRIPTION: With LM, stomatocyst 152 cannot be distinguished from stomatocysts 234 (Figs. 26C, D) and 53.

REFERENCES: Stomatocyst 152 has been observed in British Columbia, Canada (Duff & Smol, 1994; Zeeb, 1994), Minnesota, U.S.A. (Zeeb & Smol, 1993a), and Poland (Rybak, 1987 - cyst 2).

ECOLOGY: The ecological optima and tolerances of this morph are unknown.

Stomatocyst 53, Duff & Smol, 1991 (14) Fig. 29

BIOLOGICAL AFFINITY: Similar cysts from live colonies of a species of *Uroglena* have been observed by P.A. Siver (pers. commun.).

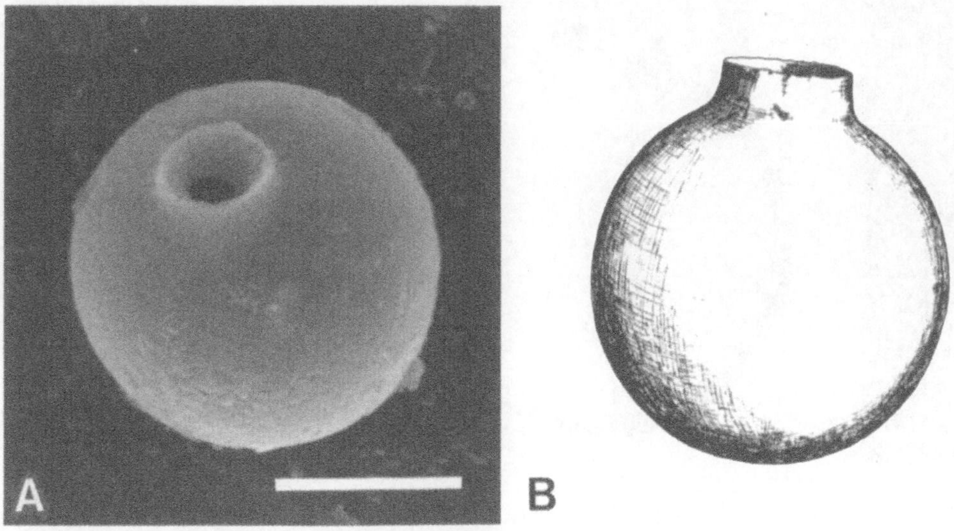

Fig. 28. Stomatocyst 152. A: SEM. B: Line drawing of SEM. Scale bar = 2 μm. See Figs. 26C, D for LM.

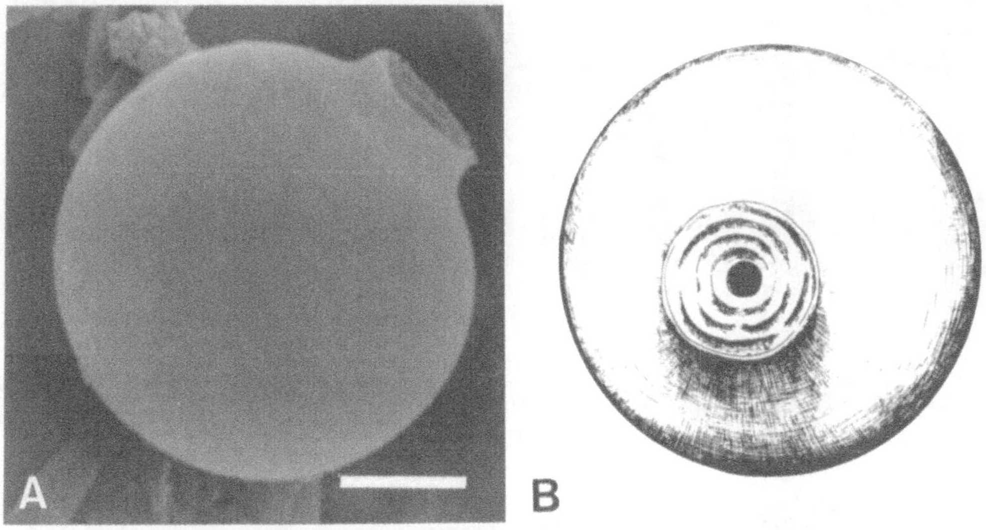

Fig. 29. Stomatocyst 53. A: SEM. B: Line drawing of SEM. Scale bar = 2 μm. See Figs. 26C, D for LM.

LOCALITY: Upper Wallface Pond (44°08.8'N; 74°03.3'W), Adirondack Park, New York, U.S.A. Gravity core, core depth 20.0-20.5 cm.

SEM DESCRIPTION: This stomatocyst is spherical (diameter 5.8-9.0 μm) with a smooth surface. The collar is low and cylindrical to slightly obconical, usually with a thickened apex (diameter 1.9-3.1 μm; height 0.3-0.7 μm). There is a flat to sloping planar annulus that always has some evidence of radial annular rings. The pore is regular (diameter 0.6-0.8 μm). The collar diameter : cyst diameter ratio is 0.29-0.39.

The characteristic specimen of stomatocyst 54 (Chapter 5) is now included within stomatocyst 53.

LM DESCRIPTION: With LM, this stomatocyst cannot be distinguished reliably from stomatocysts 234 (Figs. 26C, D) and 152.

REFERENCES: Stomatocyst 53 has been observed in British Columbia, Canada (Duff & Smol, 1994; Zeeb, 1994), Ontario, Canada (Zeeb & Smol, unpubl.), and New York, U.S.A. (Duff & Smol, 1991, unpubl.). Some specimens of *cysta minima* Nygaard and *cysta modica* Nygaard *sensu* Carney & Sandgren (1983 - Michigan, U.S.A.) evidently belong to stomatocyst 53.

ECOLOGY: This morphotype is apparently tolerant of a wide range of pH (Duff & Smol, 1991, 1994; Carney & Sandgren, 1983), but detailed autecological data are still lacking.

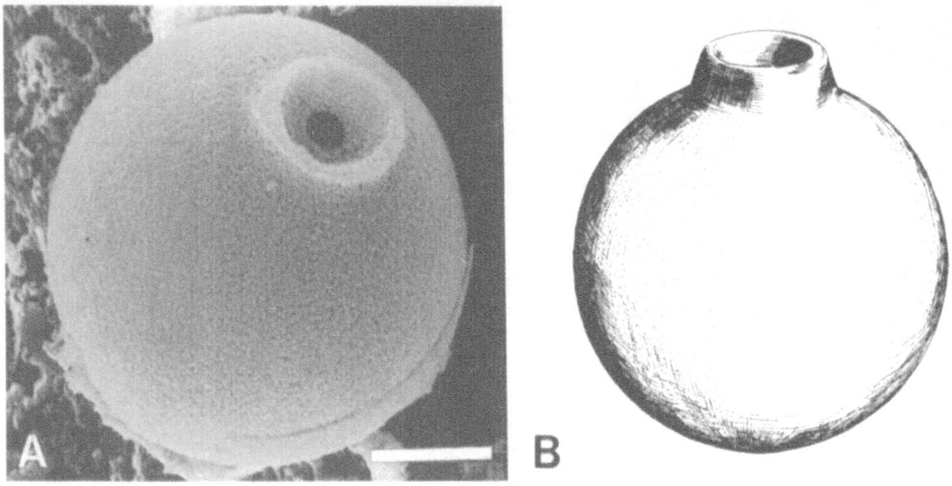

Fig. 30. Stomatocyst 146. A: SEM. B: Line drawing of SEM. Scale bar = 2 μm. See Figs. 17C, D for LM.

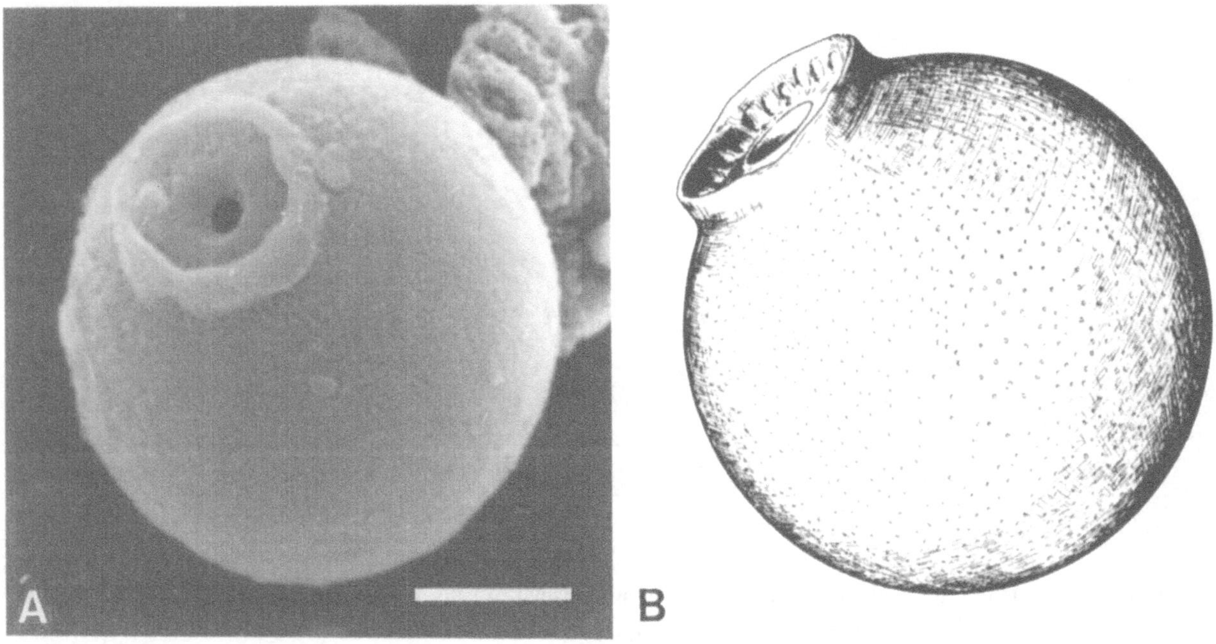

Fig. 31. Stomatocyst 198. A: SEM. B: Line drawing of SEM. Scale bar = 2 μm.

Stomatocyst 146, Zeeb & Smol in Pienitz *et al.*, 1992 (3) Fig. 30

BIOLOGICAL AFFINITY: unknown

LOCALITY: Unnamed saline lake (U60) (62°45'N; 136°38'W), near Pelly Crossing, Yukon Territory, Canada. Gravity core, core depth 25 cm.

SEM DESCRIPTION: This stomatocyst is spherical (diameter 6.6-8.6 μm). The collar is short and cylindrical, with an abrupt inner and a slightly sloping outer margin, a rounded apex, and a sloping planar annulus (collar diameter 2.2-2.7 μm; annulus diameter 1.8-2.2 μm). The pore is regular (diameter 0.6 μm). The cyst surface is smooth to psilate.

LM DESCRIPTION: Stomatocyst 146 cannot be distinguished from stomatocyst 156 (Figs. 17C, D) using LM.

REFERENCES: Stomatocyst 146 has been observed in the Yukon Territory, Canada (Pienitz *et al.*, 1992), British Columbia, Canada (Zeeb, 1994), and Minnesota, U.S.A. (Zeeb & Smol, 1993a). It was probably included within Sandgren & Carney (1983 - Michigan, U.S.A.) cyst 11 (= Carney & Sandgren, 1983 - *cysta modica*).

ECOLOGY: To date, this cyst has been found mostly in

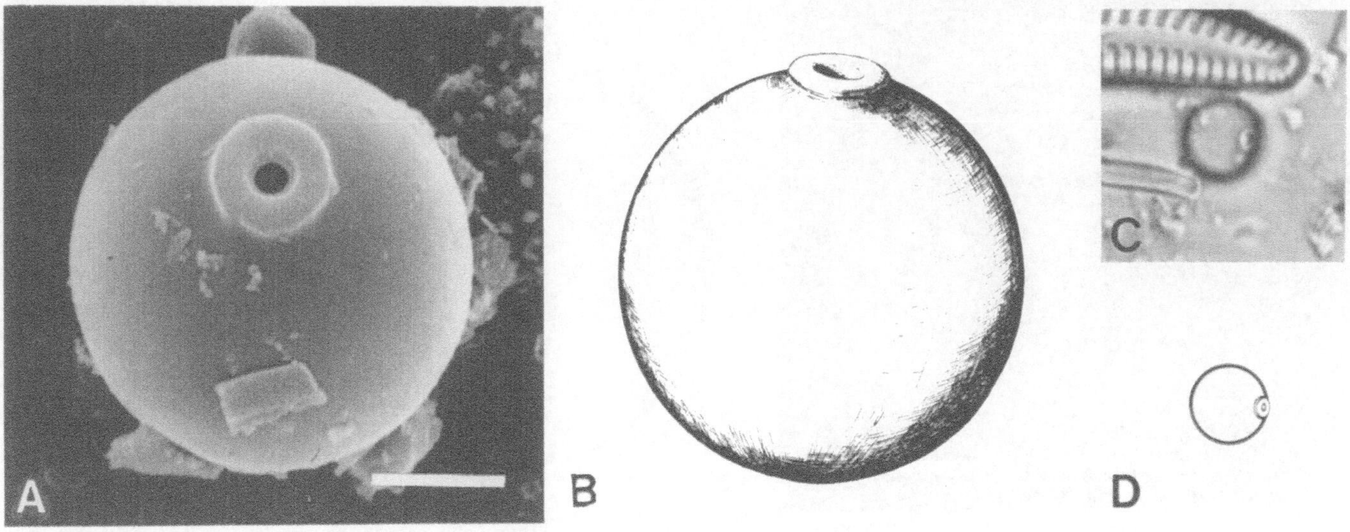

Fig. 32. Stomatocyst 125. A: SEM. B: Line drawing of SEM. C: LM. D: Line drawing of LM. Scale bar = 2 μm. LM x2000.

alkaline locations (Pienitz *et al.*, 1992; Zeeb & Smol, 1993a; Sandgren & Carney, 1983). According to Pienitz *et al.* (1992), it is intolerant of very saline (>10 mg/L) water.

Stomatocyst 198, Duff & Smol, 1994 (4) Fig. 31

BIOLOGICAL AFFINITY: unknown

LOCALITY: Elkin Lake (51°40'N; 123°45'W), British Columbia, Canada. Ekman dredge, surface (1 cm) sediment sample.

SEM DESCRIPTION: This is a spherical stomatocyst (diameter 5.6-6.2 μm) with a psilate surface. The collar is cylindrical with a rounded outer margin and a straighter inner margin (diameter 1.8-2.6 μm; height ca. 0.4 μm), a flat planar annulus (diameter 1.7-1.9 μm), and a conical pore (outer margin diameter 0.8-0.9 μm; inner margin diameter 0.5-0.6 μm). The annulus and base of the inner collar margin are ornamented with a patchy variable reticulum with roughly circular lacunae (diameter <0.1-0.2 μm).

LM DESCRIPTION: This stomatocyst has not been definitely identified with LM, but is probably indistinguishable from similar cysts (i.e. stomatocyst 53/152/234 (Figs. 26C, D) or stomatocyst 146/156 (Figs. 17C, D)).

REFERENCES: This stomatocyst has been observed in British Columbia, Canada (Duff & Smol, 1994), the Northwest Territories, Canada (Webb & Smol, unpubl.), Oregon, U.S.A. (Adam & Mehringer, 1980c - type 84), and Washington, U.S.A. (Smith & White, 1985 - Figs. 20 & 21).

ECOLOGY: No autecological details are available.

Stomatocyst 125, Duff & Smol in Duff *et al.*, 1992 (5) Fig. 32

BIOLOGICAL AFFINITY: The stomatocyst of *Paraphysomonas antarctica* Takahashi (Takahashi, 1987) resembles both stomatocysts 125 and 234.

LOCALITY: "Horseshoe Pond" (unofficial name), Cape Herschel (78°37'N; 74°42'W), Ellesmere Island, Northwest Territories, Canada. Moss sample, collected July 19, 1986.

SEM DESCRIPTION: This is a spherical, smooth-surfaced stomatocyst (diameter 6.0-7.2 μm) with a low, cylindrical to slightly obconical collar (diameter 1.3-1.8 μm; height 0.2-0.4 μm; collar diameter : cyst diameter ratio 0.22-0.29) with a thickened, flattened apex and a regular pore (diameter 0.5-0.8 μm).

LM DESCRIPTION: This stomatocyst can be distinguished from the numerous similar cysts (i.e. stomatocysts 146/156; 53/152/234; 183) by the lack of an annulus and/or the thickened collar apex, and by its cylindrical or obconical outer collar margin. However, it is difficult to distinguish from stomatocysts 181 (Figs. 16C, D, and E) and 182 (Figs. 42C, D); SEM confirmation is needed when two or more of these

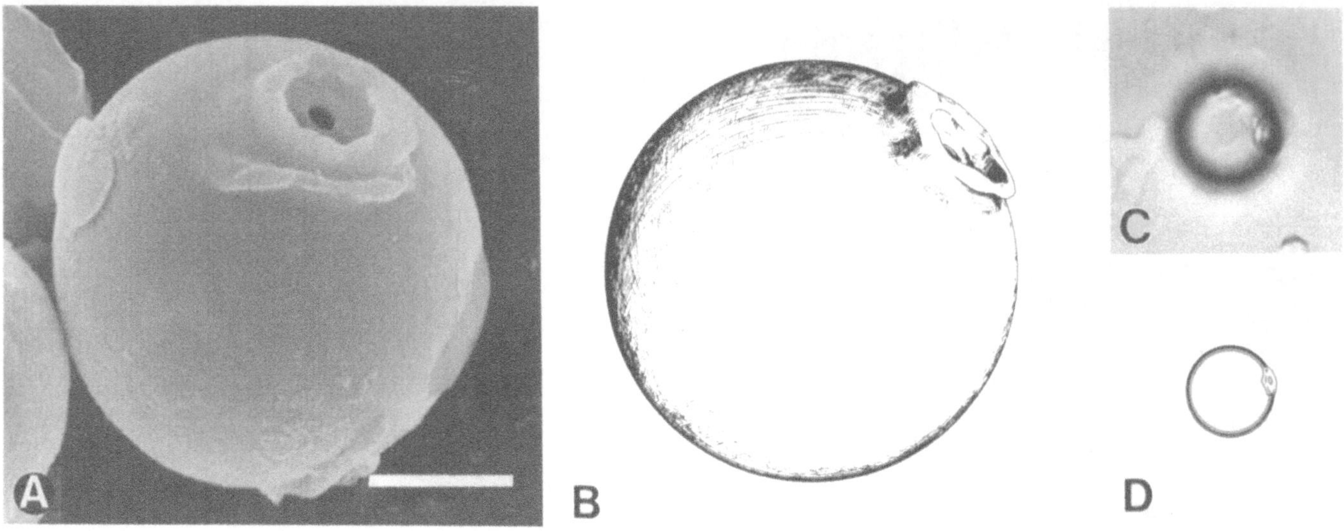

Fig. 33. Stomatocyst 183. A: SEM. B: Line drawing of SEM. C: LM. D: Line drawing of LM. Scale bar = 2 μm. LM x2000.

morphs are present in the same sample.

REFERENCES: We have only observed this cyst in the Northwest Territories, Canada (Duff *et al.*, 1992). Pienitz *et al.* (1992 - Yukon Territory, Canada) identified a similar, but degraded, morph as stomatocyst cf. 125.

ECOLOGY: This stomatocyst has been found in alkaline locations (Duff *et al.*, 1992; Pienitz *et al.*, 1992), and may be produced by a chrysophyte tolerant of high salinities (Pienitz *et al.*, 1992; Takahashi, 1987). According to Duff *et al.* (1992), it could be produced by a periphytic species.

Stomatocyst 183, Brown & Smol in Brown *et al.*, 1994 (11) Fig. 33

BIOLOGICAL AFFINITY: unknown

LOCALITY: Nordvestø (76°44'N; 73°13'W), Carey Islands, Greenland. SIPRE peat core, core depth 121-122 cm.

SEM DESCRIPTION: This stomatocyst is smooth and spherical (diameter 6.0-7.9 μm). The collar is cylindrical with a highly irregular and usually thickened apex (diameter 1.6-2.6 μm; height 0.5-0.7 μm; collar diameter : cyst diameter ratio 0.24-0.40). The outer margin is also quite irregular; usually it is acute, but it may be conical to obconical in part. There is a wide, flat planar to slightly swollen annulus (diameter 1.1-1.8 μm) surrounding a regular pore (diameter 0.3-0.5 μm).

LM DESCRIPTION: If the irregular outline of the collar is visible, stomatocyst 183 can be distinguished from similar cysts (i.e. stomatocysts 53/152/234), but is best grouped with these cysts for practical counting purposes.

REFERENCES: We have only observed this cyst in Greenland (Brown *et al.*, 1994). Adam & Mahood (1979b - California, U.S.A.) type 83 is similar but slightly smaller.

ECOLOGY: No autecological data are available.

Stomatocyst 56, Duff & Smol, 1991 (11) Fig. 34

BIOLOGICAL AFFINITY: unknown

LOCALITY: Upper Wallface Pond (44°08.8'N; 74°03.3'W), Adirondack Park, New York, U.S.A. Gravity core, core depth 0.0-0.5 cm.

SEM DESCRIPTION: This stomatocyst is spherical (10.1-13.1 μm) and has a low, cylindrical collar with an acute apex and a sloping planar annulus (collar diameter 3.1-4.3 μm; height 0.4 μm; collar diameter : cyst diameter ratio 0.31-0.44). The pore is obscured on all SEM specimens by a roughly spherical, convex siliceous plug (diameter 2.0-2.9 μm). The surface is smooth to psilate, or with a few, very low, irregularly scattered depressions.

Stomatocyst 56 is similar to stomatocyst 112, but is smaller with a lower collar diameter : cyst diameter ratio. It is distinguished from stomatocyst 241 because it is larger and lacks verrucose or reticular ornamentation. The cyst

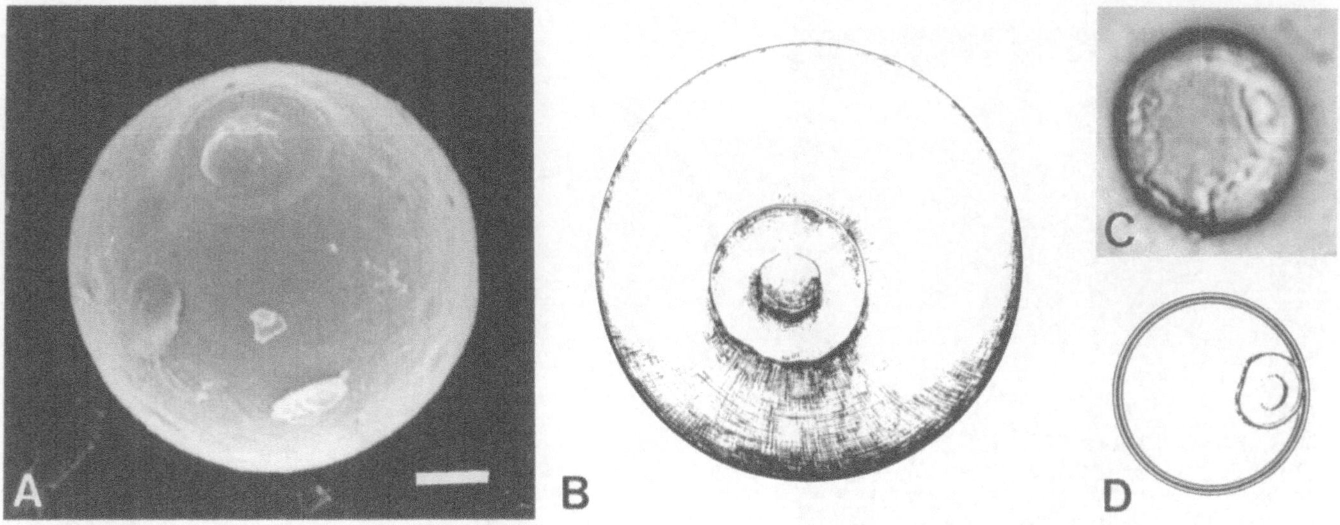

Fig. 34. Stomatocyst 56. A: SEM. B: Line drawing of SEM. C: LM. D: Line drawing of LM. Scale bar = 2 μm. LM x2000.

identified by Zeeb & Smol (1993a) as stomatocyst 56 has now been reassigned to stomatocyst 241.

LM DESCRIPTION: This cyst can be distinguished from similar morphotypes (i.e. stomatocysts 112 and 241) on the basis of size, collar diameter : cyst diameter ratio, and ornamentation.

REFERENCES: We have observed stomatocyst 56 in New York, U.S.A. (Duff & Smol, 1991, unpubl.), and California, U.S.A. (Zeeb *et al.*, subm.). The cyst identified by Rybak (1986 - Poland) as *cysta cingens* Nygaard is identical in size to stomatocyst 56, but its collar and ornamentation more closely resemble stomatocyst 112; it does not appear to be the same morph depicted by Nygaard (1956). Similarly, Rybak (1987 - Poland) cyst 5 appears to be intermediate in morphology between stomatocysts 56 and 112. Adam & Mahood (1979b - California, U.S.A.) type 85 is the same size as stomatocyst 56, but has reticular ornamentation similar to that of stomatocyst 241.

ECOLOGY: Autecological details are not yet available.

Stomatocyst 112, Zeeb *et al.*, 1990 *emend.* Duff & Smol, 1994 (4) Fig. 35

BIOLOGICAL AFFINITY: This morphotype appears to be identical to the cyst of *Ochromonas globosa* Skuja (Nygaard, 1977), but SEM confirmation is needed.

LOCALITY: Little Round Lake (44°47.5'N; 76°41.5'W), Ontario, Canada. One metre piston core, core depth 15 cm.

SEM DESCRIPTION: This is a large spherical stomatocyst (diameter 13.0-15.8 μm; up to 19.4 μm observed with LM). The collar is cylindrical with a rounded apex and gradually sloping inner margin, which may flatten into a planar annulus (collar diameter 6.8-8.0 μm; height 1.0-1.3 μm; collar diameter : cyst diameter ratio 0.45-0.51). The pore is obscured in all but one of our SEM specimens by a rounded siliceous plug, but is visible on most LM specimens (pore diameter 0.9-2.1 μm; plug diameter 2.9-3.5 μm; Fig. 35C). One SEM specimen differs slightly from the rest. Its collar has a more gradual outer margin, and it has very low verrucae (diameter 0.3-0.6 μm) ornamenting the posterior hemisphere.

Stomatocyst 112 is identical to stomatocyst 241, but is twice as large and lacks reticular forms.

LM DESCRIPTION: This stomatocyst is recognized by its size and low, wide, rounded collar.

REFERENCES: We have observed stomatocyst 112 in Ontario, Canada (Zeeb *et al.*, 1990), and British Columbia, Canada (Duff & Smol, 1994; Zeeb, 1994). It is probably identical to Sandgren & Carney (1983 - Michigan, U.S.A.) cyst 2 and Carney & Sandgren (1983 - Michigan, U.S.A.) *cysta cingens* Nygaard. The latter, however, is probably not equivalent to Nygaard's original morph, which had a distinct double collar rather than a collar and plug.

ECOLOGY: *Ochromonas globosa* has been found in acidic, oligotrophic sites, including bog lakes (Nygaard, 1977; Duthie

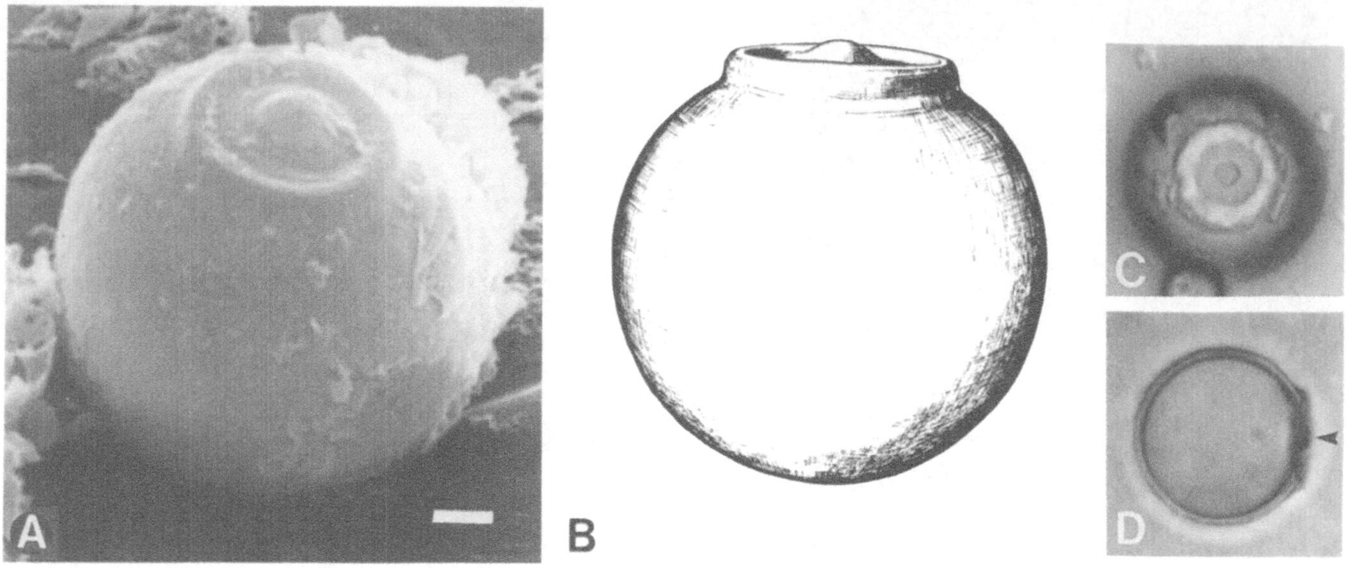

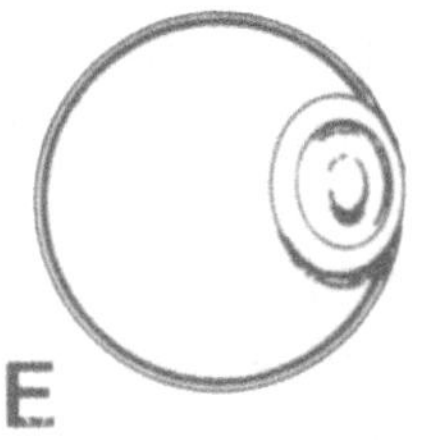

Fig. 35. Stomatocyst 112. A: SEM. B: Line drawing of SEM. C: LM, apical view. D: LM, cross-sectional view. Siliceous plug indicated by arrow. E: Line drawing of LM. Scale bar = 2 μm. LM x1500.

et al., 1976). Stomatocyst 112 has been associated with oligotrophic conditions in paleolimnological studies (Carney & Sandgren, 1983; Zeeb *et al.*, 1990), but has only been found in alkaline locations (Carney & Sandgren, 1983; Sandgren & Carney, 1983; Zeeb *et al.*, 1990; Duff & Smol, 1994).

Stomatocyst 116, Zeeb *et al.*, 1990 (38) Fig. 36

BIOLOGICAL AFFINITY: unknown

LOCALITY: Little Round Lake (44°47.5'N; 76°41.5'W), Ontario, Canada. One metre piston core, core depth 20 cm.

SEM DESCRIPTION: This stomatocyst is spherical to very widely oval (diameter 9.2-12.7 μm) with a smooth or slightly psilate surface texture. It has a cylindrical collar with abrupt inner and outer margins, a gradual base, and an acute apex (collar diameter 1.8-3.0 μm). The collar margins are never striated. The collar height is always less than or equal to the collar diameter. Two *formae* are recognized, which are differentiated on the basis of collar height.

***Forma* A** - The collar height is <1.0 μm (Fig. 36A).

***Forma* B** - The collar height varies from 1.0 to ca. 2.6 μm, but is always less than or equal to the collar diameter (Figs. 36B, C).

Stomatocyst 116 closely resembles stomatocyst 161 except that the collar margin of the latter morph is radially striated. It is distinguished from stomatocyst 153 by its shorter collar, and from stomatocysts 154 and 199 on the basis of body shape (width:length ratio ≥0.90).

LM DESCRIPTION: This cyst can be differentiated from stomatocyst 161 if the collar striations of the latter are visible; this can only be achieved with high resolution optics. It can be differentiated from stomatocysts 154 and 199 on the basis of body shape, and from stomatocyst 127 because it lacks an annulus.

REFERENCES: Stomatocyst 116 has been observed in British Columbia, Canada (Duff & Smol, 1994; Zeeb, 1994), Ontario, Canada (Zeeb *et al.*, 1990, 1994), and Minnesota, U.S.A. (Zeeb & Smol, 1993a). Rybak *et al.* (1987 - Ontario, Canada) cysts 10 and 13 may be identical, although the written descriptions for these cysts differ somewhat from ours.

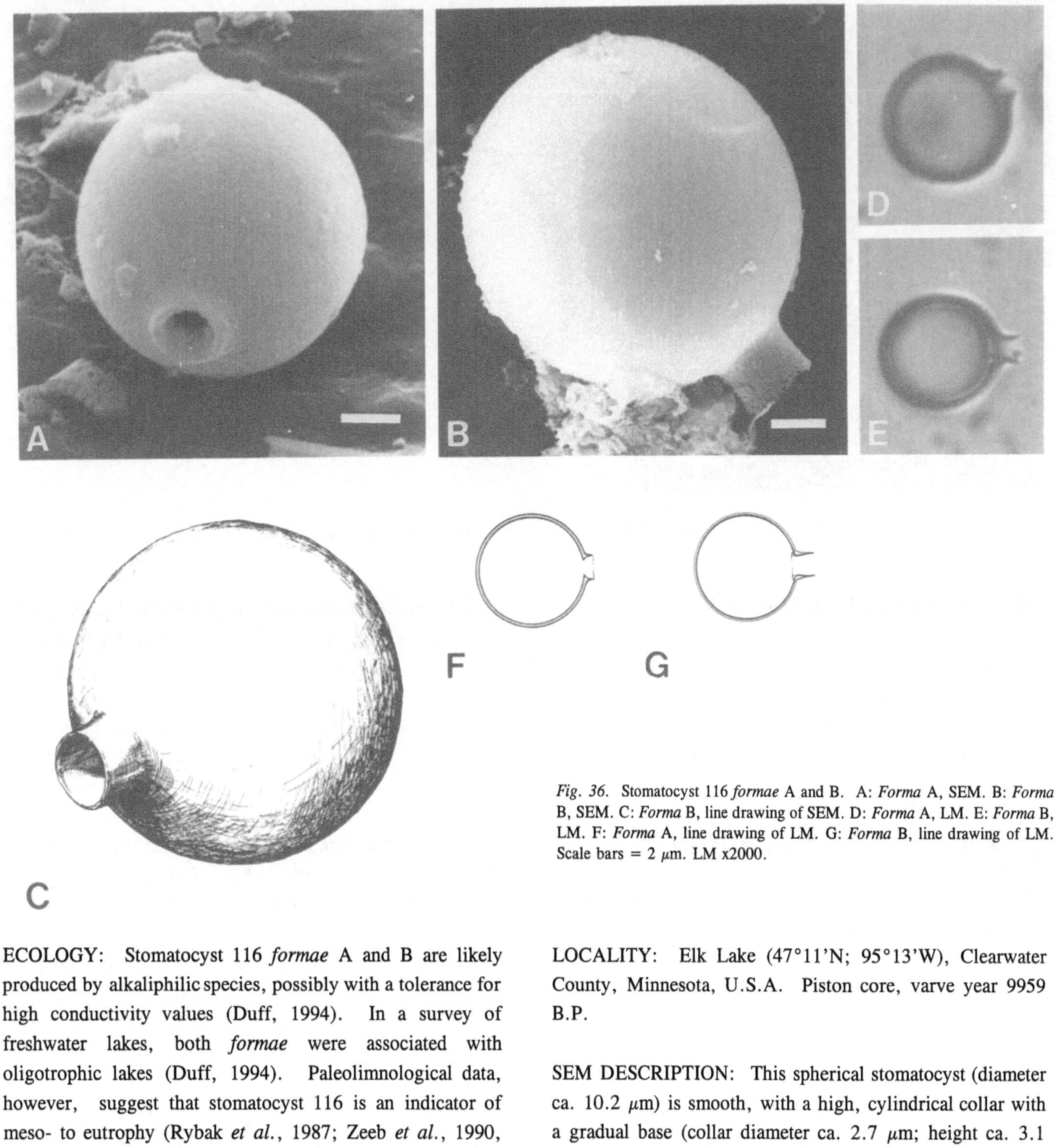

Fig. 36. Stomatocyst 116 *formae* A and B. A: *Forma* A, SEM. B: *Forma* B, SEM. C: *Forma* B, line drawing of SEM. D: *Forma* A, LM. E: *Forma* B, LM. F: *Forma* A, line drawing of LM. G: *Forma* B, line drawing of LM. Scale bars = 2 μm. LM x2000.

ECOLOGY: Stomatocyst 116 *formae* A and B are likely produced by alkaliphilic species, possibly with a tolerance for high conductivity values (Duff, 1994). In a survey of freshwater lakes, both *formae* were associated with oligotrophic lakes (Duff, 1994). Paleolimnological data, however, suggest that stomatocyst 116 is an indicator of meso- to eutrophy (Rybak *et al.*, 1987; Zeeb *et al.*, 1990, 1994; Zeeb & Smol, 1993b). In Elk Lake, Minnesota (Zeeb & Smol, 1993a), it thrived during the mid-Holocene Prairie period, when climatic conditions were warmer and drier, and lakewater was likely more saline and/or productive.

Stomatocyst 153, Zeeb & Smol, 1993a (2) Fig. 37

BIOLOGICAL AFFINITY: unknown

LOCALITY: Elk Lake (47°11'N; 95°13'W), Clearwater County, Minnesota, U.S.A. Piston core, varve year 9959 B.P.

SEM DESCRIPTION: This spherical stomatocyst (diameter ca. 10.2 μm) is smooth, with a high, cylindrical collar with a gradual base (collar diameter ca. 2.7 μm; height ca. 3.1 μm). The collar height is always greater than the collar diameter; this characteristic feature is used to distinguish stomatocyst 153 from stomatocyst 116 *forma* B.

LM DESCRIPTION: This cyst is distinguished from similar cysts (i.e. stomatocysts 116, 161) by the length of its collar and the lack of spiral collar striations.

REFERENCES: We have only observed this cyst in

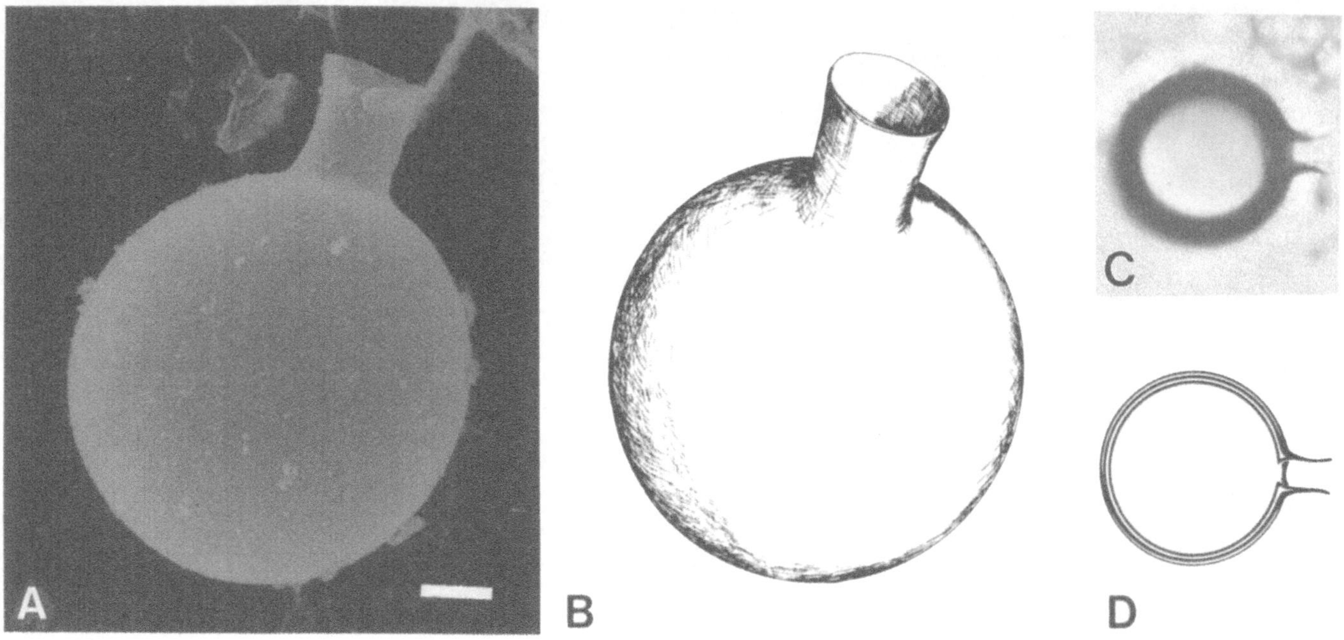

Fig. 37. Stomatocyst 153. A: SEM. B: Line drawing of SEM. C: LM. D: Line drawing of LM. Scale bar = 2 μm. LM x2000.

Minnesota, U.S.A. (Zeeb & Smol, 1993a).

ECOLOGY: In Zeeb & Smol's (1993b) paleolimnological analysis, stomatocyst 153 was most abundant (ca. 10% relative abundance) immediately following glaciation, but declined with the onset of the Holocene Prairie Period.

Stomatocyst 161, Zeeb & Smol, 1993a (13) Fig. 38

BIOLOGICAL AFFINITY: *Dinobryon divergens* Imhof (Sheath *et al.*, 1975; Sandgren, 1980a)

LOCALITY: Elk Lake (47°11'N; 95°13'W), Clearwater County, Minnesota, U.S.A. Sediment trap sample.

SEM DESCRIPTION: This is a smooth, spherical to slightly oval stomatocyst (diameter 9.6-12.7 μm) with a smooth surface. The collar is cylindrical, usually with a fairly abrupt delineation between the collar base and the cyst body. The collar margin(s) are ornamented by spiral striations, giving a "wrinkled" appearance (collar diameter 2.1-3.1 μm; height 1.0-2.8 μm).

This cyst is very similar to stomatocyst 116 except for the presence of the collar striations.

LM DESCRIPTION: This stomatocyst can only be distinguished from stomatocyst 116 *forma* B by the presence of the collar striations, and they are difficult to distinguish using LM except by careful focusing with high resolution optics (Fig. 38D). SEM confirmation is recommended.

REFERENCES: This stomatocyst has been observed in British Columbia, Canada (Duff & Smol, 1994; Zeeb, 1994); Ontario, Canada (Rybak, 1987 - cyst 6 in Fig. 6 but not Fig. 5; Rybak *et al.*, 1991 - cyst 26); the Northwest Territories, Canada (Sheath *et al.*, 1975); Minnesota, U.S.A. (Zeeb & Smol, 1993a); New York, U.S.A. (Duff & Smol, unpubl.); and Michigan, U.S.A. (Sandgren & Carney, 1983 - cyst 13).

ECOLOGY: *Dinobryon divergens* is a very common, cosmopolitan species (Kristiansen, 1980). It can be classified as a pH-indifferent or alkaliphilic species (Eloranta, 1989), and is found across a wide trophic gradient (Duthie & Ostrofsky, 1974; Moore, 1978; Eloranta, 1989). It can be considered an early spring-summer blooming species (Asmund, 1955; Moss, 1972; Davis, 1973; Duthie & Ostrofsky, 1974; Sheath *et al.*, 1975). Sheath *et al.* (1975) observed maximum cyst development in the summer.

The cyst is most common in alkaline locations (Sandgren & Carney, 1983; Zeeb & Smol, 1993a; Duff & Smol, 1994); Rybak *et al.* (1991) classified it as alkaliphilic. Its distribution was associated with deep, clear lakes in one survey (Duff, 1994).

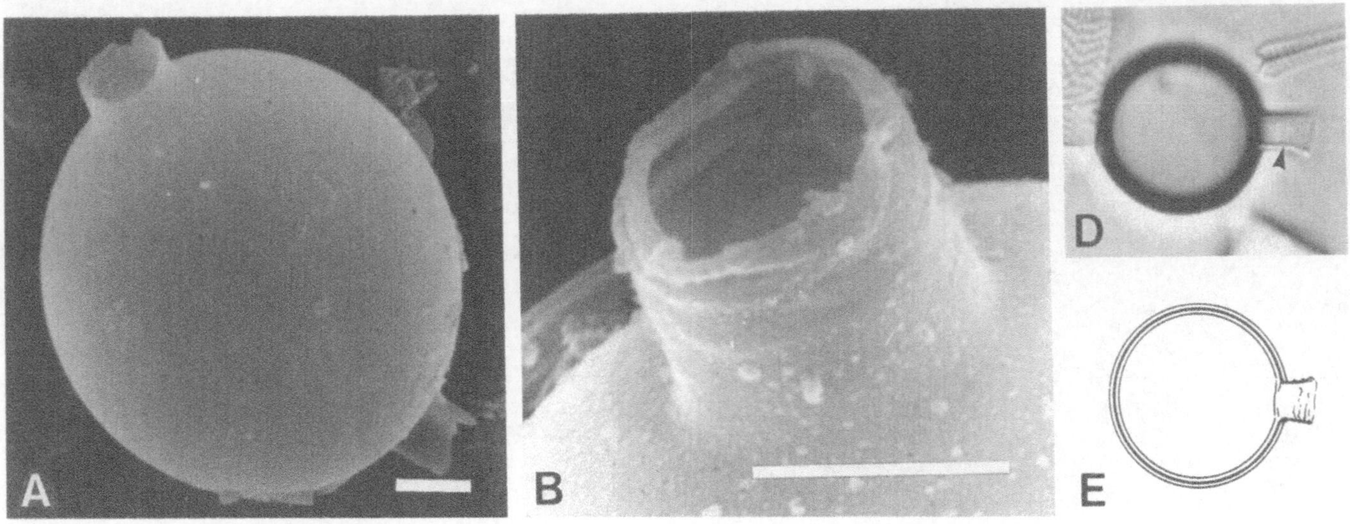

Fig. 38. Stomatocyst 161. A: SEM. B: SEM, close-up of collar. C: Line drawing of SEM. D: LM. Collar striations indicated by arrow. E: Line drawing of LM. Scale bars = 2 μm. LM x2000.

Sandgren (1989) discusses cyst development.

Stomatocyst 118, Zeeb *et al.*, 1990 (26) Fig. 39

BIOLOGICAL AFFINITY: *Mallomonas akrokomos* Ruttner in Pascher (Cronberg, 1980)

LOCALITY: Little Round Lake (44°47.5'N; 76°41.5'W), Ontario, Canada. One metre piston core, core depth 20 cm.

SEM DESCRIPTION: This is a smooth, oval stomatocyst with a considerable range in size and shape (6.1-7.6 μm x 7.8-10.2 μm; width : length ratio 0.66-0.88, median 0.78). The collar is cylindrical to obconical, may be situated centrically or acentrically on the posterior hemisphere, and often has a fluted outer margin (collar diameter 1.3-3.0 μm; height 0.3-1.1 μm). The pore is regular (pore diameter 0.4-0.6 μm).

LM DESCRIPTION: With LM, stomatocyst 118 is indistinguishable from stomatocyst 58. Stomatocyst 154 has a narrower collar than stomatocyst 118. Stomatocyst 19 is similar but lacks the collar of stomatocyst 118; however, immature specimens of stomatocyst 118 could be confused with stomatocyst 19.

REFERENCES: This common stomatocyst has been observed in British Columbia, Canada (Duff & Smol, 1994; Zeeb, 1994); Ontario, Canada (Zeeb *et al.*, 1990; Rybak *et al.*, 1987 - cyst 12); New York, U.S.A. (Duff & Smol, unpubl.); Michigan, U.S.A. (Carney & Sandgren, 1983 - *cysta teres* Nygaard; Sandgren & Carney, 1983 - cyst 47); Alaska, U.S.A. (Hilliard, 1966); Austria (Wawrik, 1980); Denmark (Nygaard, 1956 - *cysta teres*); Germany (Hickel & Maaβ, 1989); Italy (Leventhal, 1970 - *cysta teres*); Romania (Péterfi, 1967); and Argentina (Vigna, 1988). The cyst identified by Gritten (1977) as *cysta teres* is not identical to stomatocyst 118.

ECOLOGY: *Mallomonas akrokomos* is an acidophilic or pH-indifferent species with a wide tolerance for conductivity and total phosphorus; it has a clear preference for cold water (<10°C; Siver, 1991a). Paleolimnological analyses suggest that this morph tends to be found during oligotrophic to mildly eutrophic conditions (Zeeb *et al.*, 1990; Carney & Sandgren, 1983; Rybak *et al.*, 1987), although it was most

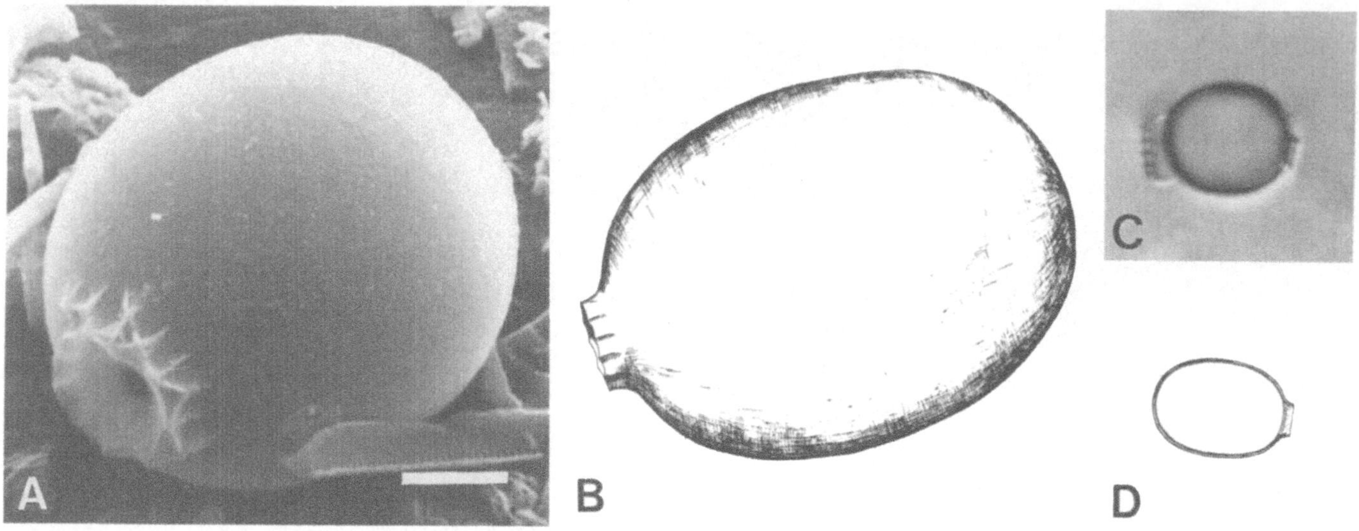

Fig. 39. Stomatocyst 118. A: SEM. B: Line drawing of SEM. C: LM. D: Line drawing of LM. Scale bar = 2 μm. LM x2000.

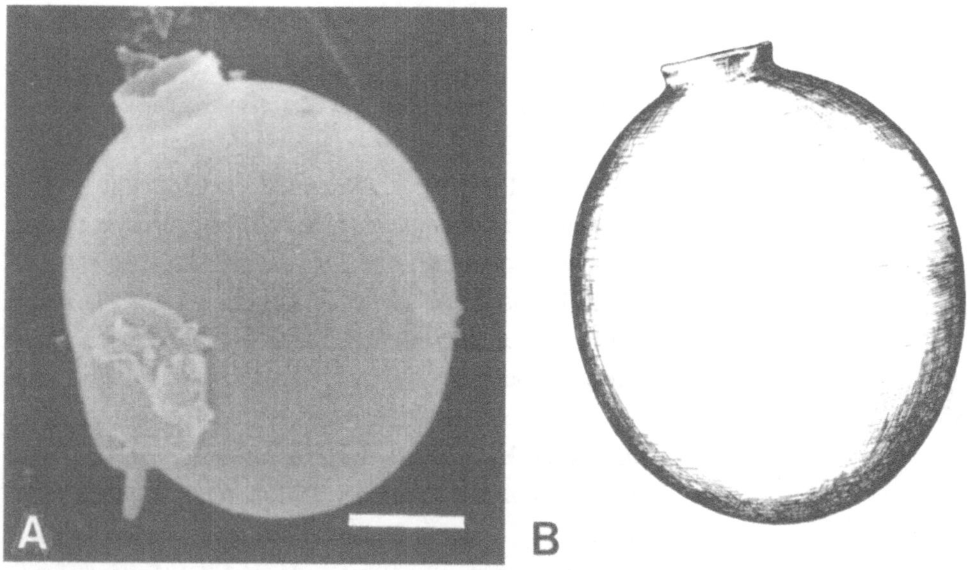

Fig. 40. Stomatocyst 58. A: SEM. B: Line drawing of SEM. Scale bar = 2 μm. See Figs. 39C, D for LM.

common in large, deep, productive freshwater lakes in British Columbia, Canada (Duff, 1994). In this survey, its distribution was also typical of a circumneutral species.

Stomatocyst 58, Duff & Smol, 1991 (3) Fig. 40

BIOLOGICAL AFFINITY: unknown

LOCALITY: Upper Wallface Pond (44°08.8'N; 74°03.3'W), Adirondack Park, New York, U.S.A. Gravity core, core depth 20.0-20.5 cm.

SEM DESCRIPTION: This is an oval stomatocyst (6.4-7.1 x 7.3-8.6 μm), with a low, fairly thinly silicified, cylindrical to obconical collar (diameter 1.9-2.3 μm; height 0.4-0.6 μm; collar diameter : cyst diameter ratio 0.28-0.40). There is a sharp delineation between the basal outer collar margin and the cyst body. The inner margin is acute. Pore details are unknown.

This stomatocyst is very similar to stomatocyst 118, but the latter has a gradual basal margin and a fluted collar. Stomatocyst 154 is larger, with a smaller collar diameter : cyst diameter ratio.

LM DESCRIPTION: With LM, this stomatocyst is indistinguishable from stomatocyst 118 (Figs. 39C, D).

REFERENCES: We have only observed stomatocyst 58 in New York, U.S.A. (Duff & Smol, 1991, unpubl.).

ECOLOGY: Stomatocyst 58 may be produced by an

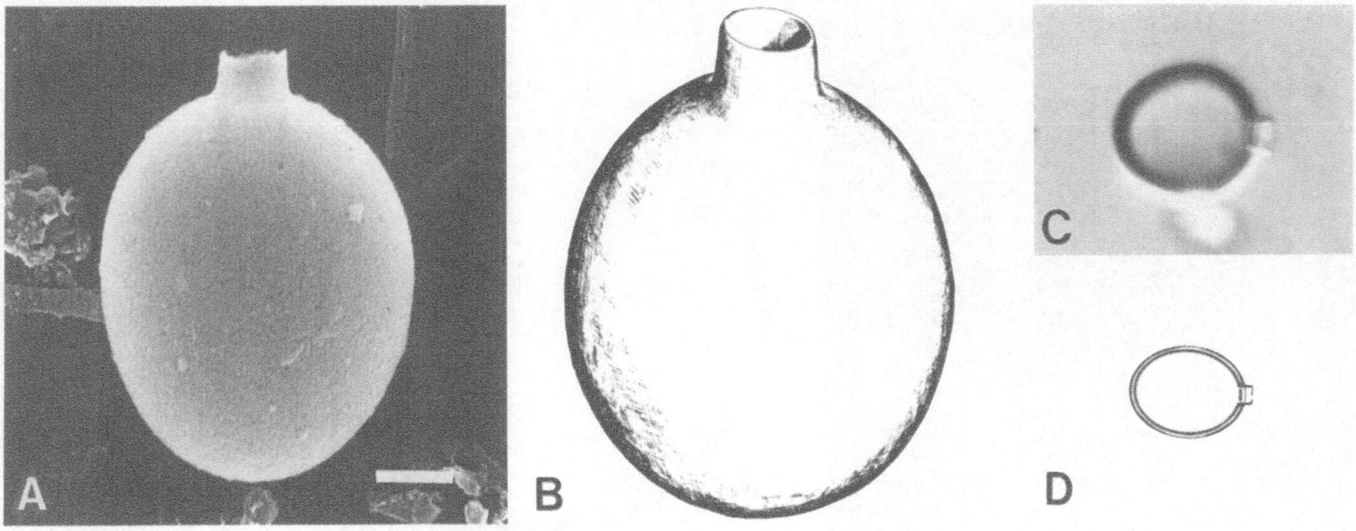

Fig. 41. Stomatocyst 154. A: SEM. B: Line drawing of SEM. C: LM. D: Line drawing of LM. Scale bar = 2 μm. LM x2000.

acidobiontic, and possibly metal-tolerant, chrysophyte (Duff & Smol, 1991).

Stomatocyst 154, Zeeb & Smol, 1993a (4) Fig. 41

BIOLOGICAL AFFINITY: unknown

LOCALITY: Elk Lake (47°11'N; 95°13'W), Clearwater County, Minnesota, U.S.A. Piston core, varve year A.D. 1695.

SEM DESCRIPTION: This is a smooth, oval stomatocyst (8.4-10.3 x 9.6-12.2 μm). The collar is cylindrical (collar diameter 1.8-2.2 μm; height to 2.0 μm; collar diameter : cyst diameter ratio 0.20-0.23). On one SEM specimen, an annulus appeared to be located approximately half-way up the inner collar margin, but this requires confirmation by additional SEM analysis.

This cyst is distinguished from stomatocyst 116 on the basis of shape alone, a somewhat arbitrary characteristic. It is differentiated from stomatocysts 118 and 58 on the basis of collar morphology.

LM DESCRIPTION: With LM, this cyst can be distinguished from stomatocyst 116 on the basis of shape. It is probably indistinguishable from stomatocyst 199.

REFERENCES: Zeeb & Smol (1993a - Minnesota, U.S.A.) is the only record of stomatocyst 154 to date.

ECOLOGY: This morph was most abundant during periods of inferred longer, wetter winters in Elk Lake, Minnesota (Zeeb & Smol, 1993b).

Spherical stomatocysts with a simple obconical collar but no ornamentation

Stomatocyst 182, Brown & Smol in Brown *et al.*, 1994 (10) Fig. 42

BIOLOGICAL AFFINITY: unknown

LOCALITY: Nordvestø (76°44'N; 73°13'W), Carey Islands, Greenland. SIPRE peat core, core depth 121-122 cm.

SEM DESCRIPTION: This is a smooth spherical (diameter 5.2-6.7 μm) stomatocyst. The collar is cylindrical with a very wide, flattened, obconical apex and an abrupt inner margin (maximum diameter 1.6-2.9 μm; height 0.3-0.5 μm; collar diameter : cyst diameter ratio 0.29-0.56). A swollen annulus surrounds a regular pore (diameter 0.3-0.4 μm).

LM DESCRIPTION: Stomatocyst 182 is difficult to distinguish from stomatocysts 125 (Figs. 32C, D) and/or 181 (Figs. 16C, D, and E). If two or more of these similar stomatocysts are present, SEM confirmation is needed.

REFERENCES: This cyst has only been found in Greenland (Brown *et al.*, 1994).

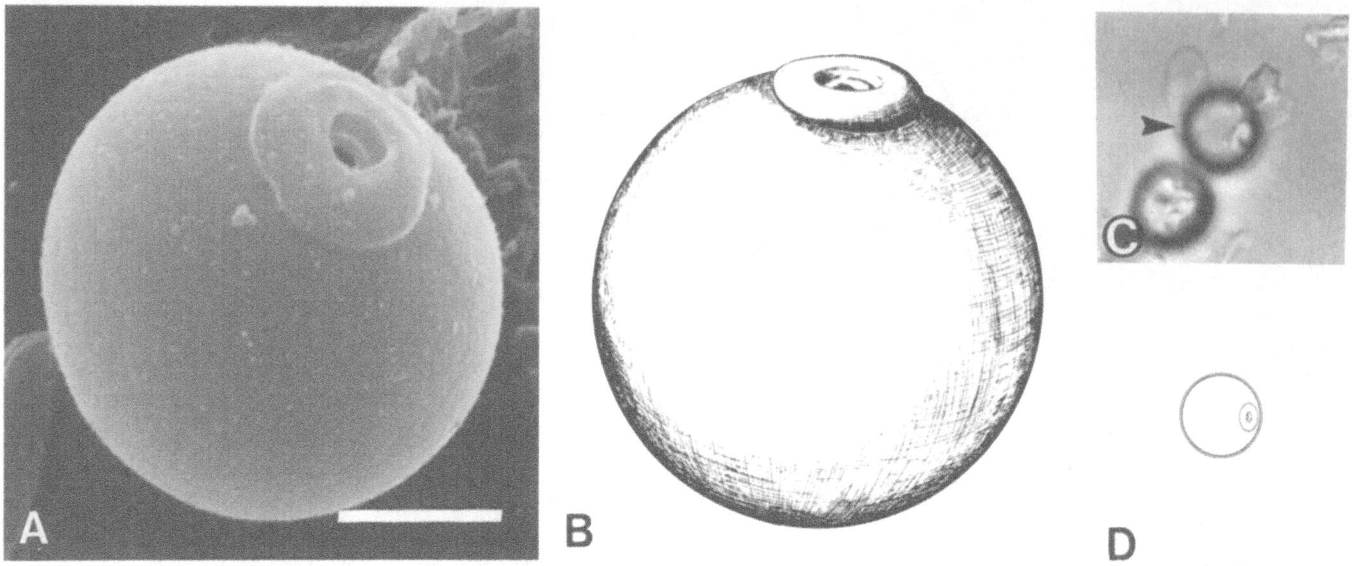

Fig. 42. Stomatocyst 182. A: SEM. B: Line drawing of SEM. C: LM. Specimen indicated by arrow. D: Line drawing of LM. Scale bar = 2 μm. LM x2000.

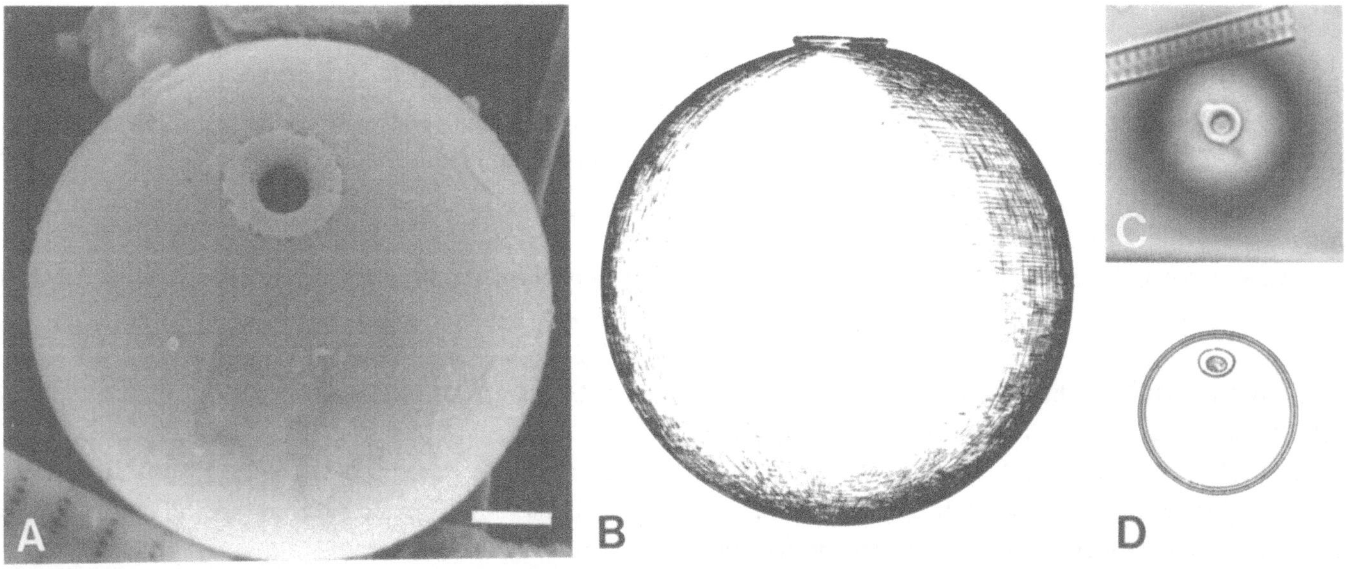

Fig. 43. Stomatocyst 151. A: SEM. B: Line drawing of SEM. C: LM. D: Line drawing of LM. Scale bar = 2 μm. LM x2000.

ECOLOGY: Autecological details are unknown.

Stomatocyst 151, Zeeb & Smol, 1993a (4) Fig. 43

BIOLOGICAL AFFINITY: unknown

LOCALITY: Elk Lake (47°11'N; 95°13'W), Clearwater County, Minnesota, U.S.A. Piston core, varve year A.D. 1786.

SEM DESCRIPTION: This is a large spherical stomatocyst (diameter 12.6-16.3 μm) with a smooth to psilate surface. The collar is very low and grossly obconical and surrounds a regular pore (collar diameter 2.8-3.7 μm; height <0.1-0.4 μm; pore diameter 1.0-1.4 μm). One questionable specimen has a sloping planar annulus (diameter 2.0 μm).

LM DESCRIPTION: This stomatocyst is distinguished from similar cysts (e.g. stomatocysts 127 and 197) by its obconical collar.

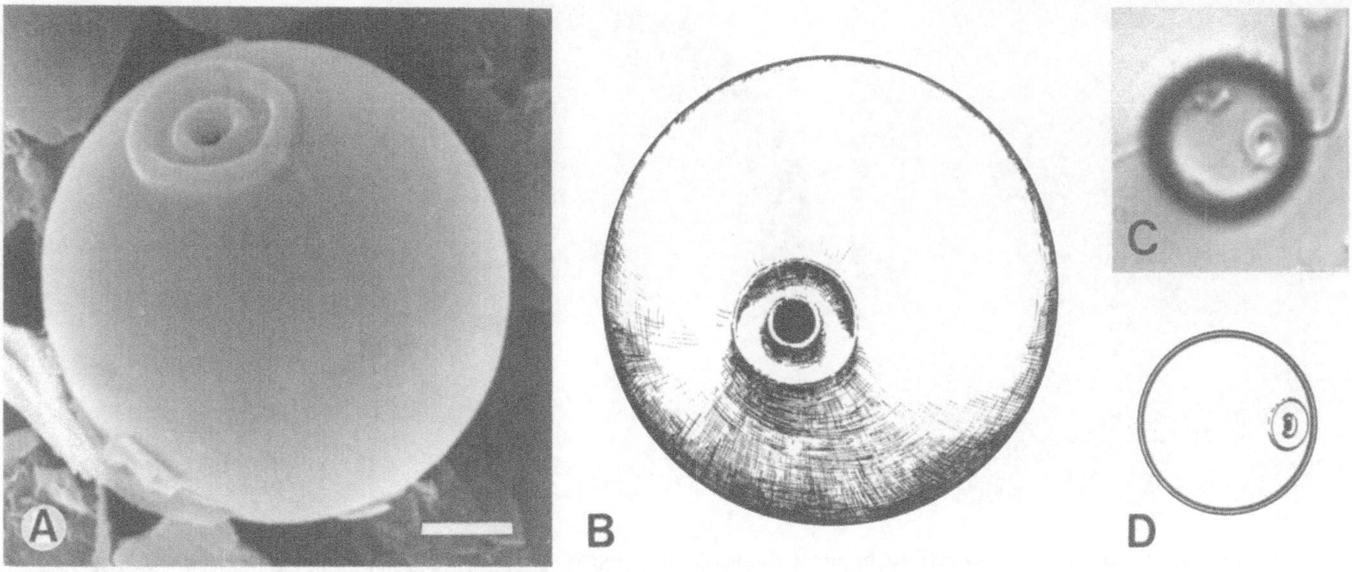

Fig. 44. Stomatocyst 16. A: SEM. B: Line drawing of SEM. C: LM. D: Line drawing of LM. Scale bar = 2 μm. LM x2000.

REFERENCES: This stomatocyst has been observed in Ontario, Canada (Zeeb & Smol, unpubl.), British Columbia, Canada (Duff & Smol, 1994; Zeeb, 1994), and Minnesota, U.S.A. (Zeeb & Smol, 1993a).

ECOLOGY: Ecological data for this rare morph are unavailable.

Spherical or oval stomatocysts with a true or false complex collar but no ornamentation

Stomatocyst 16, Duff & Smol, 1988 (8) Fig. 44

BIOLOGICAL AFFINITY: unknown

LOCALITY: "Rock Basin Lake" (unofficial name) (78°29.5'N; 76°44.8'W), Ellesmere Island, Northwest Territories, Canada. Piston core, core depth 15-16 cm.

SEM DESCRIPTION: This stomatocyst is spherical (diameter 7.5-11.2 μm) and smooth. A low, cylindrical primary collar (diameter 1.3-1.7 μm) immediately surrounds the regular pore (diameter 0.6-1.2 μm). A planar interannulus separates the primary collar from the low, cylindrical secondary collar (diameter 2.4-3.5 μm). The primary and secondary collars are approximately equal in height, and the collar apices may be rounded.

LM DESCRIPTION: Stomatocyst 16 can be distinguished from similar cysts of this size and shape (e.g. stomatocyst 127) by the double collar.

REFERENCES: This stomatocyst has been observed in the Northwest Territories, Canada (Smol, 1985 - Fig. 2B; Duff & Smol, 1988), the Yukon Territory, Canada (Brown & Smol, unpubl.), and California, U.S.A. (Mahood & Adam, 1979 - type 6; Adam, 1981 - type 6).

ECOLOGY: The presence of this cyst in arctic locations indicates that it is produced by a cold-tolerant species (Smol, 1985; Duff & Smol, 1988).

Stomatocyst 203, Duff & Smol, 1994 (3) Fig. 45

BIOLOGICAL AFFINITY: unknown

LOCALITY: King Lake West (51°45'N; 120°50'W), British Columbia, Canada. Ekman dredge, surface (1 cm) sediment sample.

SEM DESCRIPTION: This is a spherical stomatocyst (diameter 12.2-12.3 μm) with a false complex collar. The outer collar margin is very gradual and conical, the apex is acute with a central groove, and the inner collar margin protrudes beyond the outer collar margin (outer collar margin diameter 6.1-6.2 μm; height 1.4-1.8 μm; inner collar margin diameter 3.5-3.8 μm; height 1.9-2.3 μm); the pore has not

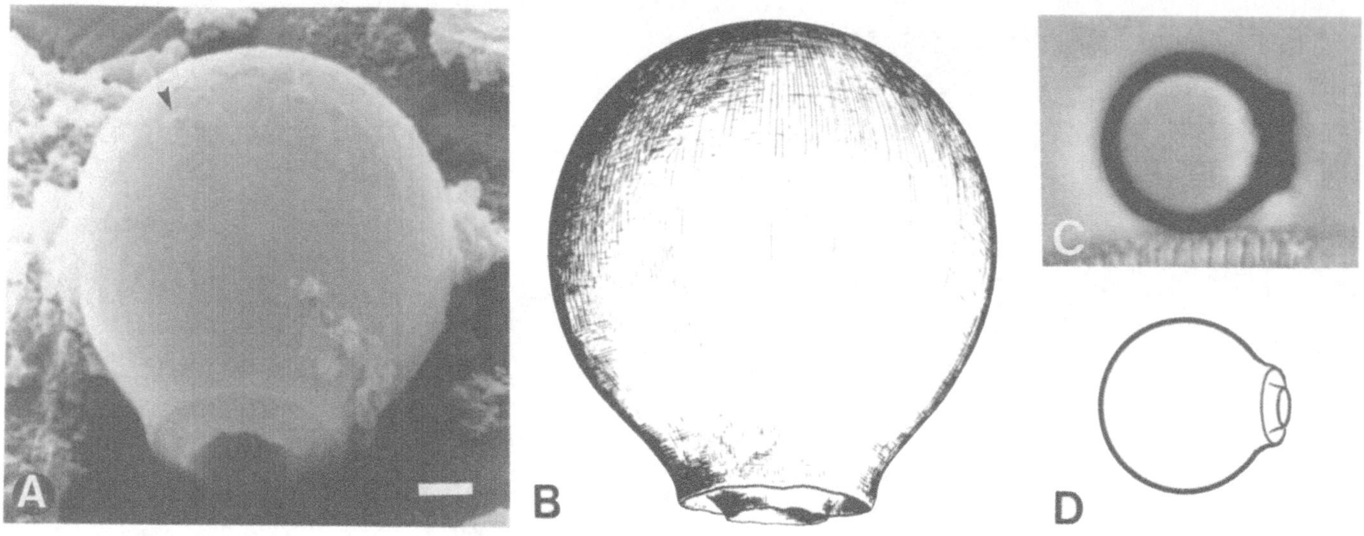

Fig. 45. Stomatocyst 203. A: SEM. Verruca indicated by arrow. B: Line drawing of SEM. C: LM. D: Line drawing of LM. Scale bar = 2 μm. LM x2000.

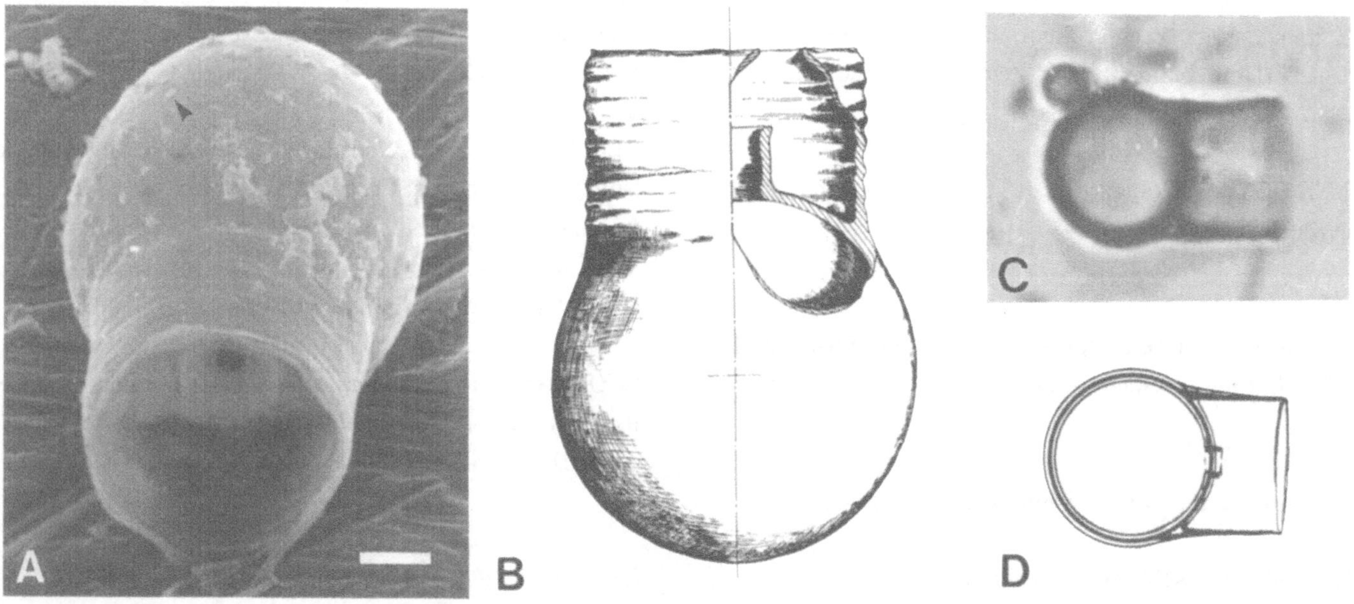

Fig. 46. Stomatocyst 204. A: SEM. Verruca indicated by arrow. B: Line drawing of SEM. Cutaway shows primary collar and primary interannulus. C: LM. D: Line drawing of LM. Scale bar = 2 μm. LM x2000.

been observed. The surface is smooth to ornamented with very low rounded verrucae (diameter 0.4-0.5 μm; Fig. 45A).

LM DESCRIPTION: This stomatocyst can be distinguished from stomatocyst 204 by the lack of an interannulus, and the gradual outer margin of the collar.

REFERENCES: We have only observed this stomatocyst in British Columbia, Canada (Duff & Smol, 1994; Zeeb, 1994), and it could not be definitely identified with descriptions in the literature.

ECOLOGY: Autecological details are not yet available.

Stomatocyst 204, Duff & Smol, 1994 (9) Fig. 46

BIOLOGICAL AFFINITY: *Uroglena volvox* Ehrenberg (Kristiansen, 1980)

LOCALITY: Durrance Lake (48°30'N; 123°25'W), British Columbia, Canada. Ekman dredge, surface (1 cm) sediment sample.

SEM DESCRIPTION: This stomatocyst is large and spherical (diameter 9.1-12.7 μm) with a true complex collar. The primary collar is cylindrical (diameter 2.1-3.2 μm; height 0.6-1.6 μm) and surrounds a regular pore (diameter 0.8-1.1 μm). The cylindrical secondary collar (diameter 4.6-9.2 μm; height 0.2-5.2 μm) is separated from the primary collar by a planar interannulus (Fig. 46B). The secondary collar is often broken, and it may be horizontally striated on more mature specimens. The cyst surface may be smooth, or ornamented with a few to many irregular verrucae (diameter 0.2-0.7 μm; Fig. 46A).

Sandgren (1980b) describes the cyst ultrastructure.

LM DESCRIPTION: This stomatocyst is distinguished by its double collar and large body size. Stomatocyst 203 is similar, but has a false double collar (i.e. it lacks an interannulus) and a more gradual outer collar margin.

REFERENCES: This stomatocyst has been observed in British Columbia, Canada (Donaldson & Stein, 1984; Duff & Smol, 1994; Zeeb, 1994); Ontario, Canada (Zeeb & Smol, unpubl.); New York, U.S.A. (Duff & Smol, unpubl.); Sweden (Cronberg & Kristiansen, 1980); and France (Andrieu, 1937 - *Clericia spectabilis* Andrieu).

ECOLOGY: *Uroglena volvox* is a widespread species (Hilliard, 1959; Kristiansen, 1976; Cronberg & Kristiansen, 1980; Kristiansen, 1980; Donaldson & Stein, 1984). In freshwater lakes in British Columbia, Canada (Duff, 1994), stomatocyst 204 demonstrated an association with slightly acidic to circumneutral, productive lakes. Cronberg & Kristiansen (1980) observed this cyst in eutrophic, polluted Lake Trummen, Sweden (pH 7.29, temperature 9.8°C). Donaldson & Stein (1984) observed encystment in August.

Stomatocyst 187, Brown & Smol in Brown *et al.*, 1994 (3) Fig. 47

BIOLOGICAL AFFINITY: unknown

LOCALITY: Nordvestø (76°44'N; 73°13'W), Carey Islands, Greenland. SIPRE peat core, core depth 50-51 cm.

SEM DESCRIPTION: This stomatocyst is oblate (4.9-5.6 x 4.1-4.7 μm) with a complex collar. The primary collar is cylindrical (apical diameter 1.2-1.6 μm), with a vertical projection which terminates in a roughly circular, flattened, mushroom-like cap (height of projection 1.5-1.6 μm; cap diameter 1.7-1.9 μm). A planar interannulus separates the primary and secondary collars. The secondary collar is thick, cylindrical to obconical with a very irregular apex (apical diameter 3.3-3.8 μm; maximum height 0.5-0.8 μm). Pore details are unknown. The cyst surface is smooth to sparsely scabrate (Fig. 47A).

Stomatocyst 187 resembles stomatocyst 135 but is smaller, lacks a tertiary collar, and differs with respect to the projection morphology. However, it is likely that the descriptions of these similar cysts will be modified with further SEM analysis.

LM DESCRIPTION: Stomatocysts 187 and 135 can only be distinguished reliably on the basis of size. Stomatocyst 187 is differentiated from stomatocyst 134 by the presence of the secondary collar.

REFERENCES: Cysts identical to stomatocyst 187 have not yet been found outside of Greenland (Brown *et al.*, 1994), although there are reports of a much larger, but similar, cyst (see discussion under stomatocyst 135).

ECOLOGY: We do not yet know the ecological optima and tolerances of this morphotype.

Stomatocyst 135, Duff & Smol in Duff *et al.*, 1992 (18) Fig. 48

BIOLOGICAL AFFINITY: The biological affinity of stomatocyst 135 is unknown, but several species of *Uroglena* are known to produce cysts with hooked collar projections (e.g. *U. notabilis* Mack (Bourrelly, 1957); *U. lindiae* Bourrelly (Bourrelly, 1957; Nygaard, 1977); *U. soniaca* Conrad (Conrad, 1938)), although these cysts are typically ornamented with short spines or verrucae.

LOCALITY: "Camp Pond" (unofficial name), Cape Herschel (78°37'N; 74°42'W), Ellesmere Island, Northwest Territories, Canada. Moss sample, collected July 20, 1986.

SEM DESCRIPTION: This stomatocyst varies in shape from widely oval to nearly spherical, and has a considerable range of size (dimensions 5.7-12.6 x 7.3-15.0 μm). It is distinguished by its complex collar. Around the regular pore (diameter 0.5 μm) is a conical primary collar (diameter 1.1-3.8 μm). A conical secondary collar with an inwardly hooked

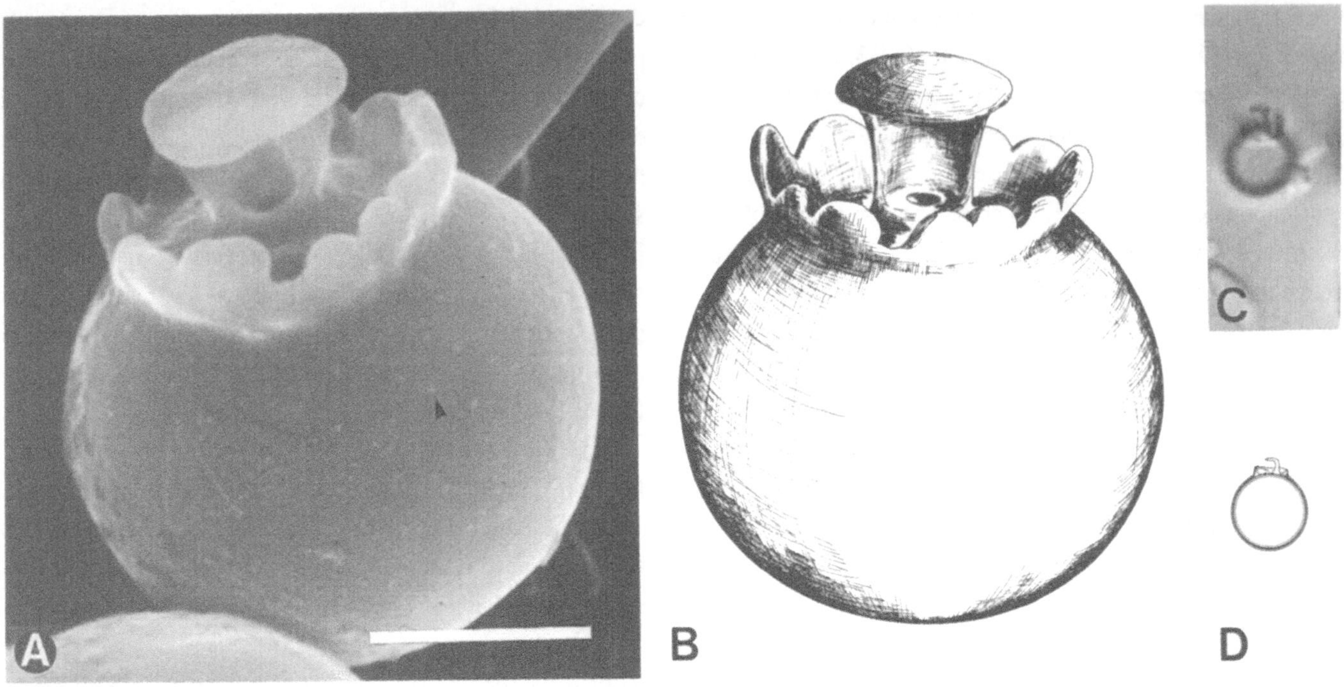

Fig. 47. Stomatocyst 187. A: SEM. Scabra indicated by arrow. B: Line drawing of SEM. C: LM. D: Line drawing of LM. Scale bar = 2 μm. LM x2000.

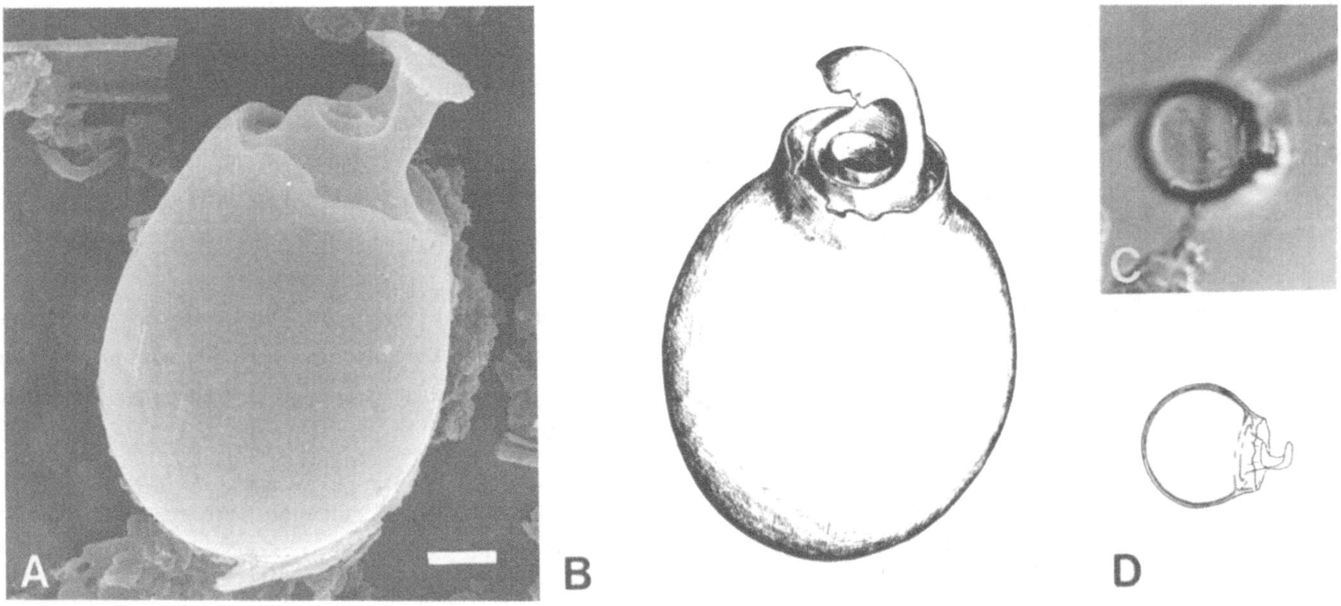

Fig. 48. Stomatocyst 135. A: SEM. B: Line drawing of SEM. C: LM. D: Line drawing of LM. Scale bar = 2 μm. LM x2000.

projection arising from one end immediately surrounds the primary collar (secondary collar diameter 2.1-3.4; height 0.9-1.5; height of projection above collar 0.7-3.0 μm). The projection varies widely in shape. The tertiary collar is separated from the secondary collar by a planar interannulus, and is wide, low, cylindrical to obconical (diameter 3.7-8.3 μm; height 0.4-1.5 μm). The apex of the tertiary collar may be irregular.

See the description of stomatocyst 187 for the characteristics used to distinguish the two morphotypes.

LM DESCRIPTION: This cyst is identified by its collar complex. It can be distinguished reliably from stomatocyst 187 only on the basis of size.

REFERENCES: Numerous cysts roughly conforming to this

type have been described in the literature. Stomatocyst 135, as defined here, has been observed in the Northwest Territories, Canada (Duff *et al.*, 1992); British Columbia, Canada (Duff & Smol, 1994; Zeeb, 1994); California, U.S.A. (Mahood & Adam, 1979 - types 1, 2, 4 and 5; Adam, 1980b - type 160; Adam, 1981 - types 2, 5 and 160; Adam & Mahood, 1980b - type 160; Adam & Mahood, 1980c - type 2; Adam & Mahood 1981b - type 160; Zeeb *et al.*, subm.); Montana, U.S.A. (Adam & Mehringer, 1980a - type 160); Connecticut, U.S.A. (Conrad, 1940 - *Carnegia operculata* Frenguelli ≠ *Carn. operculata* Frenguelli (1936); Gritten, 1977 - *Carn. frenguelli* Clerici); Washington, U.S.A. (VanLandingham, 1964 - cyst type 24); Denmark (Nygaard, 1956 - *Uroglena soniaca* Conrad (specimens without spines)); Italy (Leventhal, 1970 - *U.* cf. *soniaca* Conrad ("immature" forms), *Carn.* cf. *operculata* Frenguelli, *Carn.* cf. *frenguelli* Clerici, and *Carn.* sp. (Fig. X-12)); Spain (Rull, 1986 - *formae* 43, 44, 45, and 63); Argentina (Frenguelli, 1925 - *Clericia complexa* Frenguelli; Frenguelli, 1932 - *Trachelomonas complexa* Frenguelli, *T. complexa* var. *major* Frenguelli, *T. frenguelli* Clerici; Frenguelli, 1935b - *Outesia tecta* Frenguelli; Frenguelli, 1936 - *Carn. frenguelli* (Clerici) Deflandre); and the Kerguelen Islands, Southern Indian Ocean (Andrieu, 1936 - *Carn. frenguelli* (Clerici) Deflandre).

There are several reports of cysts that are similar to stomatocyst 135, but are ornamented with verrucae or conula: California, U.S.A. (Adam, 1980c - type 160), Spain (Rull, 1986 - *forma* 53), Denmark (Nygaard, 1956 - *Uroglena soniaca* Conrad), and Italy (Leventhal, 1970 - *U.* cf. *soniaca* Conrad).

Reticular morphs have also been found: Connecticut, U.S.A. (Conrad, 1940 - *Carnegia johannis* Andrieu; Gritten, 1977 - *Carn. johannis*, *Carn. willingtoniensis* Conrad), Michigan, U.S.A. (Sandgren & Carney, 1983 - cyst 46), and France (Andrieu, 1937 - *Carn. johannis*).

A form which is identical in size to stomatocyst 135, but has collar morphology more closely resembling stomatocyst 187, has also been described from Italy (Leventhal, 1970 - *Carnegia complexa* Conrad), Argentina (Frenguelli, 1925 - *Clericia obtecta* Frenguelli; Frenguelli, 1932 - *Trachelomonas obtecta* Frenguelli), and the Kerguelen Islands, Southern Indian Ocean (Andrieu, 1936 - *Carn. complexa* (Frenguelli) Deflandre).

ECOLOGY: This stomatocyst has been observed in shallow arctic ponds (Duff *et al.*, 1992), alpine ponds (Adam & Mahood, 1981a; Rull, 1986), and temperate lakes (Duff, 1994), and during inferred periods of low water level in Lago di Monterosi, Italy (Leventhal, 1970). It is probably produced by a species that is tolerant of cold water, and perhaps by a littoral (epiphytic or aerophilic?) species. Adam type 5 apparently prefers *Sphagnum* bogs (Adam & Mahood, 1981a). This taxon appears to be tolerant of a wide range of pH (Adam, 1980c; Duff *et al.*, 1992), although it was most common in alkaline locations in Duff's (1994) survey.

Stomatocysts with or without collar; ornamentation consists of anterior projections simulating simple or complex collar

Stomatocyst 205, Duff & Smol, 1994 (7) Fig. 49

BIOLOGICAL AFFINITY: unknown

LOCALITY: Gardom Lake (50°40'N; 119°10'W), British Columbia, Canada. Ekman dredge, surface (1 cm) sediment sample.

SEM DESCRIPTION: This thick-walled, spherical stomatocyst (diameter 9.2-14.0 μm; median 11.1 μm) has a deep, regular to conical pore (outer pore diameter 1.2-2.2 μm; pore depth 0.4-0.6 μm). There is no collar, but low, irregularly shaped and sized (roughly verrucose) siliceous projections (diameter ca. 1.0 μm) are located in a ring either immediately surrounding the pore, or up to 1.5 μm away. The cyst surface is psilate to smooth (Fig. 49A), possibly representing different developmental stages. Without the ornamentation around the pore, this stomatocyst closely resembles stomatocyst 150.

LM DESCRIPTION: This stomatocyst is identified with LM by the ornamentation surrounding the pore, the large body size, and the multiple cyst walls. If the ornamentation is undeveloped or not visible, stomatocyst 205 could be mistaken for stomatocyst 150 (Fig. 8F).

REFERENCES: This cyst has been observed in British Columbia, Canada (Duff & Smol, 1994; Zeeb, 1994); Ontario, Canada (Rybak *et al.*, 1991 - cyst 13; Zeeb & Smol, unpubl.); New York, U.S.A. (Duff & Smol, unpubl.); Poland (Rybak, 1986 - *cysta cascus* Rybak); and Argentina (Frenguelli, 1932 - *Trachelomonas volvocina* Ehrenberg var. *fibulata* Frenguelli).

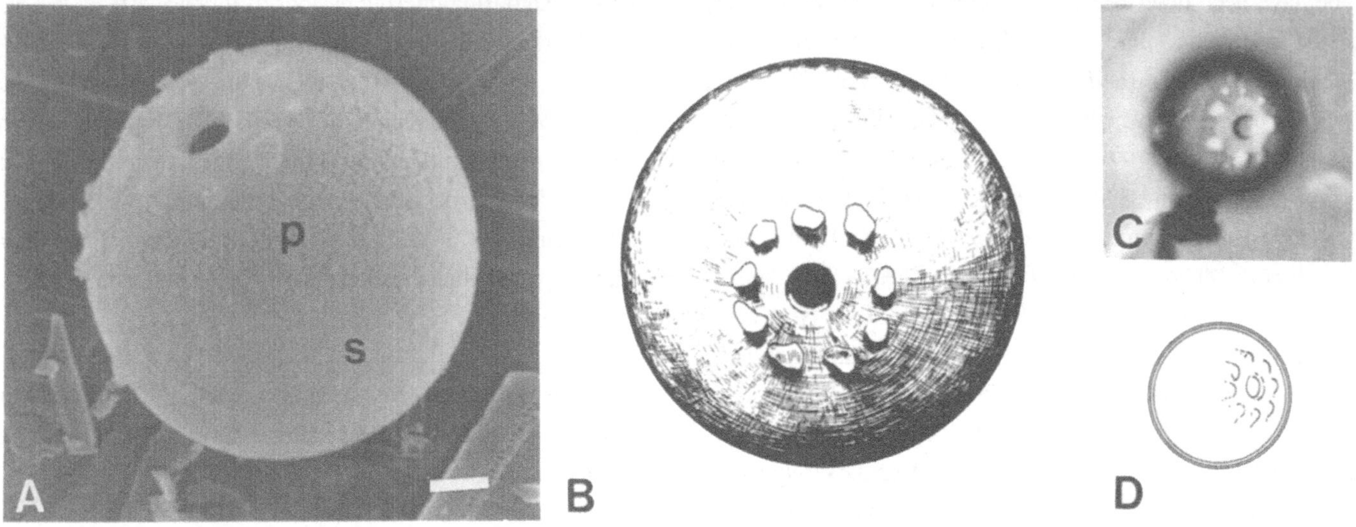

Fig. 49. Stomatocyst 205. A: SEM. p = psilate section of cyst wall; s = smooth secion of cyst wall. B: Line drawing of SEM. C: LM. D: Line drawing of LM. Scale bar = 2 μm. LM x2000.

ECOLOGY: This stomatocyst is probably produced by an alkaliphilic species (Rybak *et al.*, 1991; Duff, 1994) that is tolerant of high conductivities (Duff, 1994).

Stomatocyst 137, Duff & Smol in Duff *et al.*, 1992 (4) Fig. 50

BIOLOGICAL AFFINITY: unknown

LOCALITY: "Beach Ridge Pond" (unofficial name), Cape Herschel (78°37'N; 74°42'W), Ellesmere Island, Northwest Territories, Canada. New tow, collected July 23, 1986.

SEM DESCRIPTION: This stomatocyst is spherical (diameter 6.9-7.5 μm). The regular pore (diameter 0.7-0.8 μm) is surrounded by a sloping planar annulus (diameter 1.1-1.7 μm). There is either no collar (Fig. 50B) or a thin, low cylindrical collar (diameter 1.7-1.9 μm) with numerous irregular projections at the apex (Fig. 50A). Around the collar are many densely spaced globose, baculate or flattened projections which may have flared apices. The remainder of the cyst body is psilate or ornamented with small, shallow depressions (diameter 0.1-0.4 μm).

LM DESCRIPTION: Stomatocyst 137 can be distinguished from similar stomatocysts (e.g. stomatocyst 136) by the number and jagged appearance of the anterior projections. The depressions were not observed with LM.

REFERENCES: There are several reports of cysts with multiple projections surrounding the pore region, but they are usually much larger than stomatocyst 137. Of these, Mahood & Adam (1979 - California, U.S.A.) type 34, which is 11.1 μm in diameter, agrees most closely with our description of stomatocyst 137. Other similar but larger morphotypes have been reported from Connecticut, U.S.A. (Gritten, 1977 - *Clericia cristata* Frenguelli), Italy (Leventhal, 1970 - *C.* sp. C), and France (Andrieu, 1937 - *Carnegia arvernensis* Andrieu). Rull (1986 - Spain) *forma* 40, as originally described, is also larger than stomatocyst 137, but the SEM specimen identified by Rull (1991) as *forma* 40 is identical to stomatocyst 137. Adam & Mahood (1979b - California, U.S.A.) type 70 is also similar in size and ornamentation.

ECOLOGY: Ecological information is limited, but this stomatocyst may be produced by a littoral (periphytic?) chrysophyte (Duff *et al.*, 1992; Leventhal, 1970). In Rull's (1991) stratigraphic analysis, however, this morph was apparently less common during periods of extensive littoral zone development.

Stomatocyst 136, Duff & Smol in Duff *et al.*, 1992 (6) Fig. 51

BIOLOGICAL AFFINITY: unknown

LOCALITY: "Camp Pond" (unofficial name), Cape Herschel (78°37'N; 74°42'W), Ellesmere Island, Northwest Territories, Canada. Net tow, collected July 28, 1986.

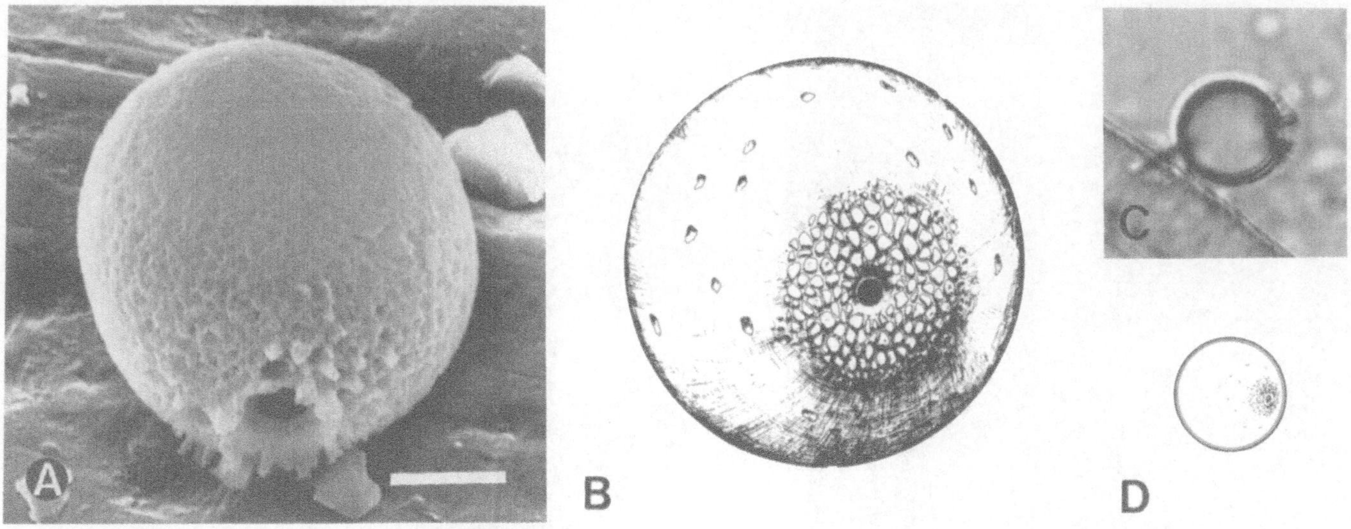

Fig. 50. Stomatocyst 137. A: SEM. B: Line drawing of SEM. C: LM. D: Line drawing of LM. Scale bar = 2 μm. LM x2000.

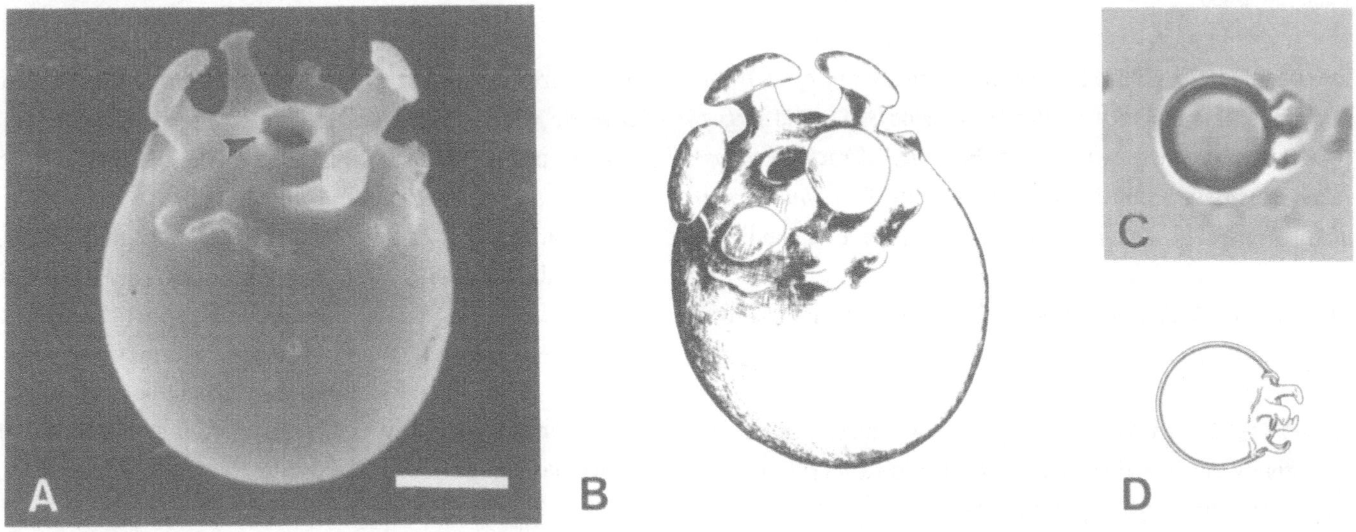

Fig. 51. Stomatocyst 136. A: SEM. Collar indicated by arrow. B: Line drawing of SEM. C: LM. D: Line drawing of LM. Scale bar = 2 μm. LM x2000.

SEM DESCRIPTION: This is a smooth, oval cyst (5.6-7.6 x 6.3-8.2 μm) which is often slightly asymmetric. The regular pore (diameter 0.6-0.9 μm) is surrounded by a low, widely conical collar (base diameter 1.9-2.0 μm; apical diameter 0.9-1.4 μm; height 0.3-0.5 μm; Fig. 51A). Around the collar are four to five projections, with inwardly flexed, flattened, circular to oval apices (height 1.1-2.5 μm), usually surrounded by additional projections. The additional projections are usually much smaller (height 0.3-0.6 μm) but occasionally equal to the central ones in length.

LM DESCRIPTION: With LM, this stomatocyst can be identified by the smooth surface and collar projections, which are fewer in number, and usually longer, than those of stomatocyst 137.

REFERENCES: This stomatocyst has been observed in the Northwest Territories, Canada (Duff *et al.*, 1992); the Yukon Territory, Canada (Brown & Smol, unpubl.); British Columbia, Canada (Zeeb, 1994); California, U.S.A. (Zeeb *et al.*, subm.; Adam, 1980b - type 157; Adam, 1981 - types 156 and 259); Connecticut, U.S.A. (Gritten, 1977 - *Carnegia arvernensis* Andrieu ≠ *C. arvernensis* Andrieu (1937)); Spain (Rull, 1986, 1991 - *formae* 36 and 37); Italy (Leventhal, 1970 - *Clericia* sp. A); and the Kerguelen Islands, Southern Indian Ocean (Andrieu, 1936 - *Carnegia deflandrei* Andrieu).

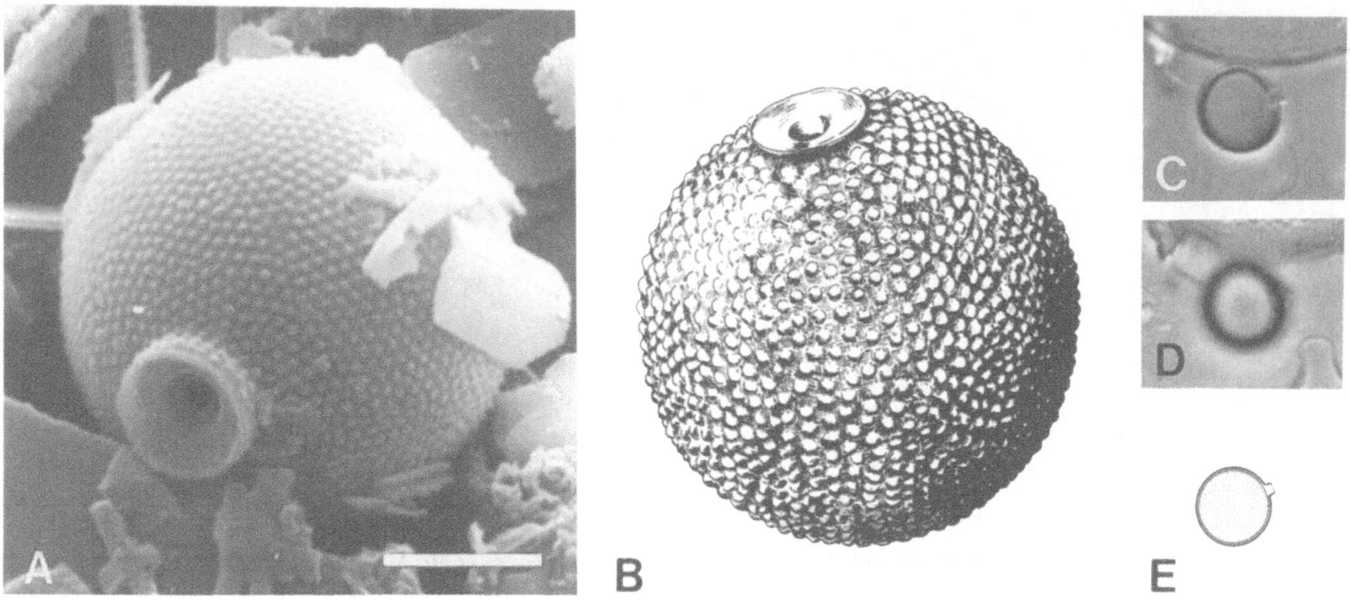

Fig. 52. Stomatocyst 117. A: SEM. B: Line drawing of SEM. C: LM, cross-sectional view. D: LM, surface view. E: Line drawing of LM. Scale bar = 2 μm. LM x2000.

Frenguelli (1925, 1932) has reported a number of cysts from Argentina, that are very similar to stomatocyst 136 but larger (*Clericia cristata* var. *paucicostata* Frenguelli, *Trachelomonas (Clericia) cristata* var. *cylindrica* Frenguelli, *T. cristata forma minor* Frenguelli, and *T. complexa* var. *minor* Frenguelli).

ECOLOGY: This cyst appears to be associated with shallow water (Duff *et al.*, 1992; Rull, 1986; Leventhal, 1970) and may be produced by a littoral (periphytic?) species. In Duff *et al.*'s (1992) survey of arctic ponds, it was most common in a relatively nutrient-rich site. According to Rull (1991), it is associated with elevated productivity.

Stomatocysts ornamented with scabrae

Stomatocyst 117, Zeeb *et al.*, 1990 (19) Fig. 52

BIOLOGICAL AFFINITY: unknown

LOCALITY: Little Round Lake (44°47.5'N; 76°41.5'W), Ontario, Canada. One metre piston core, core depth 15 cm.

SEM DESCRIPTION: This spherical stomatocyst (diameter 4.8-6.4 μm) has a short cylindrical to slightly obconical collar (apical diameter 1.2-2.0 μm; height 0.3-0.8 μm) with a conical pore (diameter 0.3-0.5 μm). The inner collar margin may be sloping, or may be steep with a flat planar annulus. The entire cyst surface and the basal portion of the collar are covered with rounded scabrae arranged equidistantly in a regular pattern (scabrae diameter ≤0.1-0.2 μm; distance between scabrae is always 0.2 μm).

Stomatocyst 117 resembles stomatocyst 210, but the latter has conula, not scabrae, and the ornamentation density is less.

LM DESCRIPTION: With LM, the scabrae are visible (Figs. 52D, E). However, this stomatocyst cannot be distinguished from stomatocyst 178, which is reticular, or stomatocyst 210, which has conula.

REFERENCES: This cyst has been observed in Ontario, Canada (Zeeb *et al.*, 1990), British Columbia, Canada (Duff & Smol, 1994), Minnesota, U.S.A. (Zeeb & Smol, 1993a), and California, U.S.A. (Mahood & Adam, 1979 - type 44; Adam & Mahood, 1979b - types 54, 98 and 99; Adam & Mahood, 1981b - type 54). Adam & Mahood (1979b) type 97 is very similar to stomatocyst 117 but has several echinate equatorial spines. Adam & Mahood (1979b) type 98 is not identical to the morph identified as type 98 in their other publications (Adam & Mahood, 1980b; Adam & Mehringer, 1980c), which is equivalent to our stomatocyst 210.

ECOLOGY: Stomatocyst 117 has been classified as an oligotrophic indicator (Zeeb *et al.*, 1990). However, as a collective category, with stomatocysts 178 and 210, it was most common in slightly acidic to circumneutral, productive lakes in British Columbia, Canada (Duff, 1994).

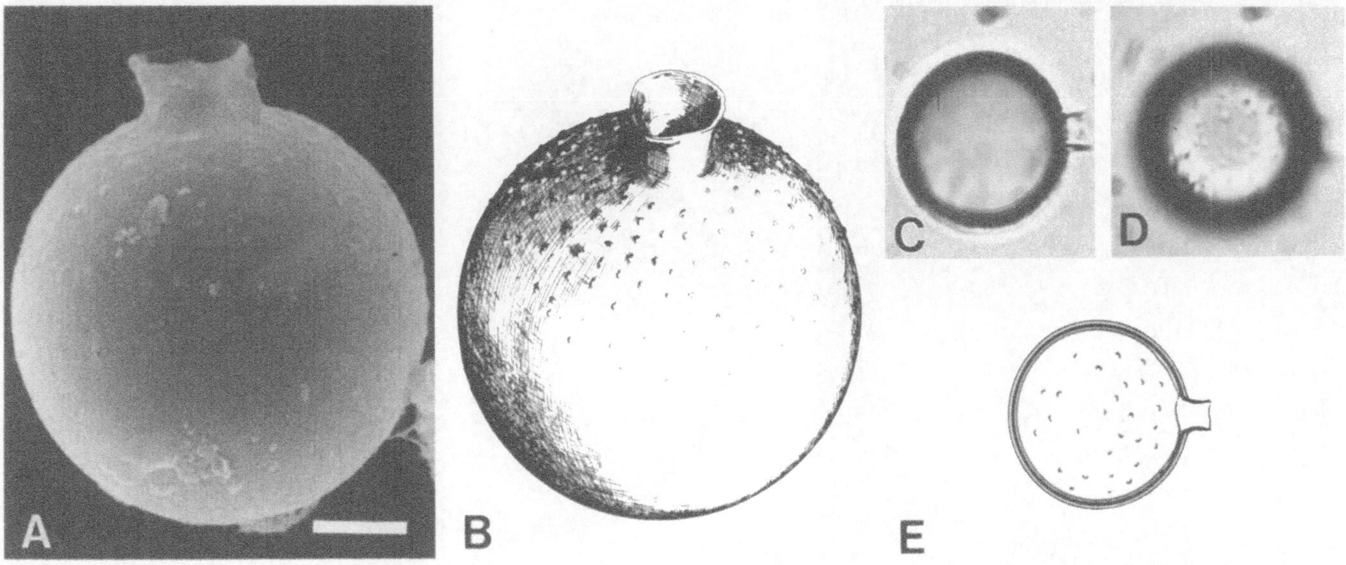

Fig. 53. Stomatocyst 129. A: SEM. B: Line drawing of SEM. C: LM, cross-sectional view. D: LM, surface view. E: Line drawing of LM. Scale bar = 2 μm. LM x2000.

Stomatocyst 129, Duff & Smol in Duff *et al.*, 1992 (2) Fig. 53

BIOLOGICAL AFFINITY: unknown

LOCALITY: "Elison Pond" (unofficial name), Cape Herschel (78°37'N; 74°42'W), Ellesmere Island, Northwest Territories, Canada. Epilithon, collected August 3, 1986.

SEM DESCRIPTION: This spherical stomatocyst (diameter 8.2-9.0 μm) has a cylindrical collar with a slightly flaring apex (collar diameter 2.1-2.3 μm; height 1.1-1.3 μm); the inner collar margin is not striated. The pore is regular (diameter 0.8 μm). The cyst surface, particularly in the anterior region surrounding the collar, is slightly scabrate.

Stomatocyst 129 is similar to stomatocyst 79, but stomatocyst 79 has collar striations, is ornamented with verrucae, and usually has a more obconical collar.

LM DESCRIPTION: The scabrae of this stomatocyst are visible with LM when focusing on the surface (Fig. 53D), but are not visible in cross-sectional view (Fig. 53C). Stomatocyst 129 can be distinguished from stomatocyst 116 *forma* B by the presence of ornamentation, and from stomatocyst 79 by the relatively reduced ornamentation and the more cylindrical collar.

REFERENCES: Identical cysts have not been found elsewhere, although stomatocyst 13 (Chapter 5 - Northwest Territories, Canada), and Rybak *et al.* (1987 - Ontario, Canada) cyst 18, are similar.

ECOLOGY: Autecological details are not yet available for this rare morphotype.

Stomatocyst 233, Zeeb *et al.*, 1994 (6) Fig. 54

BIOLOGICAL AFFINITY: *Dinobryon cylindricum* Imhof produces a very similar stomatocyst (Donaldson & Stein, 1984; Sandgren, 1980b, 1981, 1983a, b, 1989), but scabrate forms have not been recorded.

LOCALITY: Lake 227 (49°42'N; 93°42'W), Experimental Lakes Area, Ontario, Canada. Freeze core, varve year A.D. 1974.

SEM DESCRIPTION: This stomatocyst is large and spherical (diameter 12.3-14.4 μm). The collar has a conical base that narrows gradually into a cylindrical, hooked apical portion (basal diameter ca. 3.6 μm; apical diameter 1.6-2.6 μm; height to flexure 2.8-3.2 μm; length to 5.5 μm), and is radially striated. Dense scabrae with occasional smooth patches are found over the cyst surface. The scabrae may be elongated (diameter 0.2-0.4 μm; height 0.2-0.3 μm) and extend up the length of the collar.

Stomatocyst 233 resembles both stomatocysts 41 and 78; these three morphs are likely produced by closely related chrysophyte species.

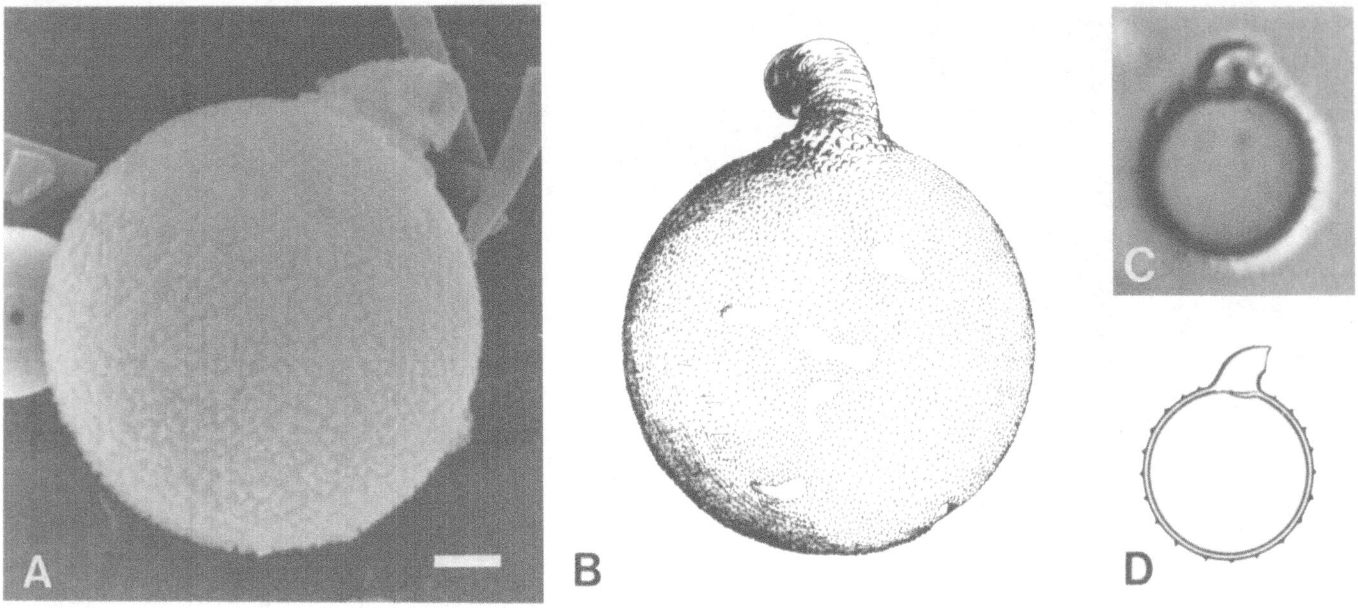

Fig. 54. Stomatocyst 233. A: SEM. B: Line drawing of SEM. C: LM. D: Line drawing of LM. Scale bar = 2 μm. LM x2000.

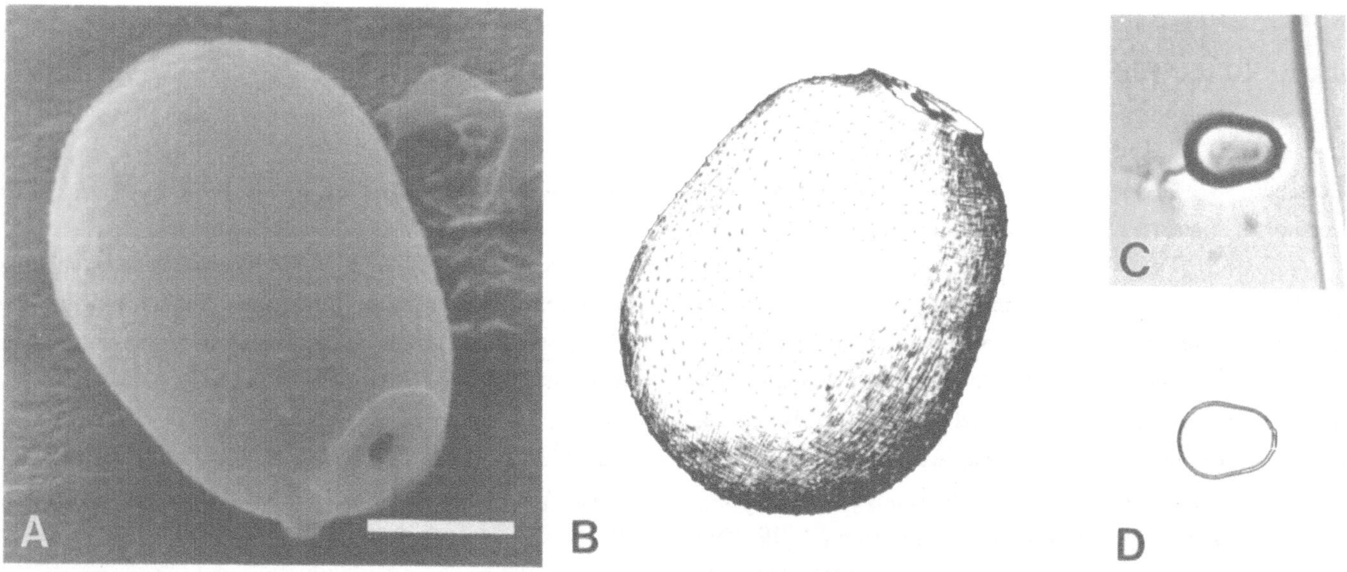

Fig. 55. Stomatocyst 235. A: SEM. B: Line drawing of SEM. C: LM. D: Line drawing of LM. Scale bar = 2 μm. LM x2000.

LM DESCRIPTION: Stomatocyst 233 can be distinguished from stomatocyst 41 (Figs. 20D, E) by the presence of scabrae. It is differentiated from stomatocyst 78 (Figs. 62D, E, and F) by the lack of a bipartite collar and the absence of verrucae.

REFERENCES: We have recorded this stomatocyst from Ontario, Canada (Zeeb *et al.*, 1994), and New York, U.S.A. (Duff & Smol, 1991 - misidentified as stomatocyst 78).

ECOLOGY: In the manipulated Lake 227, stomatocyst 233 increased in relative abundance after fertilization with phosphorus and nitrogen, but declined when the N:P ratio was reduced (Zeeb *et al.*, 1994).

Stomatocyst 235, Zeeb & Smol in Duff *et al.* (this volume) (3) Fig. 55

BIOLOGICAL AFFINITY: unknown

NEGATIVE NUMBER: J.P. Smol 575

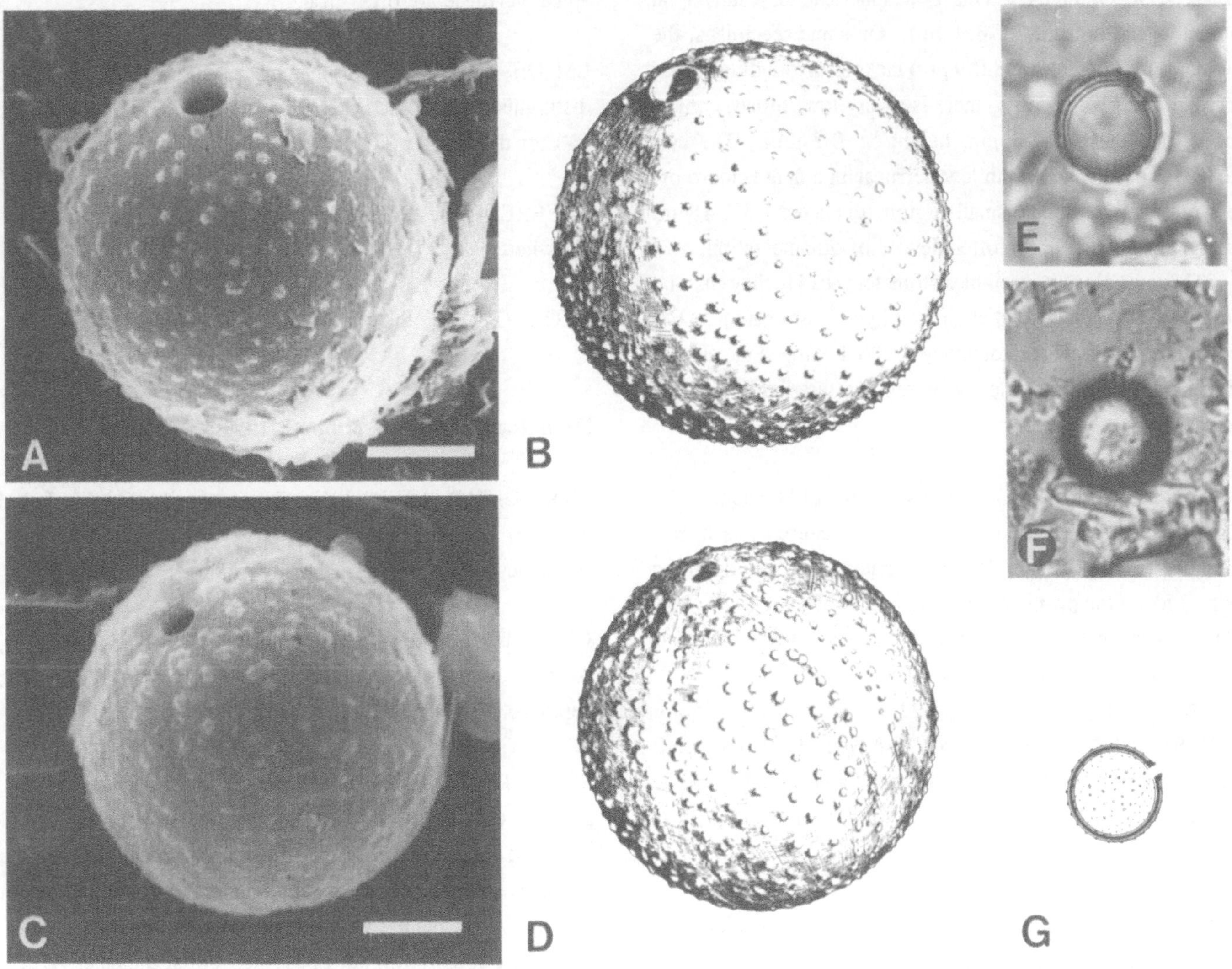

Fig. 56. Stomatocysts 140 and 141. A: Stomatocyst 140, SEM. B: Stomatocyst 140, line drawing of SEM. C: Stomatocyst 141, SEM. D: Stomatocyst 141, line drawing of SEM. E: Stomatocysts 140/141, LM, cross-sectional view. F: Stomatocysts 140/141, LM, surfical view. G: Stomatocysts 140/141, line drawing of LM. Scale bars = 2 μm. LM x2000.

LOCALITY: Lake 227 (49°42'N; 93°42'W), Experimental Lakes Area, Ontario, Canada. Freeze core, varve year A.D. 1989.

SEM DESCRIPTION: This stomatocyst is oval to ovate (4.5-4.9 x 5.3-6.3 μm) with a low, wide obconical collar (diameter 1.7-2.2 μm; height 0.1-0.2 μm) surrounding a flat, planar annulus (diameter 1.5-1.6 μm) and a regular pore (diameter 0.4-0.7 μm). The entire cyst surface is densely scabrate.

LM DESCRIPTION: Stomatocyst 235 can be recognized by its distinctive body shape and the wide, low collar. The ornamentation is not visible.

REFERENCES: We have found no other references to this rare cyst.

ECOLOGY: Autecological details are not yet available.

Stomatocysts ornamented with verrucae only

Stomatocyst 140, Duff & Smol in Duff *et al.*, 1992 (7) Figs. 56A, B

BIOLOGICAL AFFINITY: unknown

LOCALITY: "Horseshoe Pond" (unofficial name), Cape Herschel (78°37'N; 74°42'W), Ellesmere Island, Northwest Territories, Canada. Surface sediment sample, collected July 23, 1986.

SEM DESCRIPTION: This is a spherical to widely oval stomatocyst (diameter 7.0-8.9 μm). On some specimens, the regular pore (diameter 0.8-0.9 μm) lacks a surrounding collar; on others (e.g. Fig. 56A), there is a very low, broad, conical collar (diameter 0.9-1.1 μm; height ca. 0.2 μm). The cyst surface is ornamented with low verrucae in a dense, more-or-less regular pattern. A small region (diameter 1.3-1.4 μm) surrounding the collar is often devoid of ornamentation.

This cyst closely resembles stomatocyst 141; they may be produced by closely related chrysophytes. Stomatocyst 140 is differentiated from stomatocyst 206 by the conical, not obconical collar, and the more regular distribution of the verrucae.

LM DESCRIPTION: Stomatocyst 140 cannot be distinguished from stomatocyst 141. Immature cysts in this category (i.e. specimens lacking verrucae) can be identified using LM by the prominent pore and multiple cyst walls (Fig. 56E). These features are also used to distinguish stomatocysts 140/141 from stomatocyst 206.

REFERENCES: We have not observed stomatocyst 140 outside of the Northwest Territories, Canada (Duff *et al.*, 1992), but Rybak *et al.* (1991 - Ontario, Canada) cyst 28 is very similar.

ECOLOGY: Stomatocysts 140/141 were very abundant in a relatively large arctic pond with a substantial moss bank around part of its perimeter (Duff *et al.*, 1992); these cysts may be produced by epiphytic species.

Stomatocyst 141, Duff & Smol in Duff *et al.*, 1992 (3) Figs. 56C, D

BIOLOGICAL AFFINITY: unknown

LOCALITY: "Elison Pond" (unofficial name), Cape Herschel (78°37'N; 74°42'W), Ellesmere Island, Northwest Territories, Canada. Epilithon, collected August 3, 1986.

SEM DESCRIPTION: This spherical stomatocyst (diameter 7.5-8.2 μm) has a regular pore (diameter 0.6-0.8 μm) with no collar. The cyst surface is densely ornamented with low verrucae except for six to eight evenly spaced, narrow (width ca. 0.3-0.4 μm), unornamented, strip-like regions extending longitudinally from the pore, and an unornamented circular region (diameter 1.0-1.4 μm) around the pore. The most apical verrucae are often fused.

LM DESCRIPTION: Stomatocysts 140 and 141 cannot be distinguished with LM. See the discussion under stomatocyst 140 for details.

REFERENCES: This cyst has not yet been observed elsewhere.

ECOLOGY: See the discussion under stomatocyst 140.

Stomatocyst 75, Duff & Smol, 1991 (14) Fig. 57

BIOLOGICAL AFFINITY: *Dinobryon bavaricum* Imhof produces a very similar, but larger (diameter 9-12 μm), stomatocyst (G. Cronberg, pers. commun.).

LOCALITY: Upper Wallface Pond (44°08.8'N; 74°03.3'W), Adirondack Park, New York, U.S.A. Gravity core, core depth 10.0-10.5 cm.

SEM DESCRIPTION: All the SEM specimens of this stomatocyst that we studied are spherical (diameter 5.8-8.3 μm), but oblate specimens have been observed with LM. The collar is cylindrical to grossly obconical; the inner collar margin is ornamented with a series of low, highly irregular, concentric or spiralling siliceous rings (collar diameter 1.7-2.8 μm; height 0.1-0.7 μm). The pore is regular (diameter 0.5-0.9 μm). Numerous, irregularly scattered verrucae that vary greatly in size (<0.2-0.5 μm) are found over the entire cyst surface.

Stomatocyst 75 is nearly identical to stomatocyst 53, except for the presence of ornamentation.

LM DESCRIPTION: This stomatocyst can usually be identified by its highly irregular surface pattern of ornamentation, and the low, wide collar. The collar spirals are generally not visible.

REFERENCES: This stomatocyst has been observed in Ontario, Canada (Smol, 1988a - Plate 1A; Rybak *et al.*, 1991 - cyst 38), and New York, U.S.A. (Duff & Smol, 1991).

ECOLOGY: *Dinobryon bavaricum* is a cosmopolitan species (Moore, 1978; Dürrschmidt, 1982b; Eloranta, 1989). It is often very abundant in clear, acidic to circumneutral, oligotrophic, softwater lakes (Duthie & Ostrofsky, 1974;

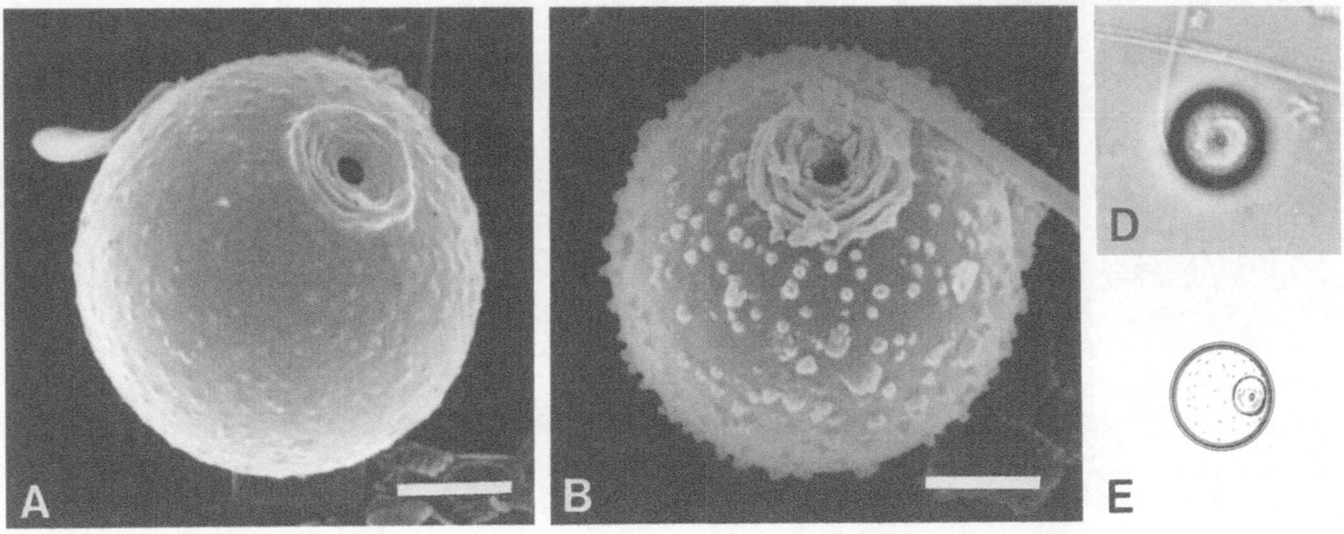

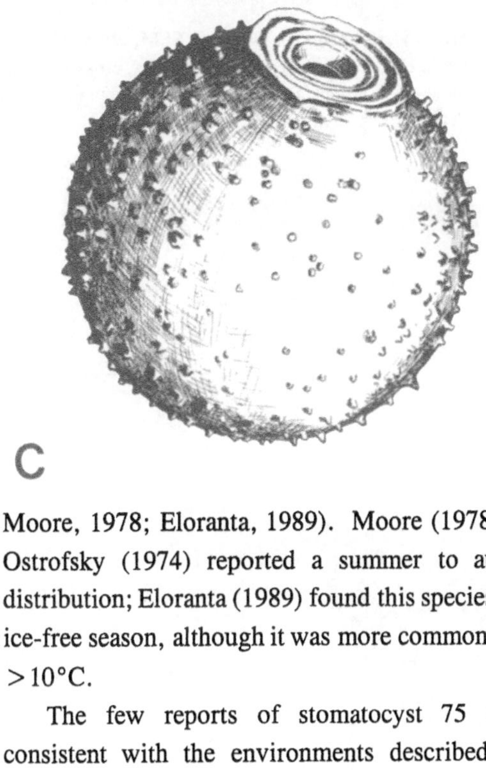

Fig. 57. Stomatocyst 75. A: SEM. Specimen with less developed ornamentation. B: SEM. Highly developed specimen. C: Line drawing of SEM. D: LM. E: Line drawing of LM. Scale bars = 2 μm. LM x2000.

Moore, 1978; Eloranta, 1989). Moore (1978) and Duthie & Ostrofsky (1974) reported a summer to autumn seasonal distribution; Eloranta (1989) found this species throughout the ice-free season, although it was more common at temperatures >10°C.

The few reports of stomatocyst 75 are from sites consistent with the environments described above (Smol, 1988a; Duff & Smol, 1991; Rybak *et al.*, 1991). Rybak *et al.* (1991) classified the cyst as acidophilic.

Stomatocyst 79, Duff & Smol, 1991 (18) Fig. 58

BIOLOGICAL AFFINITY: This stomatocyst is probably formed by *Dinobryon sociale* var. *americanum* (Brunthaler) Bachmann (G. Cronberg, pers. commun.).

LOCALITY: Upper Wallface Pond (44°08.8'N; 74°03.3'W), Adirondack Park, New York, U.S.A. Gravity core, core depth 0.0-0.5 cm.

SEM DESCRIPTION: This cyst is spherical (diameter 7.7-13.1 μm). The collar is long and gradually obconical with a conical base; the apex may be more widely flared (collar diameter 1.9-3.8 μm; height 1.5-4.4 μm). The collar margins are radially striated. The ornamentation is highly variable. It consists of randomly scattered protrusions, usually irregular verrucae but sometimes low, thin, baculate spines with splayed apices (verrucae diameter $\leq$0.1-0.7 μm; height <0.1-0.8 μm). Ornamentation may be found on the outer collar margin.

LM DESCRIPTION: Stomatocyst 79 may be confused with stomatocyst 129 with LM, but the latter cyst is ornamented more heavily in the anterior than the posterior hemisphere, and has a more cylindrical collar. Also, the ornamentation of stomatocyst 129 is not visible in cross-sectional view.

REFERENCES: This cyst has been observed in British Columbia, Canada (Duff & Smol, 1994; Zeeb, 1994), and New York, U.S.A. (Smol, 1988a - Plate 1B; Duff & Smol, 1991).

ECOLOGY: This stomatocyst may be produced by an acidophilic species (Duff & Smol, 1991; Smol, 1988a).

Stomatocyst 206, Duff & Smol, 1994 (3) Fig. 59

BIOLOGICAL AFFINITY: unknown

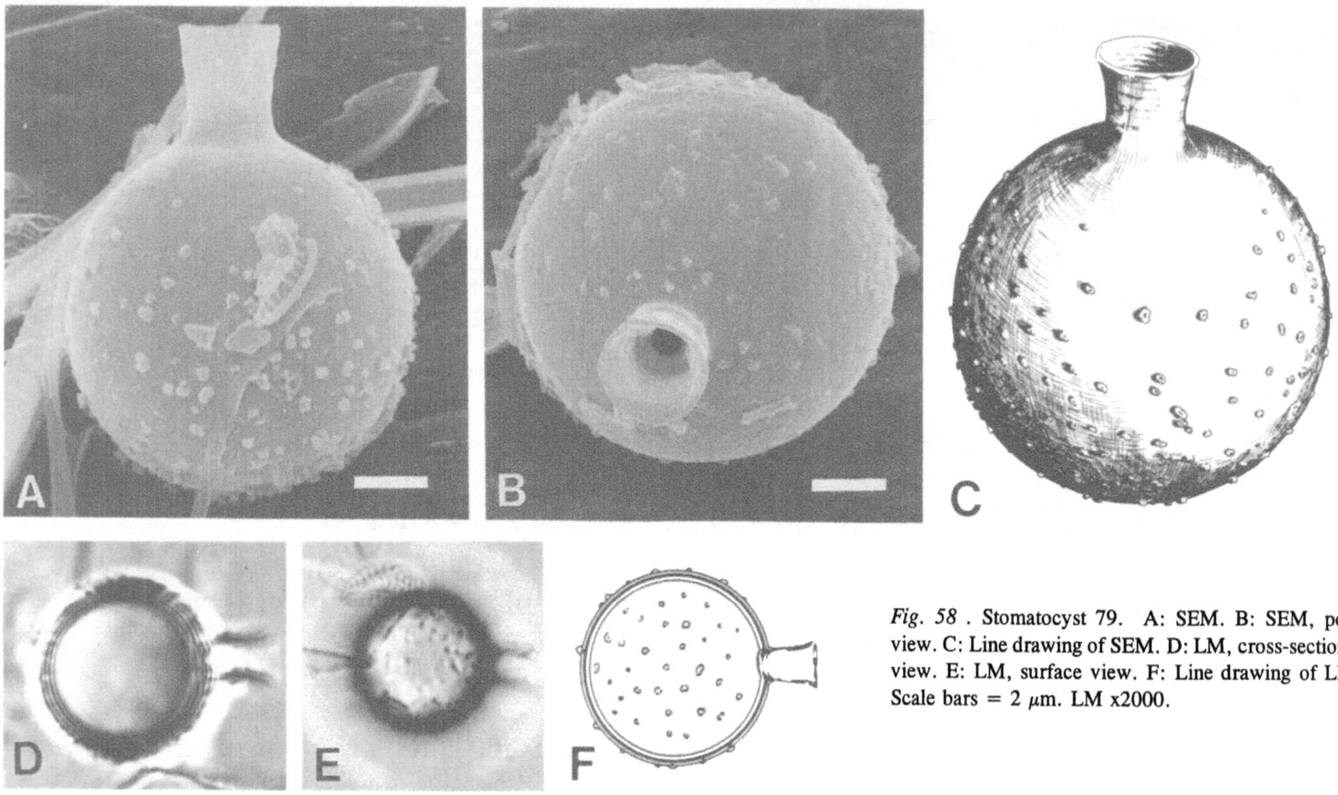

Fig. 58 . Stomatocyst 79. A: SEM. B: SEM, pore view. C: Line drawing of SEM. D: LM, cross-sectional view. E: LM, surface view. F: Line drawing of LM. Scale bars = 2 μm. LM x2000.

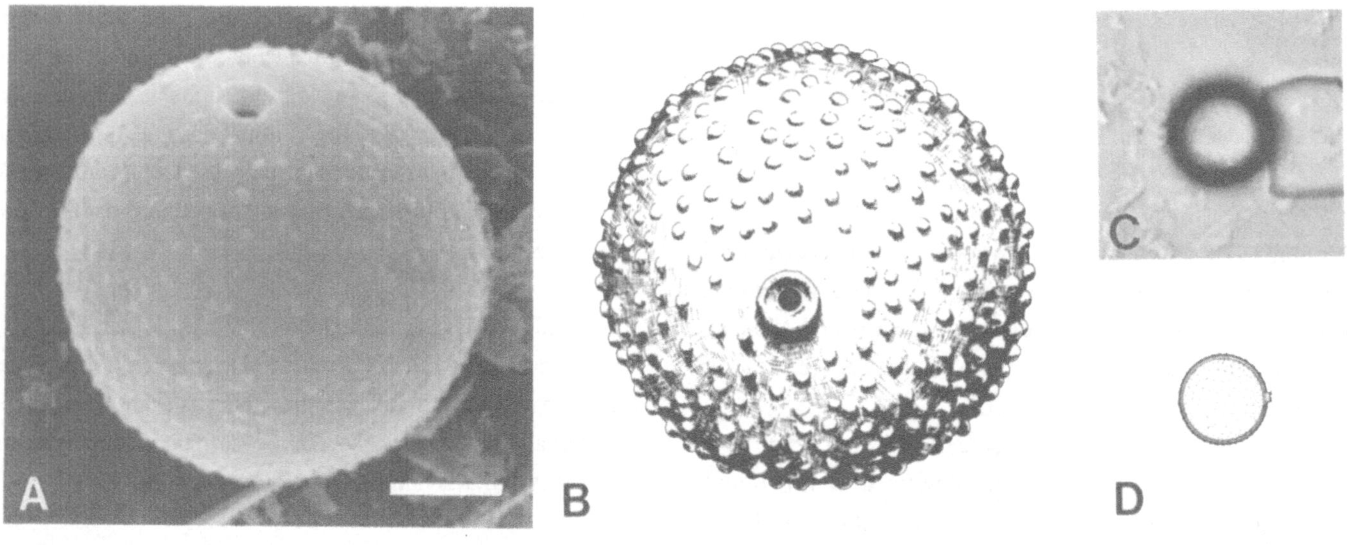

Fig. 59. Stomatocyst 206. A: SEM. B: Line drawing of SEM. C: LM. Collar is not visible. D: Line drawing of LM. Scale bar = 2 μm. LM x2000.

LOCALITY: Green Lake (52°30'N; 121°00'W), British Columbia, Canada. Ekman dredge, surface (1 cm) sediment sample.

SEM DESCRIPTION: This stomatocyst is spherical (diameter 7.4-8.1 μm). The regular pore (diameter 0.3-0.5 μm) is surrounded by a low, obconical collar (diameter 1.0-1.7 μm; height ca. 0.2 μm) with a sloping inner margin and a rounded apex. The ornamentation consists of randomly spaced verrucae, which are usually low and widely rounded (diameter 0.1-0.5 μm; height up to 0.2 μm), but occasionally resemble blunt conula. The ornamentation density is high but variable over the cyst surface.

LM DESCRIPTION: The verrucae are faint but visible with LM (Fig. 59C). This cyst is similar in size, shape, and ornamentation to stomatocysts 140/141, and possibly the two could be confused; however, stomatocysts 140/141 have

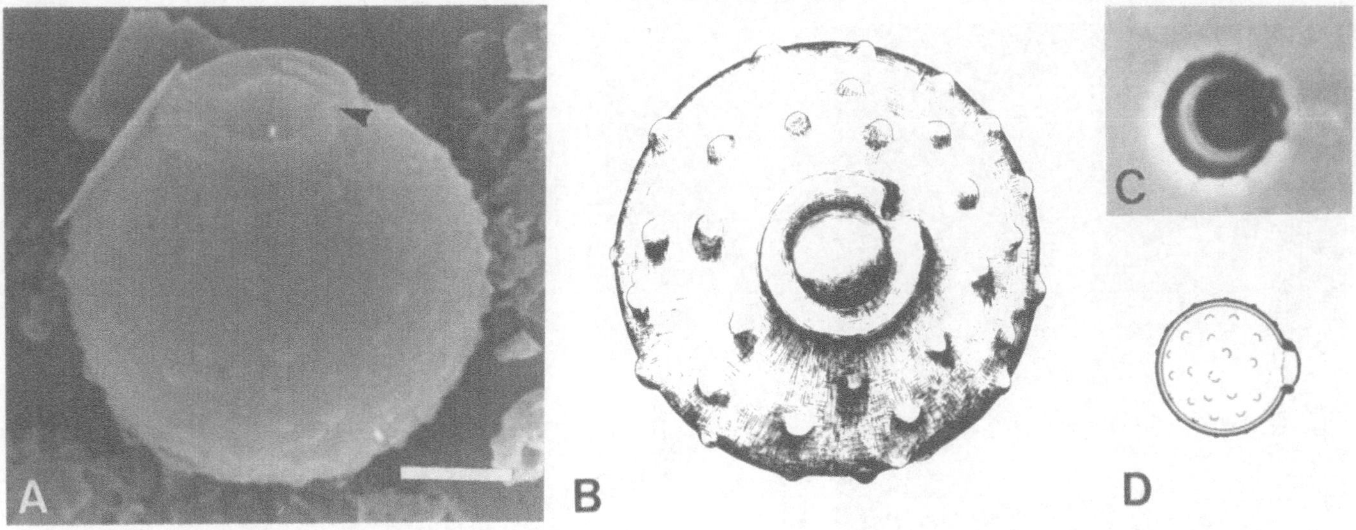

Fig. 60. Stomatocyst 158. A: SEM. Arrow indicates notch-like indentation in the collar. B: Line drawing of SEM. C: LM. D: Line drawing of LM. Scale bar = 2 μm. LM x2000.

multiple walls and a deep pore. Stomatocyst 206 is smaller than stomatocyst 207, another similar morph.

REFERENCES: We have observed this cyst in British Columbia, Canada (Duff & Smol, 1994; Zeeb, 1994), Ontario, Canada (Zeeb & Smol, unpubl.), and the Yukon Territory, Canada (Brown & Smol, unpubl.). Rybak *et al.* (1991 - Ontario, Canada) cyst 36 is probably identical.

ECOLOGY: Rybak *et al.* (1991) cyst 36 was classified as pH-indifferent.

Stomatocyst 158, Zeeb & Smol, 1993a (11) Fig. 60

BIOLOGICAL AFFINITY: unknown

LOCALITY: Elk Lake (47°11'N; 95°13'W), Clearwater County, Minnesota, U.S.A. Piston core, varve year A.D. 1864.

SEM DESCRIPTION: This is a spherical stomatocyst (diameter 6.1-8.5 μm; median 7.0 μm). The collar is very wide, low, conical to cylindrical with a rounded apex (collar diameter 2.4-3.7 μm; height ca. 0.6 μm; collar diameter : cyst diameter ratio 0.36-0.55, median 0.50). On many of our specimens, we observed a notch-like indentation in the collar (Fig. 60A). Although the taxonomic significance of this character is debatable, it is interesting to note that an identical indentation was observed in the collar of stomatocyst 241 (Fig. 126B), a similar cyst with reticular ornamentation. The pore of stomatocyst 158 is usually obscured by a rounded siliceous cap (diameter 1.6-2.2 μm). The ornamentation consists of numerous, fairly large verrucae (diameter 0.2-1.2 μm; height usually equal to diameter).

LM DESCRIPTION: This stomatocyst can be differentiated from stomatocyst 164, with both SEM and LM, by its smaller size, lower and wider collar, denser verrucae, and higher collar diameter : cyst diameter ratio, although there is some overlap in all of these features. It is distinguished from stomatocyst 241 on the basis of ornamentation, although immature (unornamented) forms could be confused.

REFERENCES: This stomatocyst has been observed in British Columbia, Canada (Duff & Smol, 1994; Zeeb, 1994), and Minnesota, U.S.A. (Zeeb & Smol, 1993a). The cyst identified by Rybak (1986 - Poland) as *cysta crassicollis* Nygaard is slightly larger with more pointed verrucae, but Nygaard's (1956 - Denmark) specimens are not the same as our stomatocyst 158.

ECOLOGY: Zeeb & Smol (1993b) found stomatocyst 158 during periods of greater lake productivity, but Rybak's (1986) morph was believed to be an indicator of oligotrophy.

Stomatocyst 164, Zeeb & Smol, 1993a (14) Fig. 61

BIOLOGICAL AFFINITY: unknown

LOCALITY: Elk Lake (47°11'N; 95°13'W), Clearwater

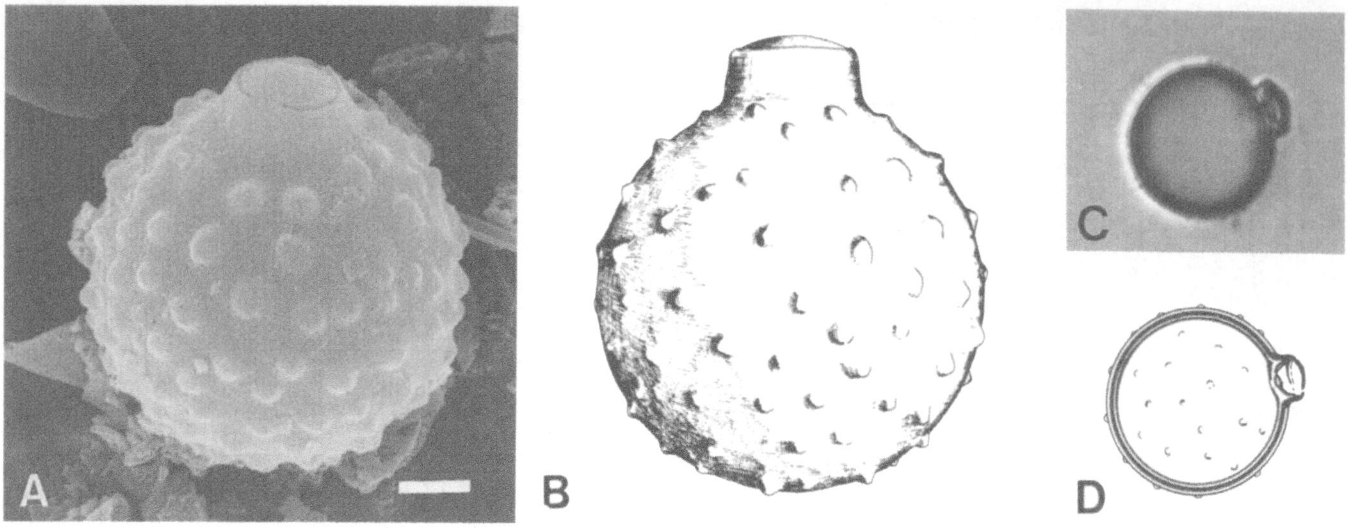

Fig. 61. Stomatocyst 164. A: SEM. B: Line drawing of SEM. C: LM. D: Line drawing of LM. Scale bar = 2 μm. LM x2000.

County, Minnesota, U.S.A. Piston core, varve year A.D. 1851.

SEM DESCRIPTION: This is a spherical to slightly oblate stomatocyst (diameter 6.6-10.8 μm; median 9.7 μm). The collar is conical to cylindrical (apical diameter 2.0-4.1 μm; height 0.5-1.6 μm) with a regular pore (diameter 0.8-0.9 μm; depth 0.2-0.3 μm); it is usually covered by a convex siliceous cap (diameter 2.2-3.2 μm; height 0.3-0.8 μm). The collar diameter : cyst diameter ratio is 0.29-0.40 (median 0.33). The cyst is smooth when immature, but when mature is covered with widely spaced verrucae or blunt conula in a more-or-less irregular pattern (diameter 0.5-1.2 μm; height 0.2-0.9 μm).

LM DESCRIPTION: See stomatocyst 158 for the features used to distinguish these two morphs.

REFERENCES: This stomatocyst has been found in British Columbia, Canada (Duff & Smol, 1994; Zeeb, 1994), Ontario, Canada (Rybak *et al.*, 1987 - cyst 20 in Fig. 23 but not Fig. 24; Rybak *et al.*, 1991 - cyst 41), Minnesota, U.S.A. (Zeeb & Smol, 1993a), and Poland (Rybak, 1986 - cyst 17).

ECOLOGY: This cyst is apparently produced by a species with a wide tolerance for pH, but a preference for alkaline water (Rybak, 1986; Rybak *et al.*, 1991; Duff, 1994).

Stomatocyst 78, Duff & Smol, 1991 (7) Fig. 62

BIOLOGICAL AFFINITY: This stomatocyst may be produced by a *Dinobryon* species; *D. cylindricum* Imhof produces a similar stomatocyst (Donaldson & Stein, 1984; Sandgren, 1980b, 1981, 1983a, b, 1989), but without a bipartite collar or scabrae.

LOCALITY: Upper Wallface Pond (44°08.8'N; 74°03.3'W), Adirondack Park, New York, U.S.A. Gravity core, core depth 20.0-20.5 cm.

SEM DESCRIPTION: This is a spherical stomatocyst (diameter 12.2-14.1 μm; median 12.9 μm). The collar consists of a low, wide, cylindrical basal portion with an abrupt outer margin and a flattened apex, which connects abruptly with a narrow, hooked, cylindrical distal portion (base diameter 4.3-5.4 μm; base height 0.9-1.6 μm; apical diameter 1.7-2.3 μm; length to 5.7 μm; height to flexure 2.5-4.3 μm). The cyst surface is usually scabrate, and usually possesses several scattered protrusions or "false" verrucae, formed by a clump of slightly elongated scabrae (diameter 0.6-2.2 μm; height 0.3-0.7 μm).

Stomatocyst 78 resembles both stomatocysts 41 and 233; these three morphs are likely produced by closely related chrysophyte taxa.

LM DESCRIPTION: Stomatocyst 78 can be distinguished from stomatocysts 41 (Figs. 20D, E) and 233 (Figs. 54C, D) by the bipartite, hooked collar and the surface texture, which is visible with LM (Fig. 62D).

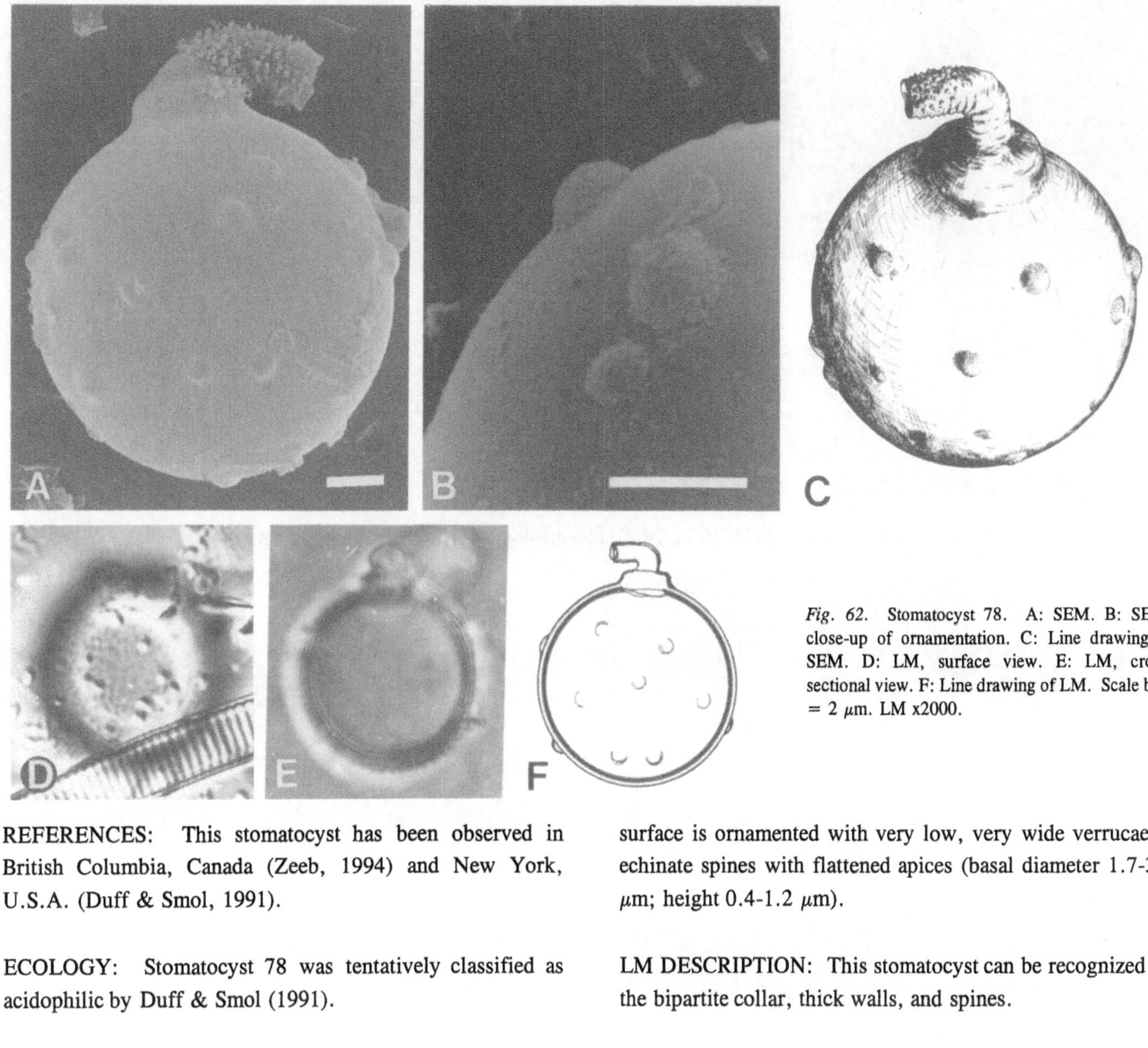

Fig. 62. Stomatocyst 78. A: SEM. B: SEM, close-up of ornamentation. C: Line drawing of SEM. D: LM, surface view. E: LM, cross-sectional view. F: Line drawing of LM. Scale bars = 2 μm. LM x2000.

REFERENCES: This stomatocyst has been observed in British Columbia, Canada (Zeeb, 1994) and New York, U.S.A. (Duff & Smol, 1991).

ECOLOGY: Stomatocyst 78 was tentatively classified as acidophilic by Duff & Smol (1991).

Stomatocyst 209, Duff & Smol, 1994 (3) Fig. 63

BIOLOGICAL AFFINITY: unknown

LOCALITY: Florence Lake (48°30'N; 123°30'W), British Columbia, Canada. Ekman dredge, surface (1 cm) sediment sample.

SEM DESCRIPTION: This is a spherical stomatocyst (diameter 12.4-15.3 μm). The collar is bipartite. The basal portion of the collar is conical to cylindrical (diameter 5.1-7.2 μm; height 1.7-3.8 μm) with an irregular outline. The apical portion of the collar is cylindrical and hooked (apical diameter 1.7-3.4 μm; height 1.3-1.4 μm), and was broken or undeveloped in all our SEM specimens (Fig. 63A). The surface is ornamented with very low, very wide verrucae to echinate spines with flattened apices (basal diameter 1.7-3.5 μm; height 0.4-1.2 μm).

LM DESCRIPTION: This stomatocyst can be recognized by the bipartite collar, thick walls, and spines.

REFERENCES: This morph has been observed in British Columbia, Canada (Duff & Smol, 1994), and Ontario, Canada (Rybak *et al.*, 1987 - cyst 20). Rybak *et al.* (1991 - Ontario, Canada) cyst 40 is also very similar to our stomatocyst 209.

ECOLOGY: This cyst may be produced by a eutrophic (Rybak *et al.*, 1987) taxon. Rybak *et al.* (1991) cyst 40 was classified as alkaliphilic.

Stomatocysts ornamented with either verrucae or conula

Stomatocyst 210, Duff & Smol, 1994 (10) Fig. 64

BIOLOGICAL AFFINITY: unknown

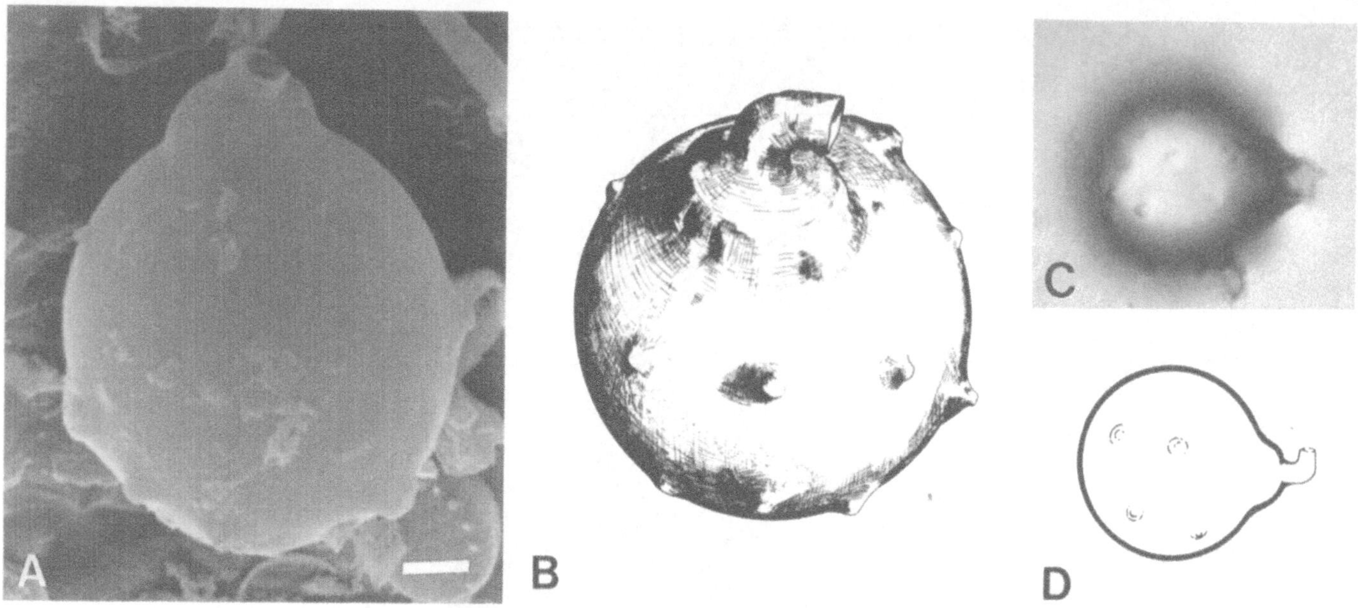

Fig. 63. Stomatocyst 209. A: SEM. B: Line drawing of SEM. C: LM. D: Line drawing of LM. Scale bar = 2 μm. LM x2000.

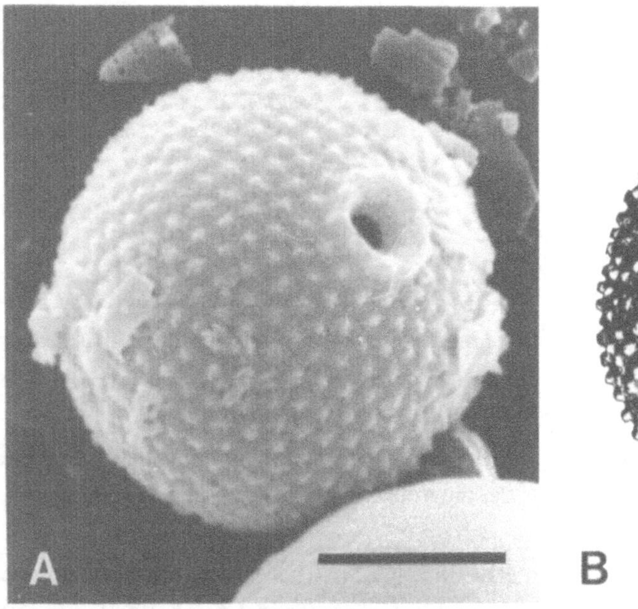

Fig. 64. Stomatocyst 210. A: SEM. B: Line drawing of SEM. Scale bar = 2 μm. See Figs. 52C, D, and E for LM.

LOCALITY: Champion Lake #3 (49°15'N; 117°30'N), British Columbia, Canada. Ekman dredge, surface (1 cm) sediment sample.

SEM DESCRIPTION: This is a spherical cyst (diameter 5.1-6.3 μm) with a regular pore (diameter 0.5-0.6 μm) and a short cylindrical to conical collar (diameter 1.0-1.7 μm; height 0.2-0.6 μm). The surface ornamentation consists of very short, blunt conula or verrucae (diameter 0.1-0.4 μm; height 0.1-0.2 μm) arranged equidistantly in a regular pattern (distance between conula 0.3-0.4 μm); they may also be found on the outer collar margin and/or the collar apex.

Stomatocyst 210 resembles both stomatocysts 117 and 31, but the ornamentation density is intermediate between these two morphotypes.

LM DESCRIPTION: With LM, stomatocyst 210 cannot be distinguished from stomatocysts 117 (Figs. 52C, D, and E) and 178. The surface ornamentation of this group of cysts can be observed only with high resolution optics.

REFERENCES: This cyst has been observed in British Columbia, Canada (Duff & Smol, 1994), Ontario, Canada (Rybak *et al.*, 1991 - cyst 34), Oregon, U.S.A. (Adam &

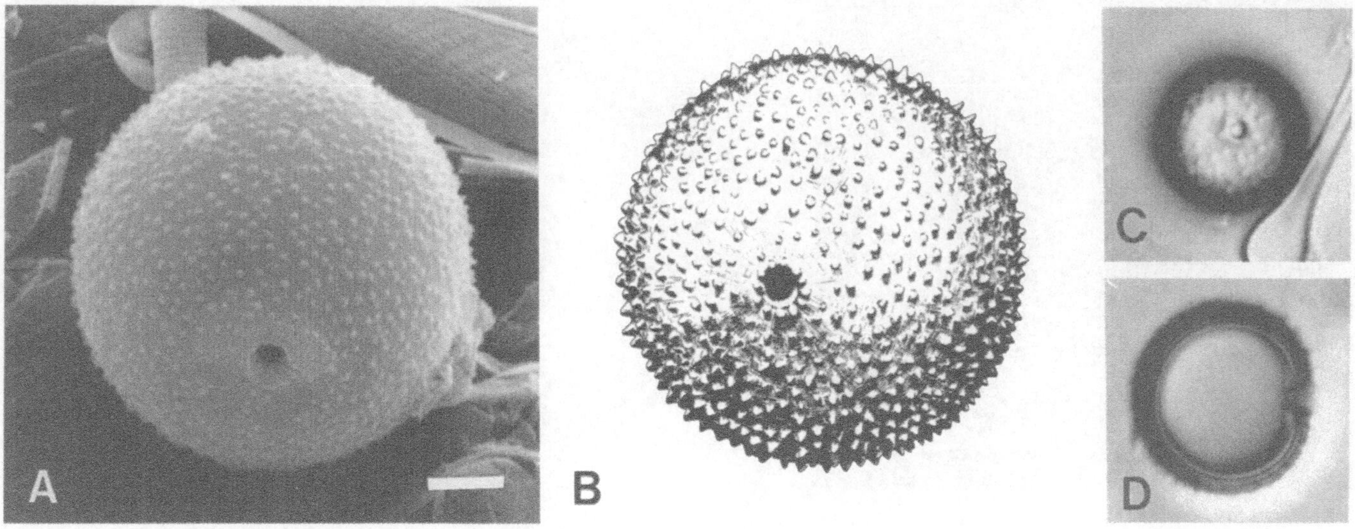

Fig. 65. Stomatocyst 207. A: SEM. B: Line drawing of SEM. C: LM, apical surface view. D: LM, cross-sectional view. E: Line drawing of LM. Scale bar = 2 μm. LM x2000.

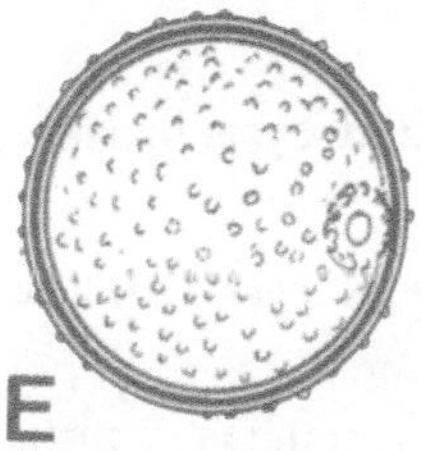

Mehringer, 1980c - type 98 ≠ Adam & Mahood 1979b - type 98), and California, U.S.A. (Adam & Mahood, 1979b - type 96; Adam & Mahood, 1980b - type 98; Carney *et al.*, 1992 - cyst 6). Rybak (1987 - Poland) cysts 7 and 8 are similar but smaller (diameter ca. 3-4 μm).

ECOLOGY: Generally, this cyst is reported to be an oligotrophic, acidophilic, or pH-indifferent indicator (Adam & Mahood, 1979, 1980; Adam & Mehringer, 1980; Rybak *et al.*, 1991; Carney *et al.*, 1992;). Rybak (1987) cysts 7 and 8, however, were inferred to be alkaliphilic and eutrophic indicators.

Stomatocyst 207, Duff & Smol, 1994 (6) Fig. 65

BIOLOGICAL AFFINITY: unknown

LOCALITY: Durrance Lake (48°30'N; 123°25'W), British Columbia, Canada. Ekman dredge, surface (1 cm) sediment sample.

SEM DESCRIPTION: This stomatocyst is spherical (diameter 8.3-10.8 μm), and ornamented with small, densely spaced conula or verrucae (diameter 0.2-0.4 μm; height to 0.3 μm), which vary in shape even on the same specimen. The deep, conical pore (diameter 0.6-0.8 μm) is surrounded by a small conical collar (diameter 1.6-1.9 μm), which is formed by the fusion of a ring of verrucae around the outer pore margin.

This stomatocyst is distinguished from stomatocyst 206 with SEM on the basis of size and collar morphology, and by the presence of conula which tend to be more densely spaced than on stomatocyst 206. Stomatocyst 142 (Chapter 5) is similar, but larger (diameter 11.4-12.5 μm), with more rounded verrucae, and a large concave pore surrounded by a low conical collar.

LM DESCRIPTION: This cyst is thick-walled with a prominent pore when viewed with LM; the ornamentation is visible.

REFERENCES: This stomatocyst has been observed in British Columbia, Canada (Duff & Smol, 1994; Zeeb, 1994), and New York, U.S.A. (Duff & Smol, unpubl.). Adam & Mehringer (1980c - Oregon, U.S.A.) type 326 resembles our morphotype.

ECOLOGY: Autecological details are not yet available.

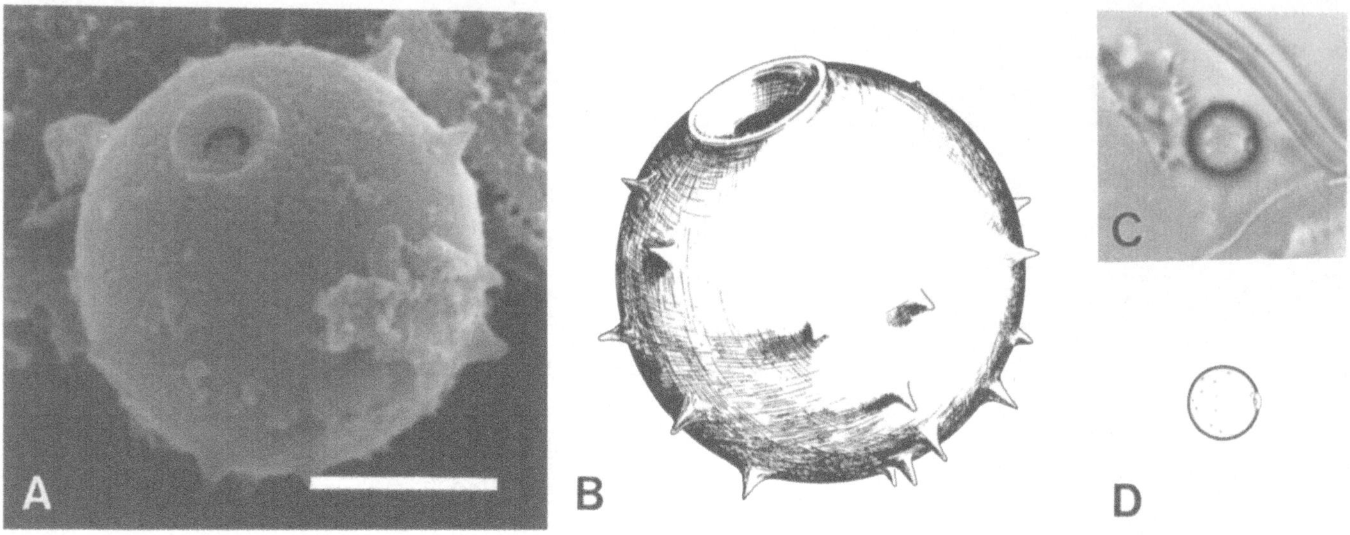

Fig. 66. Stomatocyst 170. A: SEM. B: Line drawing of SEM. C: LM. D: Line drawing of LM. Scale bar = 2 μm. LM x2000.

Stomatocysts ornamented with conula only

Stomatocyst 170, Zeeb & Smol, 1993a (5) Fig. 66

BIOLOGICAL AFFINITY: unknown

LOCALITY: Elk Lake (47°11'N; 95°13'W), Clearwater County, Minnesota, U.S.A. Sediment trap sample.

SEM DESCRIPTION: This is a small, spherical stomatocyst (diameter 3.9-5.5 μm) with a low, cylindrical to obconical collar (diameter 1.0-1.9 μm; height 0.2-0.5 μm), no annulus, and a regular pore (diameter 0.4-0.7 μm); one questionable SEM specimen has poorly developed collar struts. Irregularly scattered sharp conula are found mostly in the posterior hemisphere (diameter 0.4-1.1 μm; height 0.2-0.6 μm).

LM DESCRIPTION: High resolution optics must be used to distinguish stomatocyst 170 from similar cysts with ridges (i.e. stomatocyst 33/222, 91 and 113). Stomatocyst 5 *forma* A could be confused with stomatocyst 170 if the anterior depressions of the former are not visible.

REFERENCES: This stomatocyst has been observed in British Columbia, Canada (Duff & Smol, 1994; Zeeb, 1994), Minnesota, U.S.A. (Zeeb & Smol, 1993a), and California, U.S.A. (Zeeb *et al.*, subm.).

ECOLOGY: This stomatocyst was more common in deep, oligotrophic lakes in British Columbia, Canada (Duff, 1994).

Stomatocyst 139, Duff & Smol in Duff *et al.*, 1992 (6) Fig. 67

BIOLOGICAL AFFINITY: unknown

LOCALITY: "Horseshoe Pond" (unofficial name), Cape Herschel (78°37'N; 74°42'W), Ellesmere Island, Northwest Territories, Canada. Moss sample, collected July 19, 1986.

SEM DESCRIPTION: This is a spherical to slightly oblate stomatocyst (diameter 8.0-10.6 μm) ornamented with irregularly and widely scattered conula (base diameter 0.4-0.9 μm; height 0.2-0.3 μm), sometimes with connecting basal ridges. The collar is low and broadly conical surrounding a regular pore (collar diameter 1.0-2.1 μm; pore diameter 0.6-0.9 μm).

LM DESCRIPTION: This stomatocyst can be distinguished by the wide collar and conula.

REFERENCES: We have not definitely identified this cyst elsewhere.

ECOLOGY: This stomatocyst was relatively common in high arctic ponds with substantial moss banks (Duff *et al.*, 1992), and may be produced by an epiphytic species.

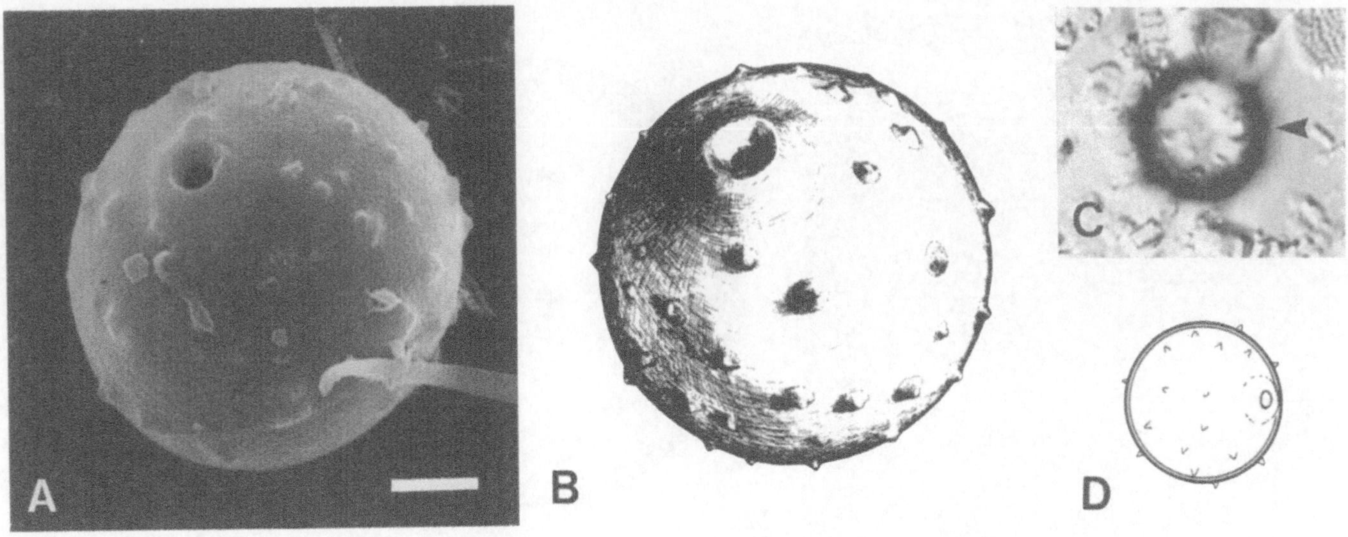

Fig. 67. Stomatocyst 139. A: SEM. B: Line drawing of SEM. C: LM. Collar indicated by arrow. D: Line drawing of LM. Scale bar = 2 μm. LM x2000.

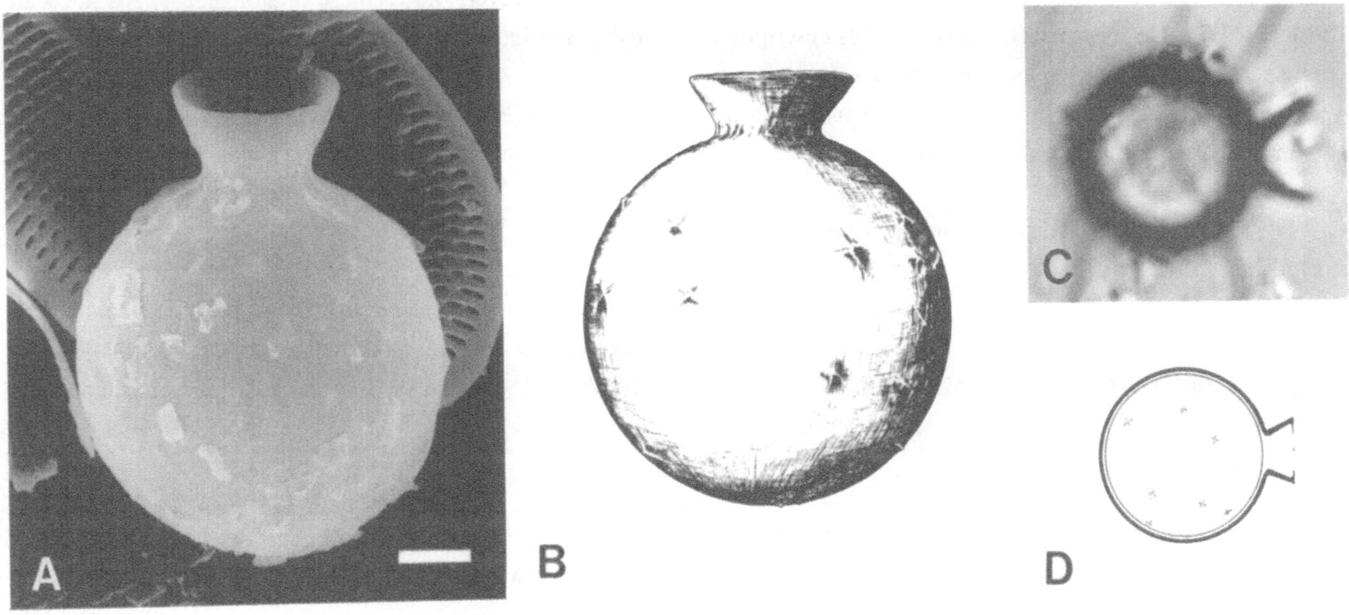

Fig. 68. Stomatocyst 131. A: SEM. B: Line drawing of SEM. C: LM. D: Line drawing of LM. Scale bar = 2 μm. LM x2000.

Stomatocyst 131, Duff & Smol in Duff *et al.*, 1992 (4) Fig. 68

BIOLOGICAL AFFINITY: unknown

LOCALITY: "Horseshoe Pond" (unofficial name), Cape Herschel (78°37'N; 74°42'W), Ellesmere Island, Northwest Territories, Canada. Moss sample, collected July 19, 1986.

SEM DESCRIPTION: This stomatocyst is spherical (diameter 9.5-11.1 μm) with a widely flared obconical collar (basal diameter 2.5-3.4 μm; apical diameter 3.5-4.4 μm; height 1.0-2.6 μm). The apex is occasionally irregular. The collar base may possess vertical struts. Inner collar details are unknown. The cyst surface has very low, wide, irregularly scattered conula, which sometimes appear to arise from the junction of two low ridges, suggesting that the ornamentation is incompletely developed.

LM DESCRIPTION: The wide, obconical collar is very

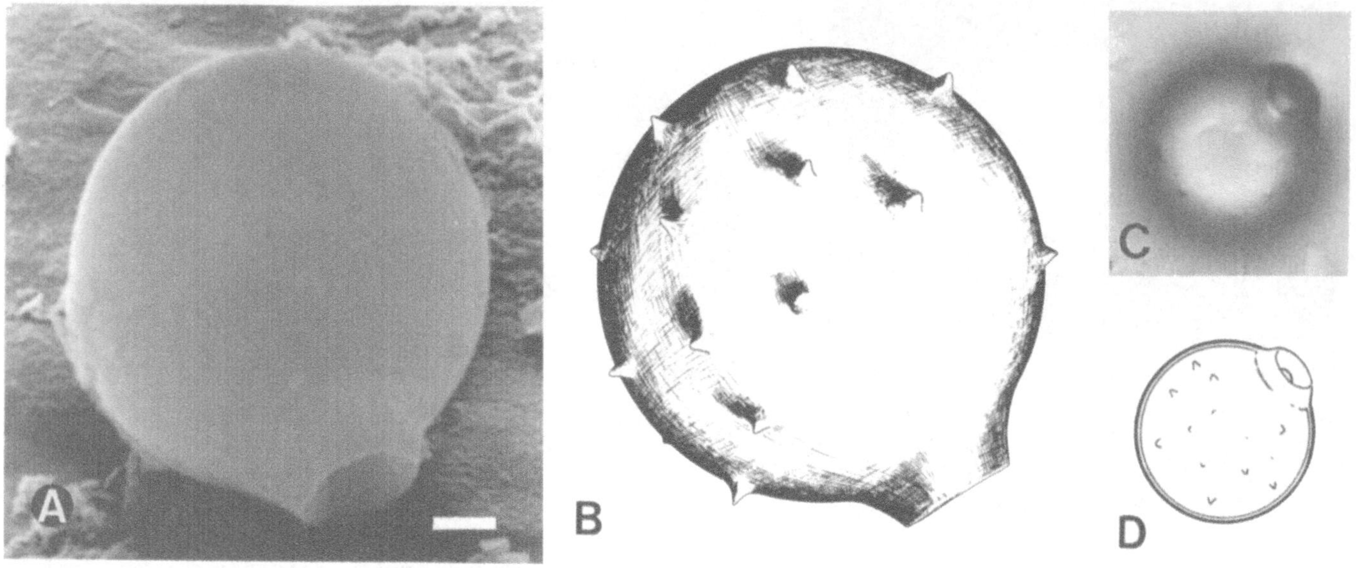

Fig. 69. Stomatocyst 208. A: SEM. B: Line drawing of SEM. C: LM. D: Line drawing of LM. Scale bar = 2 μm. LM x2000.

distinctive with LM.

REFERENCES: No cysts exactly matching this description have been found elsewhere; *cysta infundibularis* Nygaard (1956 - Denmark) is similar in size and collar morphology, but is more densely ornamented with spines.

ECOLOGY: Autecological data for this rare morph are not yet available.

Stomatocyst 208, Duff & Smol, 1994 (4) Fig. 69

BIOLOGICAL AFFINITY: unknown

LOCALITY: Davis Lake (49°50'N; 120°45'W), British Columbia, Canada. Ekman dredge, surface (1 cm) sediment sample.

SEM DESCRIPTION: This spherical stomatocyst (diameter 11.9-12.9 μm) possesses a very wide, conical collar (apical diameter 3.4-3.9 μm; height 1.8-2.1 μm) with no sharp delineation between the collar basal margin and the cyst body. The apex varies from fairly acute to quite rounded; the inner margin is gently sloping and the pore is unknown. A few to several low conula (diameter 0.6-1.5 μm; height 0.3-0.6 μm) are scattered widely on the cyst surface.

LM DESCRIPTION: This stomatocyst can be distinguished by its size and wide collar. It could be confused with specimens of stomatocyst 209 that lack the distal collar portion; however, the verrucae of stomatocyst 209 are wider and less pointed than the conula of stomatocyst 208.

REFERENCES: We have only observed this stomatocyst in British Columbia, Canada (Duff & Smol, 1994; Zeeb, 1994). Rybak *et al.* (1991 - Ontario, Canada) cyst 30 is probably identical.

ECOLOGY: Stomatocyst 208 was most common in oligotrophic, alkaline, montane lakes in Duff's (1994) survey. Rybak *et al.* (1991) cyst 30 was classified as alkalibiontic.

Stomatocyst 31, Duff & Smol, 1989 (19) Fig. 70

BIOLOGICAL AFFINITY: unknown

LOCALITY: Tasikutaaq Lake (66°23'N; 65°57'W), Baffin Island, Northwest Territories, Canada. Percussion core, core depth 26.5 cm.

SEM DESCRIPTION: This is a small, spherical stomatocyst (diameter 4.0-6.6 μm) with a short, slightly conical to cylindrical collar (diameter 0.8-1.8 μm; height 0.1-0.8 μm), a sloping inner margin, and a regular pore (diameter 0.3-0.7 μm). Duff & Smol (1989) incorrectly reported the presence of an annulus. The surface is ornamented with short, blunt to pointed conula (diameter 0.3-0.8 μm; height 0.1-0.4 μm) arranged equidistantly in a regular pattern (distance between

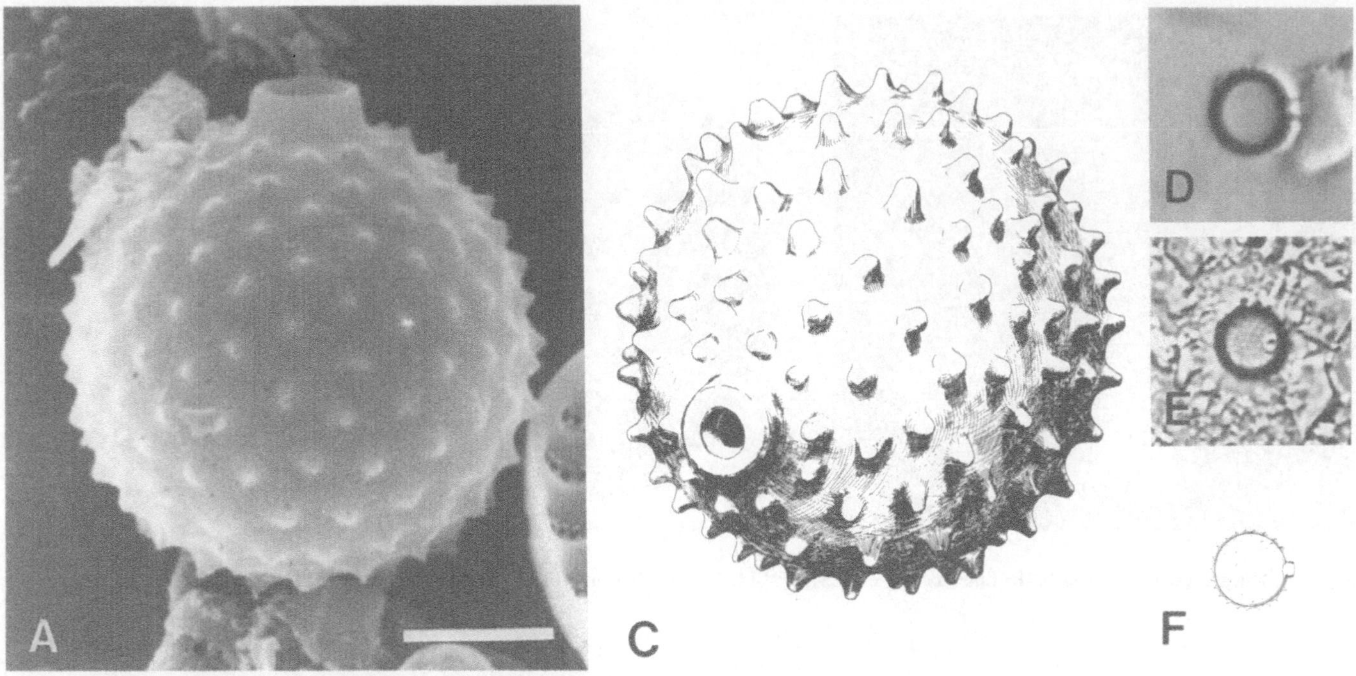

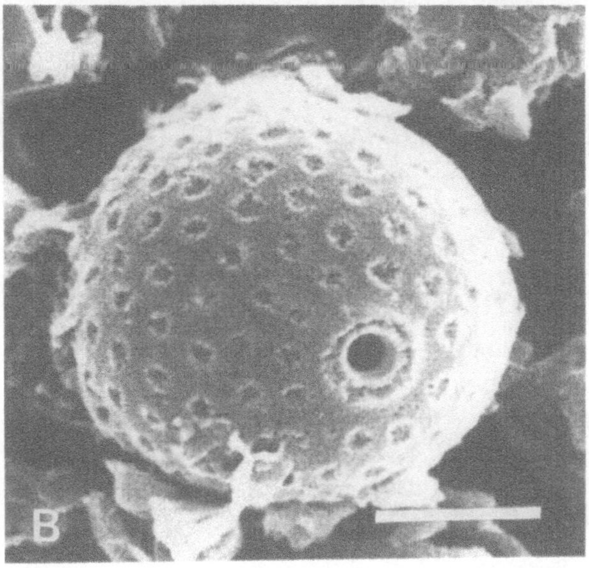

Fig. 70. Stomatocyst 31. A: SEM, mature specimen. B: SEM, immature specimen. C: Line drawing of SEM. D: LM, cross-sectional view. E: LM, surface view. F: Line drawing of LM. Scale bars = 2 μm. LM x2000.

spines 0.3-1.0 μm). An immature specimen is shown in Fig. 70B.

This cyst is similar to stomatocyst 157, but is smaller. The ornamentation density is less than that of stomatocyst 210.

LM DESCRIPTION: Stomatocyst 31 can be identified by its size, ornamentation and short, thick collar. It is distinguished from stomatocysts 117/178/210 on the basis of ornamentation density.

REFERENCES: This stomatocyst has been observed in the Northwest Territories, Canada (Duff & Smol, 1988 - stomatocyst 26 (invalidly described; Chapter 5); Duff & Smol, 1989); Ontario, Canada (Zeeb *et al.*, 1990); British Columbia, Canada (Duff, 1994; Zeeb, 1994); New York, U.S.A. (Duff & Smol, unpubl.); Minnesota, U.S.A. (Zeeb & Smol, 1993a); Michigan, U.S.A. (Sandgren and Carney, 1983 - cyst 19); and California, U.S.A. (Adam, 1980a - type 301). It is probably identical to *cysta geometrica* Nygaard (1956 - Denmark).

ECOLOGY: Stomatocyst 31 is most likely produced by a species that is tolerant of cold water (Adam , 1980a; Duff & Smol, 1988, 1989), but has a wide tolerance for trophic status (Duff & Smol, 1988, 1989; Zeeb *et al.*, 1990; Duff, 1994) and pH (Nygaard, 1956; Zeeb & Smol, 1993a; Duff, 1994).

Stomatocyst 157, Zeeb & Smol, 1993a (4) Fig. 71

BIOLOGICAL AFFINITY: unknown

LOCALITY: Elk Lake (47°11'N; 95°13'W), Clearwater County, Minnesota, U.S.A. Piston core, varve year A.D. 1851.

SEM DESCRIPTION: This is a spherical stomatocyst (diameter 5.6-7.8 μm). It has a thickly silicified, cylindrical to slightly conical collar (apical diameter 1.5-2.9 μm; height 0.5-0.8 μm), with an abrupt outer margin and a sloping inner

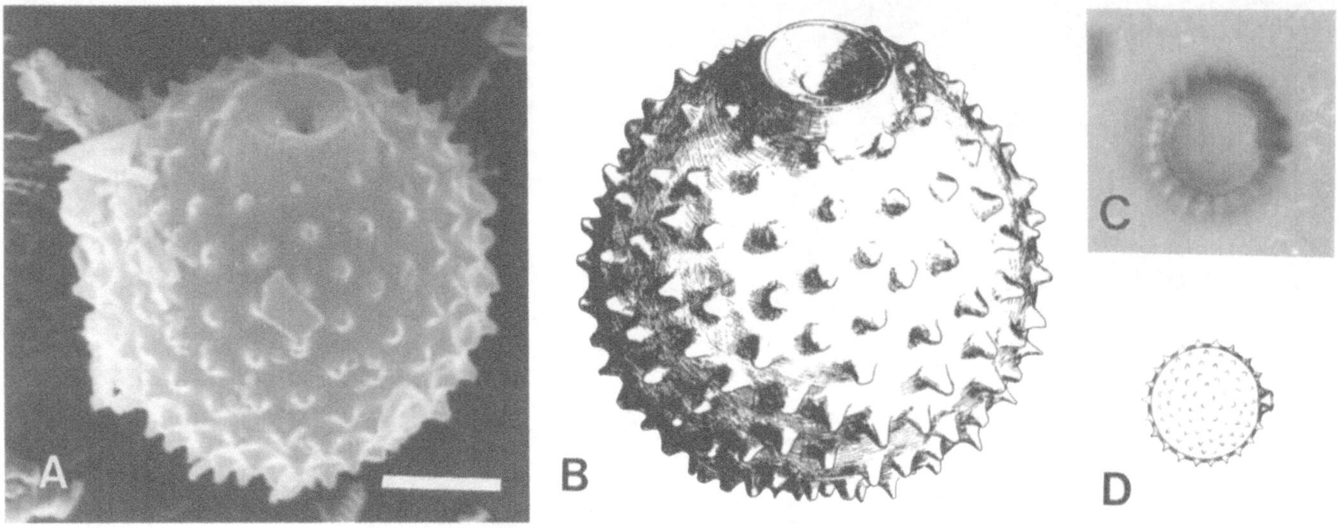

Fig. 71. Stomatocyst 157. A: SEM. B: Line drawing of SEM. C: LM. D: Line drawing of LM. Scale bar = 2 μm. LM x2000.

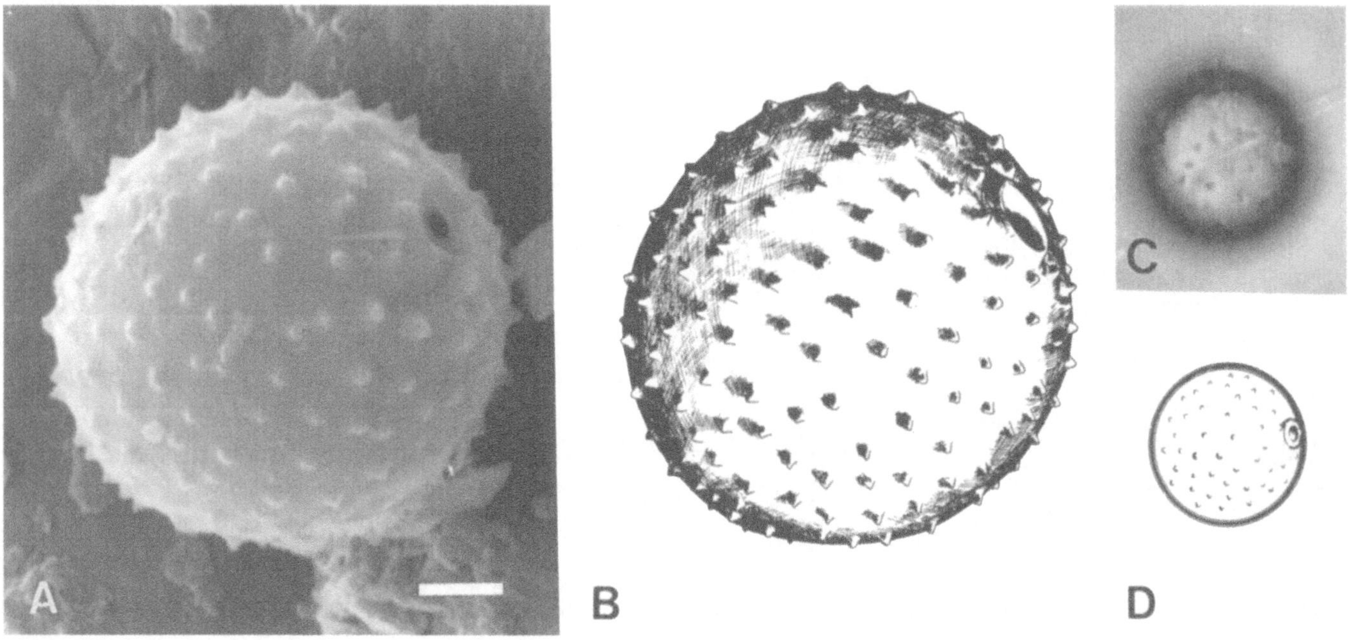

Fig. 72. Stomatocyst 211. A: SEM. B: Line drawing of SEM. C: LM. D: Line drawing of LM. Scale bar = 2 μm. LM x2000.

margin, and a conical pore (inner pore diameter 0.4-0.6 μm). The cyst surface is ornamented with conula (basal diameter 0.3-0.7 μm; height 0.4-0.5 μm) equidistantly arranged in a regular pattern (distance between conula 0.7-0.8 μm).

This stomatocyst closely resembles stomatocyst 31 but is usually larger, with a wider and thicker collar.

LM DESCRIPTION: This cyst can be separated from stomatocyst 31 by its size and collar morphology. The regular pattern of ornamentation is very distinctive.

REFERENCES: Stomatocyst 157 has been observed in Minnesota, U.S.A. (Zeeb & Smol, 1993a), and New York, U.S.A. (Duff & Smol, unpubl.).

ECOLOGY: Ecological details are as yet unknown.

Stomatocyst 211, Duff & Smol, 1994 (3) Fig. 72

BIOLOGICAL AFFINITY: unknown

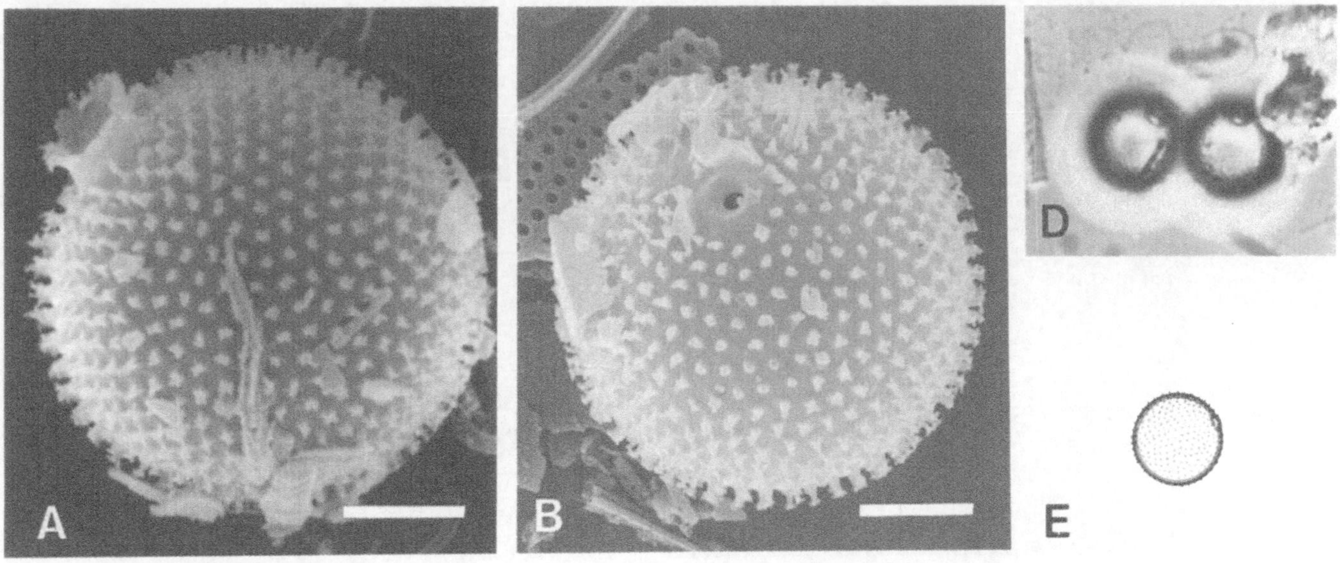

Fig. 73. Stomatocyst 73. A: SEM, mature specimen. B: SEM, immature specimen. C: Line drawing of SEM. D: LM of two specimens. E: Line drawing of LM. Scale bars = 2 μm. LM x2000.

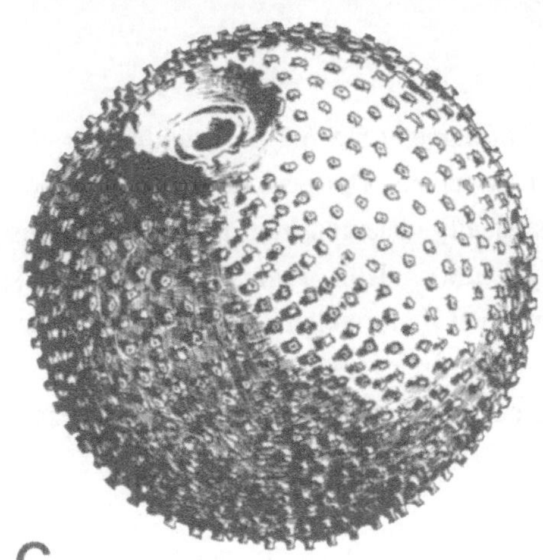

LOCALITY: Hawkins Lake (51°55'N; 120°55'W), British Columbia, Canada. Ekman dredge, surface (1 cm) sediment sample.

SEM DESCRIPTION: This stomatocyst is spherical to widely oval (9.8-10.7 x 10.2-12.1 μm; width : length ratio 0.88-0.96). The collar is low and broadly conical (apical diameter 1.6-2.1 μm; height 0.3-0.8 μm) with an acute apex, a sloping inner margin or a sloping planar annulus and a regular pore (diameter 0.7-0.9 μm). The surface is ornamented with short, broad conula (diameter 0.6-1.1 μm; height 0.3-0.8 μm; distance between conula 0.9-1.8 μm).

LM DESCRIPTION: With LM, stomatocyst 211 cannot be distinguished from stomatocysts 212, 167 and small ($\leq$14.0 μm) specimens of stomatocyst 166.

REFERENCES: We were not able to link this cyst with previously published morphotypes.

ECOLOGY: As part of a collective category, stomatocyst 211 was most common in large, slightly acidic to circumneutral lakes in British Columbia, Canada (Duff, 1994).

Stomatocysts ornamented with short, thin spines that are roughly equidistant and cover the entire cyst surface

Stomatocyst 73, Duff & Smol, 1991 (11) Fig. 73

BIOLOGICAL AFFINITY: unknown

LOCALITY: Upper Wallface Pond (44°08.8'N; 74°03.3'W), Adirondack Park, New York, U.S.A. Gravity core, core depth 10.0-10.5 cm.

SEM DESCRIPTION: All SEM specimens of this stomatocyst are spherical (diameter 5.6-7.8 μm), but oblate specimens have been observed with LM. The conical pore (inner diameter 0.2-0.5 μm) is usually, but not always, surrounded by a small conical primary collar with an acute apex (diameter 0.7-1.0 μm). A narrow, flat, planar interannulus (diameter 1.0-1.5 μm) separates the primary collar from a thinly silicified, obconical secondary collar with an irregular apex (diameter 1.8-2.5 μm; height ca. 0.7 μm). Concentric rings of regularly-spaced echinate spines cover the entire cyst body (spine diameter $\leq$ 1 μm; height 0.2-0.4 μm;

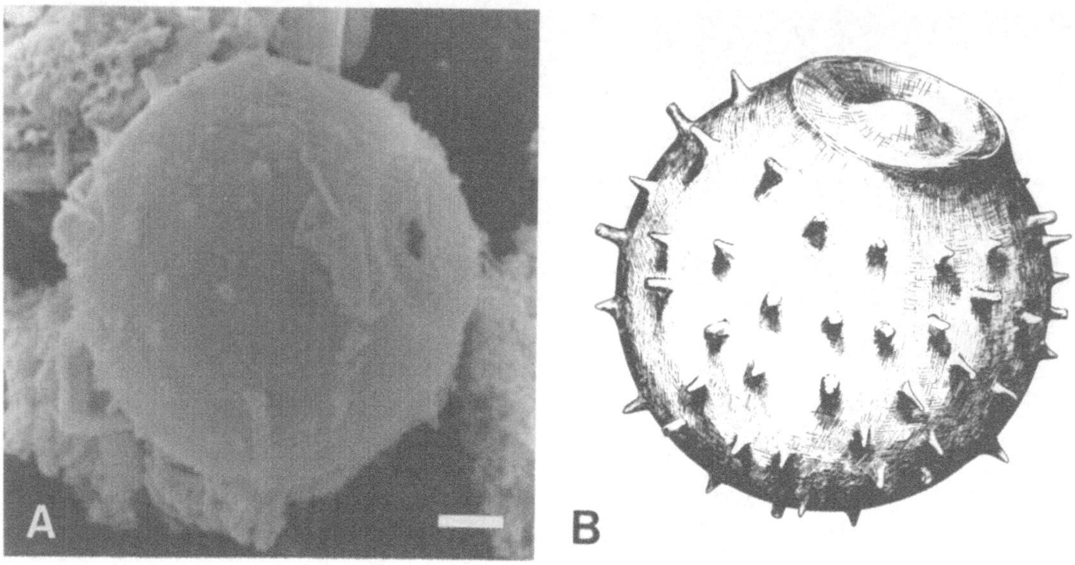

Fig. 74. Stomatocyst 194. A: SEM. B: Line drawing of SEM. Scale bar = 2 μm.

distance between spines 0.4-0.5 μm); fully developed spines have one to four finger-like apical projections. On one immature specimen (Fig. 73B), the collar appears to form from the fusion of the most anterior ring of spines.

LM DESCRIPTION: The ornamentation of this morphotype can be distinguished with high resolution optics. It is usually more thickly silicified than stomatocysts 117/178/210, the collar is longer and more obconical, and the body shape may vary from spherical to oblate.

REFERENCES: We have found this stomatocyst in the Yukon Territory, Canada (Brown & Smol, unpubl.), and New York, U.S.A. (Duff & Smol, 1991, unpubl.).

ECOLOGY: This stomatocyst is likely produced by a species tolerant of very low pH (<5.0) and possibly of elevated aluminum concentrations (Duff & Smol, 1991).

Stomatocyst 194, Zeeb & Smol in Zeeb *et al.*, subm. (5) Fig. 74

BIOLOGICAL AFFINITY: unknown

LOCALITY: Sonoma Volcanics, Napa County (38°N; 122°W), California, U.S.A. Pliocene (2-3 m.y.) sample

SEM DESCRIPTION: This stomatocyst is large and spherical (diameter 12.5-14.7 μm). The collar is low, very wide and conical (diameter 5.5-6.2 μm), with a flat planar to slightly swollen annulus. The pore is regular (diameter ca. 1.5 μm) but is usually obscured by a rounded siliceous plug (diameter 1.7-1.9 μm). Thin, echinate and baculate spines (basal diameter 0.5-0.7 μm; height 0.4-0.9 μm) are scattered randomly over the entire cyst surface.

This stomatocyst is very similar to stomatocyst 56 but is ornamented.

LM DESCRIPTION: We have not yet observed this cyst with LM.

REFERENCES: We have only observed this stomatocyst in California, U.S.A. (Zeeb *et al.*, subm.). *Cysta crassicollis* Nygaard (*sensu* Rybak, 1986 but not Nygaard, 1956) is similar but is smaller and ornamented more densely with verrucae or conula, not spines.

ECOLOGY: Autecological details are unknown.

Stomatocyst 167, Zeeb & Smol, 1993a (4) Fig. 75

BIOLOGICAL AFFINITY: unknown

LOCALITY: Elk Lake (47°11'N; 95°13'W), Clearwater County, Minnesota, U.S.A. Piston core, varve year 10 159 B.P.

SEM DESCRIPTION: This stomatocyst is spherical

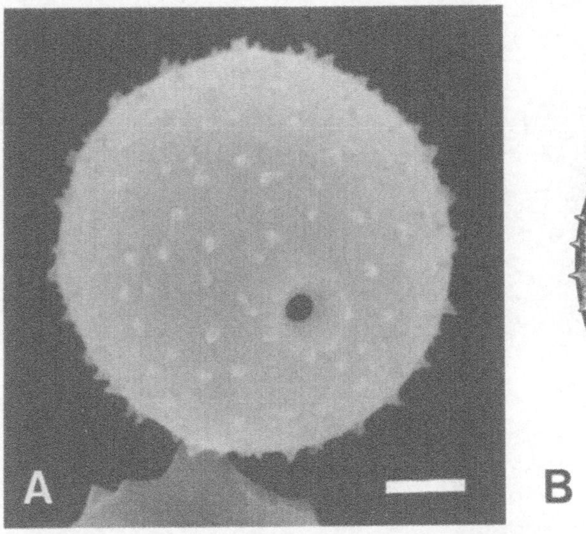

Fig. 75. Stomatocyst 167. A: SEM. B: Line drawing of SEM. Scale bar = 2 μm. See Figs. 72C, D for LM.

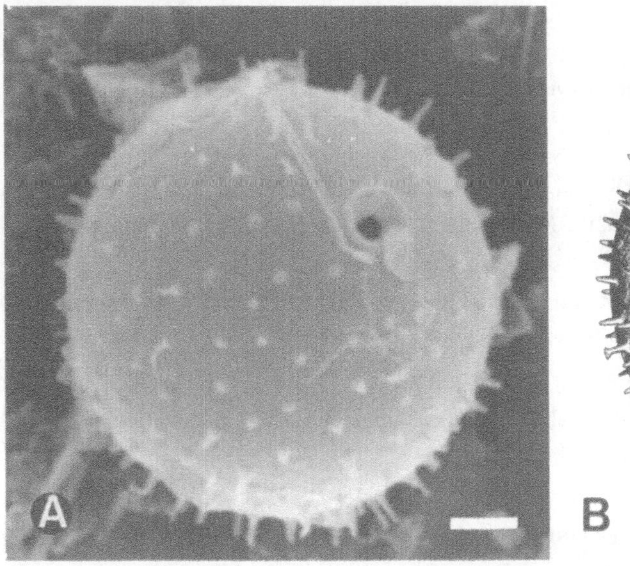

Fig. 76. Stomatocyst 212. A: SEM. B: Line drawing of SEM. Scale bar = 2 μm. See Figs. 72C, D for LM.

(diameter 9.3-10.4 μm) with a short cylindrical collar (diameter 1.3-2.0 μm; collar diameter : cyst diameter ratio 0.13-0.21) surrounding a conical pore (inner pore diameter 0.6-0.7 μm). The collar appeared to be degraded or undeveloped on all our SEM specimens. The cyst surface is covered with short, narrow, echinate spines (basal diameter 0.3-0.4 μm; height 0.1-0.6 μm; distance between spines 0.5-2.0 μm), which may have flattened or hooked apices but not corona-like apices.

LM DESCRIPTION: With LM, stomatocyst 167 cannot be distinguished from stomatocysts 211 (Figs. 72C, D), 212, and small (≤14.0 μm) specimens of stomatocyst 166.

REFERENCES: This stomatocyst has been recorded in Ontario, Canada (Zeeb & Smol, unpubl.), British Columbia, Canada (Duff & Smol, 1994), and Minnesota, U.S.A. (Zeeb & Smol, 1993a).

ECOLOGY: As a member of a collective category, this stomatocyst was most common in large, acidic to circumneutral lakes (Duff, 1994).

Stomatocyst 212, Duff & Smol, 1994 (4) Fig. 76

BIOLOGICAL AFFINITY: unknown

LOCALITY: Dahl Lake (53°50'N; 123°15'W), British Columbia, Canada. Ekman dredge, surface (1 cm) sediment sample.

SEM DESCRIPTION: This spherical stomatocyst (diameter 13.4-13.9 μm) has a low, cylindrical collar (diameter 2.2-2.5 μm; height 0.3-0.8 μm; collar diameter : cyst diameter ratio 0.16-0.18) with a gradually sloping inner margin and a regular pore (diameter 0.8-1.0 μm). The ornamentation

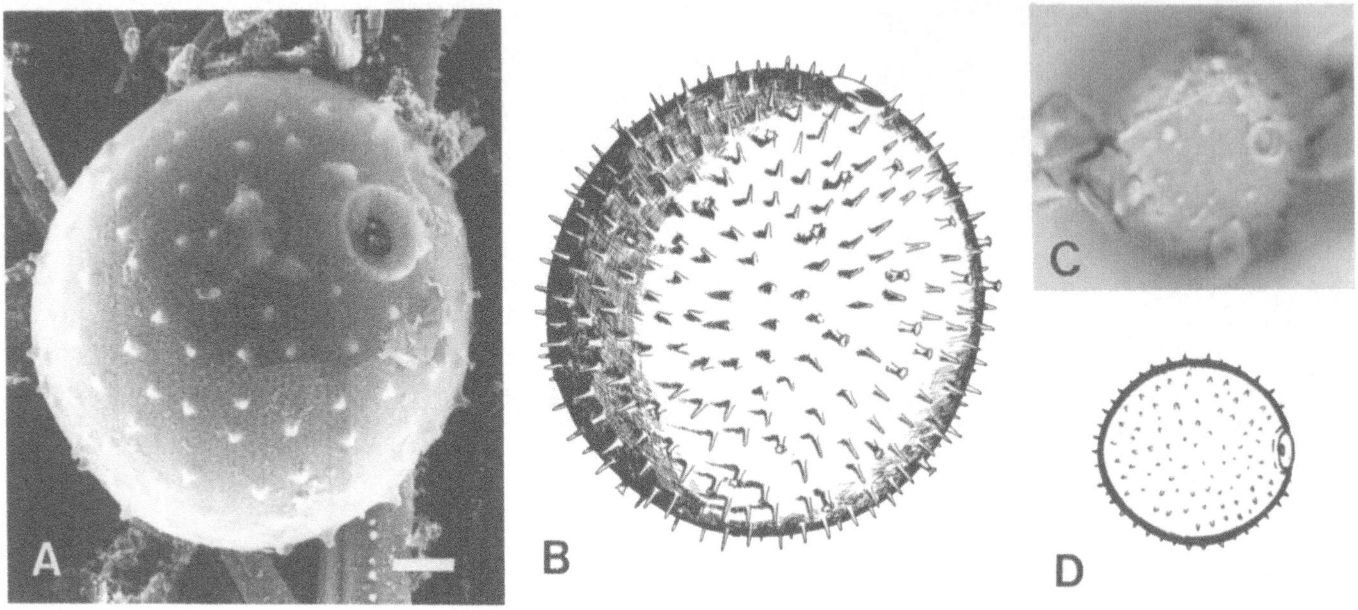

Fig. 77. Stomatocyst 166. A: SEM, showing immature honeycomb reticulum. B: Line drawing of SEM. C: LM, specimens > 14.0 μm. D: Line drawing of LM, specimens > 14.0 μm. Scale bar = 2 μm. LM x2000.

consists of short, narrow, echinate spines (basal diameter 0.2-0.8 μm; height 0.5-1.3 μm; distance between spines 1.2-3.4 μm).

Stomatocyst 212 is very similar to stomatocyst 167, but is larger with slightly different collar morphology. It is identical to stomatocyst 213, except that it is smaller, and does not have an annulus. Nevertheless, it is possible that these three morphotypes are produced by the same species.

LM DESCRIPTION: With LM, this stomatocyst cannot be distinguished from stomatocysts 167, 211 (Figs. 72C, D), and small (≤14.0 μm) specimens of 166.

REFERENCES: We have recorded stomatocyst 212 in British Columbia, Canada (Duff & Smol, 1994), Ontario, Canada (Zeeb & Smol, unpubl.), and the Yukon Territory, Canada (Brown & Smol, unpubl.).

ECOLOGY: As a member of a collective category, stomatocyst 212 was most common in large, slightly acidic to circumneutral lakes (Duff, 1994).

Stomatocyst 166, Zeeb & Smol, 1993a (9) Fig. 77

BIOLOGICAL AFFINITY: This stomatocyst is probably produced by *Mallomonas crassisquama* (Asmund) Fott (Gretz *et al.*, 1979), although we have extended the published size range.

LOCALITY: Elk Lake (47°11'N; 95°13'W), Clearwater County, Minnesota, U.S.A. Sediment trap sample.

SEM DESCRIPTION: This stomatocyst is spherical to widely oval (diameter 12.6-17.2 μm, or 11.4-16.0 x 12.9-18.4 μm; width : length ratio >0.85 (0.87-1.0)). The collar is conical to obconical (diameter 1.4-3.2 μm; height 0.3-0.9 μm) with an acute apex, a sloping inner margin, no annulus, and a regular pore (diameter 0.9-1.3 μm). The inner and outer collar margins may be vertically strutted. The collar diameter : cyst width ratio is 0.13-0.19; although this is generally smaller than that of Smol (1984) stomatocyst 1, produced by *Mallomonas pseudocoronata* Prescott, there is some overlap.

The immature honeycomb reticulum described by Skogstad (1984) and Cronberg (1988) was observed on five SEM specimens (e.g. Fig. 77A). When mature, the cyst surface is ornamented with broadly echinate spines (basal diameter 0.4-1.0 μm; height 0.2-0.7 μm) which may have hooked or corona-like apices. The spines are scattered but fairly regularly spaced (distance between spines 0.8-3.5 μm).

LM DESCRIPTION: This stomatocyst can be distinguished from Smol (1984) stomatocyst 1 and stomatocyst 236 by its width : length ratio. Small (≤14.0 μm) specimens are indistinguishable from stomatocysts 167, 211, and 212, whereas large (>14.0 μm) specimens are indistinguishable from stomatocyst 213.

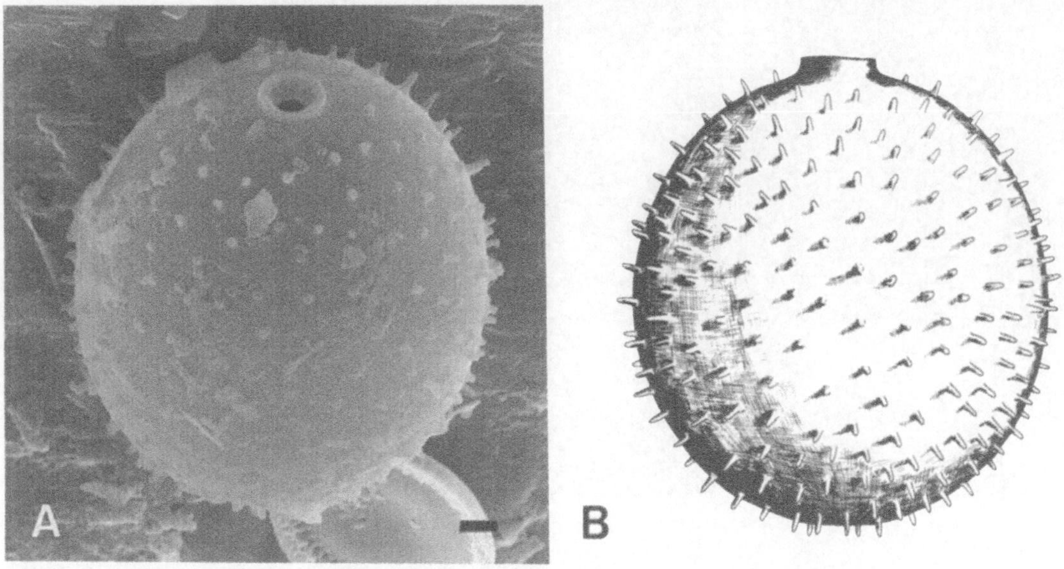

Fig. 78. Stomatocyst 213. A: SEM. B: Line drawing of SEM. Scale bar = 2 μm. See Figs. 77C, D for LM.

REFERENCES: We have observed stomatocyst 166 in British Columbia, Canada (Duff & Smol, 1994), the Yukon Territory, Canada (Brown & Smol, unpubl.), Minnesota, U.S.A. (Zeeb & Smol, 1993a), and New York, U.S.A. (Duff & Smol, unpubl.). Sandgren & Carney (1983 - Michigan, U.S.A.) cyst 45 and Smith & White (1985 - Fig. 19, Washington, U.S.A.) are probably identical.

ECOLOGY: *M. crassisquama* is a common species with a wide tolerance for many environmental variables (e.g. pH, specific conductance, trophic status, temperature), although it does not occur in very acidic (pH <5.0) sites (Siver, 1991a).

Stomatocyst 213, Duff & Smol, 1994 (2) Fig. 78

BIOLOGICAL AFFINITY: This stomatocyst is possibly produced by *Mallomonas acaroides* Perty *emend.* Ivanoff (Cronberg, 1980, as *M. acaroides* var. *striatula* Asmund).

LOCALITY: Spanish Lake (52°35'N; 121°20'W), British Columbia, Canada. Ekman dredge, surface (1 cm) sediment sample.

SEM DESCRIPTION: This stomatocyst is large, spherical to widely oval (diameter 16.0-19.4 μm). The collar is cylindrical and low (diameter 2.7-2.9 μm; height 0.2 μm; collar diameter : cyst diameter ratio 0.15-0.17) with a narrow, flat to sloping planar annulus (diameter ca. 1.5 μm) surrounding the regular pore (diameter 1.1-1.2 μm). The ornamentation consists of narrow, echinate spines (basal diameter 0.4-1.1 μm; height 0.5-1.1 μm) which are scattered but fairly regularly spaced over the cyst surface (distance between spines 1.3-2.5 μm). Spines were observed with flattened or hooked, but not corona-like, apices. Signs of developmental immaturity (the honeycomb reticulum and a vertically strutted inner collar margin) were observed on one SEM specimen.

Stomatocyst 213 is larger than stomatocyst 212, and has an annulus. It is distinguished from stomatocyst 166 with SEM by the narrow, stick-like appearance of the spines, although this feature may not be significant.

LM DESCRIPTION: With LM, stomatocyst 213 cannot be distinguished reliably from large (>14.0 μm) specimens of stomatocyst 166 (Figs. 77C, D).

REFERENCES: We have only found stomatocyst 213 in British Columbia, Canada (Duff & Smol, 1994; Zeeb, 1994), and we have not identified it definitely with other published morphs.

ECOLOGY: *Mallomonas acaroides* is an alkaliphilic, meso- to eutrophic species with a preference for cool water (Siver, 1991a). Cyst formation has been observed in the autumn (Cronberg, 1980).

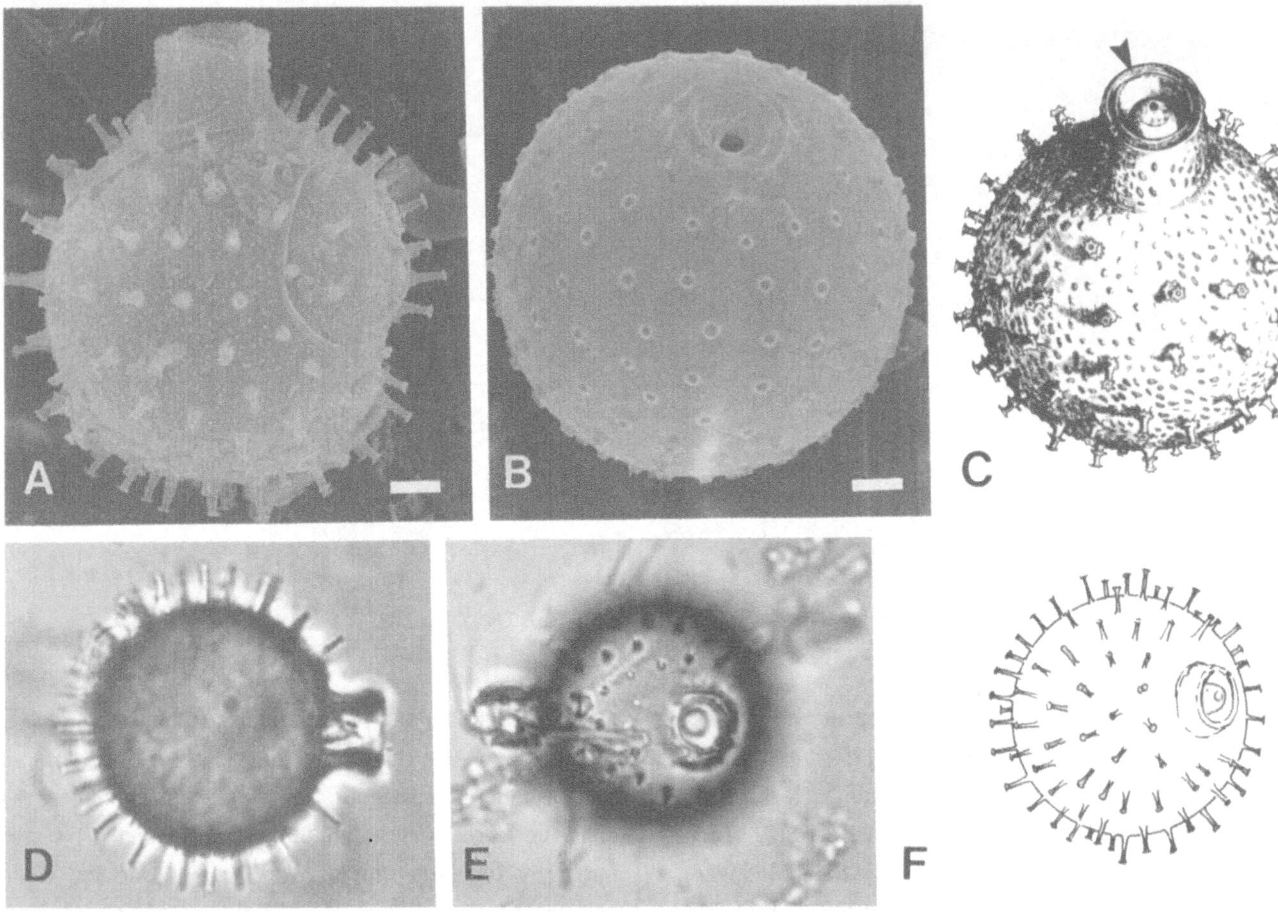

Fig. 79. Stomatocyst 72. A: SEM, mature specimen. B: SEM, immature specimen showing annulus. C: Line drawing of SEM. Apical collar groove indicated by arrow. D: LM, cross-sectional view. E: LM, surface view. F: Line drawing of LM. Scale bars = 2 μm. LM x1500.

Stomatocyst 72, Duff & Smol, 1991 (13) Fig. 79

BIOLOGICAL AFFINITY: Stomatocyst 72 is produced by an undescribed rhizopodial chrysophyte (Sandgren, 1989).

LOCALITY: Upper Wallface Pond (44°08.8'N; 74°03.3'W), Adirondack Park, New York, U.S.A. Gravity core, core depth 0-0.5 cm.

SEM DESCRIPTION: This is a very large, spherical stomatocyst (diameter 19.1-22.0 μm) with a complex collar. The regular pore (diameter 1.3-1.8 μm) is immediately surrounded by a narrow, conical to cylindrical primary collar (diameter 1.3-2.3 μm), which is separated from the secondary collar by a very narrow, flat planar annulus (Fig. 79B). The secondary collar is large and cylindrical (outer apical diameter 4.7-8.0 μm; height 2.5-6.0 μm), with an apical groove that separates the inner and outer collar margins, forming a false complex collar (apical groove width 0.6-1.6 μm; Fig. 79C). The cyst surface is ornamented with baculate spines (base diameter 0.3-2.0 μm; height 0.8-2.9 μm) set approximately equidistantly (distance between spines 1.7-4.4 μm). The spine apices are ornamented with a corona of eight to ten finger-like projections, and a ring or additional row of projections may be located approximately half-way up the spine. On immature specimens (Fig. 79B), the spines are seen to be hollow. Scabrae, which may be slightly elongated, are found on the cyst surface between the spines of mature cysts (Figs. 79A, C).

Sandgren (1989) describes the cyst development.

LM DESCRIPTION: This stomatocyst is easily identified by its size, collar morphology, and spines. The scabrae are sometimes, but not always, visible with LM.

REFERENCES: This stomatocyst has been observed in Ontario, Canada (Rybak *et al.*, 1991 - cyst 60; Zeeb *et al.*, 1994), Wisconsin, U.S.A. (Sandgren, 1989), and New York, U.S.A. (Duff & Smol, 1991, unpubl.; P.A. Siver, pers. commun.).

ECOLOGY: Stomatocyst 72 is probably produced by an

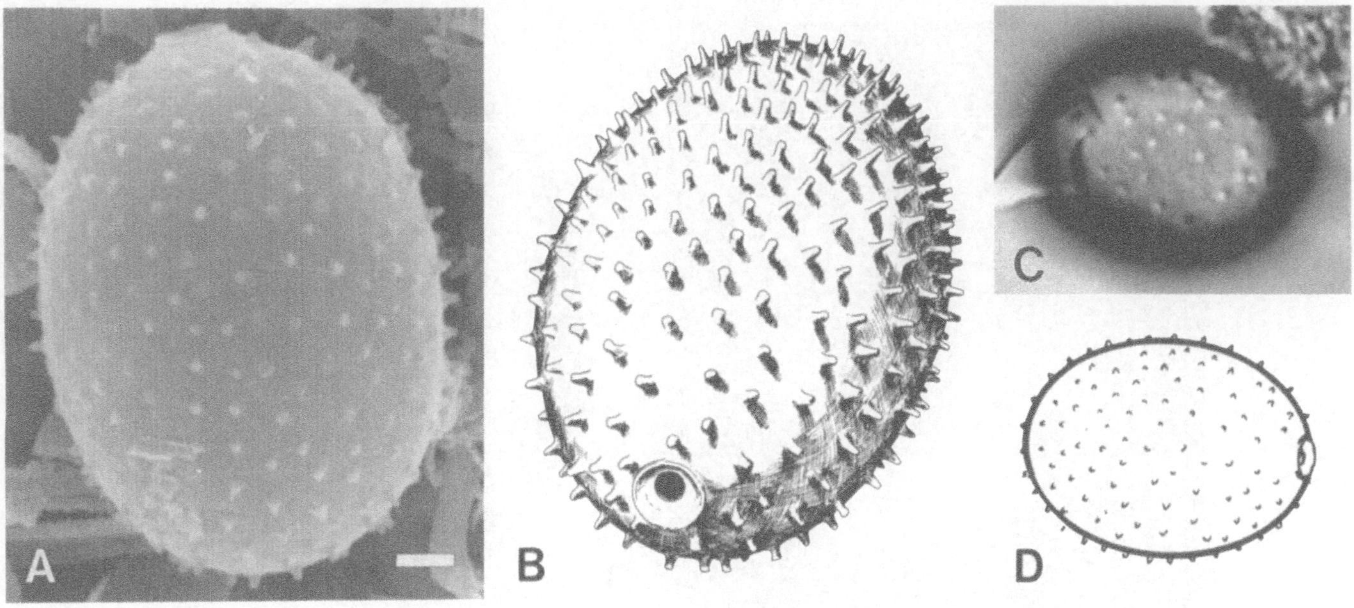

Fig. 80. Smol (1984) stomatocyst 1. A: SEM. B: Line drawing of SEM. C: LM. Collar not visible. D: Line drawing of LM. Scale bars = 2 μm. LM x2000.

acidobiontic or acidophilic species (Rybak *et al.*, 1991; Duff & Smol, 1991, unpubl.). Although it has been found in oligotrophic locations (Rybak *et al.*, 1991; Duff & Smol, 1991, unpubl.), it increased in relative abundance in an artificially eutrophied lake (Zeeb *et al.*, 1994).

Stomatocyst 1, Smol, 1984 (17) Fig. 80

BIOLOGICAL AFFINITY:
Mallomonas pseudocoronata Prescott (Smol, 1984)

LOCALITY: Lake Opinicon (44°33'N, 76°26'W), Ontario, Canada. Plankton sample, collected September 29, 1979.

SEM DESCRIPTION: This stomatocyst is oval (11.2-15.6 x 13.6-23.5 μm) with a width : length ratio ≤0.85 (0.53-0.84). The collar is short, fairly wide, conical to slightly obconical (diameter 2.1-3.6 μm; height 0.3-0.8 μm) with an acute apex, a sloping inner margin, and a regular pore (diameter ca. 1.4 μm). The thin annulus noted by Smol (1984) on his live specimens was not observed on any of our other specimens recovered from sediment samples, and may be easily dissolved. The collar diameter : cyst width ratio ranges from 0.17-0.22 (one SEM specimen has a ratio of 0.29).

The honeycomb reticulum commonly observed on immature *Mallomonas* stomatocysts (Skogstad, 1984; Cronberg, 1988) was observed on only one of our SEM specimens. The rest of our specimens are smooth surfaced, with short echinate spines (basal diameter 0.4-0.9 μm; height 0.3-0.8 μm) which may be bent or have hooked apices, but not corona-like apices. The spines are scattered but fairly regularly spaced across the entire cyst surface (distance between spines 1.0-2.7 μm).

LM DESCRIPTION: This stomatocyst is probably not distinguishable from stomatocyst 236 with LM. It can be distinguished from stomatocysts 166, 212, and 213 on the basis of body shape.

REFERENCES: This stomatocyst has been observed in Ontario, Canada (Smol, 1984; Rybak *et al.*, 1991 - cyst 41); British Columbia, Canada (Duff & Smol, 1994; Zeeb, 1994); Minnesota, U.S.A. (Zeeb & Smol, 1993a); California, U.S.A. (Adam & Mahood, 1979 - type 32); and Switzerland (Smol, 1988b). The published micrographs of Carney *et al.* (1992 - Figs. 4, 48, California, U.S.A.) cyst 45 are identical, but the written description indicates that their morph is a collective category.

ECOLOGY: *M. pseudocoronata* is an alkaliphilic species with a relatively high temperature optimum (18.1°C; Siver, 1991a). Cyst formation has been observed in the autumn (Smol, 1984). It has a fairly wide tolerance for lake productivity but is excluded from hypereutrophic systems (Siver, 1991a). The cyst has been found in large, productive lakes (Duff, 1994).

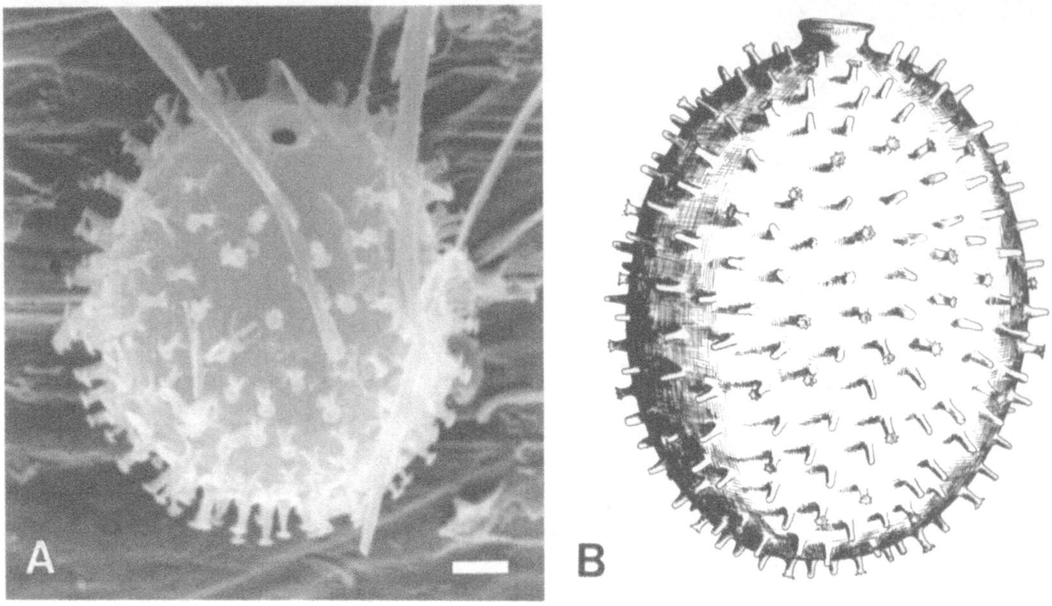

Fig. 81. Stomatocyst 236. A: SEM. B: Line drawing of SEM. Scale bar = 2 μm. See Fig. 80C, D for LM.

Stomatocyst 236, Duff & Smol in Duff *et al.* (this volume) (3) Fig. 81

BIOLOGICAL AFFINITY: Stomatocyst 236 is probably produced by *Mallomonas acaroides* var. *muskokana* Nicholls (Siver, 1991c), although our specimens are more oval than previously described.

NEGATIVE NUMBER: J.P. Smol 610

LOCALITY: "Unnamed Pond" (unofficial name) (43°11'N; 74°37'W), Adirondack Park, New York, U.S.A. Gravity core, surface (0-0.25 cm) sediment sample.

SEM DESCRIPTION: This is an oval stomatocyst (11.1-12.8 x 13.0-15.9 μm) with a width : length ratio of 0.78-0.85. There is a low, obconical collar (diameter 1.7-2.4 μm; height 0.5-0.6 μm; collar diameter : cyst width ratio 0.15-0.19) with a slightly rounded apex, a sloping inner margin, no annulus, and a regular pore (diameter 0.7-0.9 μm). The cyst body is ornamented with baculate spines (basal diameter 0.3-0.9 μm; height 0.9-1.4 μm) with flattened to splayed apices which may be corona-like. These are fairly regularly spaced (distance between spines 1.0-2.1 μm).

LM DESCRIPTION: We have not observed this stomatocyst definitely with LM; it is probably indistinguishable from Smol (1984) stomatocyst 1 (Fig. 80C, D).

REFERENCES: This stomatocyst has been observed in the Yukon Territory, Canada (Brown & Smol, unpubl.), New York, U.S.A. (Duff *et al.*, this volume; Siver, 1991b), and Connecticut, U.S.A. (Siver, 1991b). Adam & Mahood (1979b - California, U.S.A.) type 47 is similar.

ECOLOGY: *Mallomonas acaroides* var. *muskokana* is an acidobiontic species (estimated optimum pH of 5.4) found primarily in oligotrophic, softwater (specific conductance <40 μS) lakes; it is also common in acid bogs (Siver, 1991a). It is most abundant in late spring to early summer, and has a relatively high temperature optimum of 17.5°C (Siver, 1991a). The few reports of this cyst are from sites similar to those described above (Siver, 1991b; Duff & Smol, unpubl.).

Stomatocysts with short spines that are usually concentrated in the posterior hemisphere

Stomatocyst 185, Brown & Smol in Brown *et al.*, 1994 (11) Fig. 82

BIOLOGICAL AFFINITY: unknown

LOCALITY: Nordvestø (76°44'N; 73°13'W), Carey Islands, Greenland. SIPRE peat core, core depth 50-51 cm.

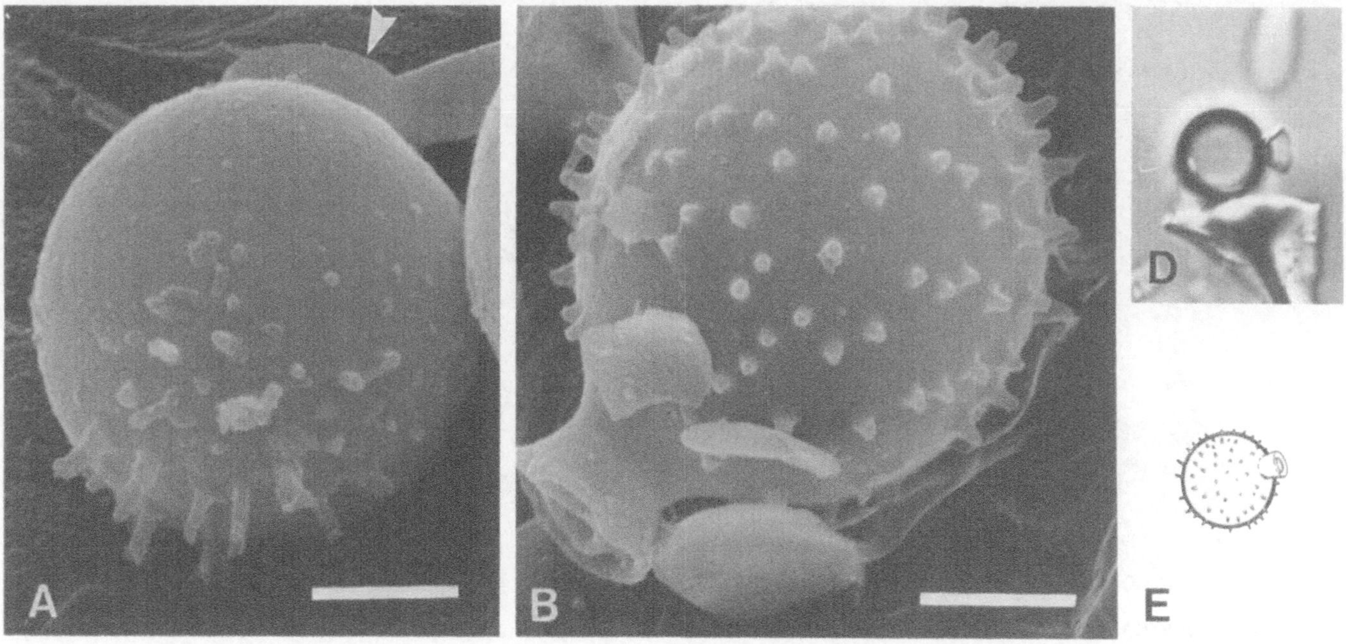

Fig. 82. Stomatocyst 185 *formae* A and B. A: *Forma* A, SEM. Collar indicated by arrow. B: *Forma* B, SEM. C: *Forma* A, line drawing of SEM. D: *Forma* A, LM. Ornamentation not visible. D: *Forma* B, line drawing of LM. Scale bars = 2 μm. LM x2000.

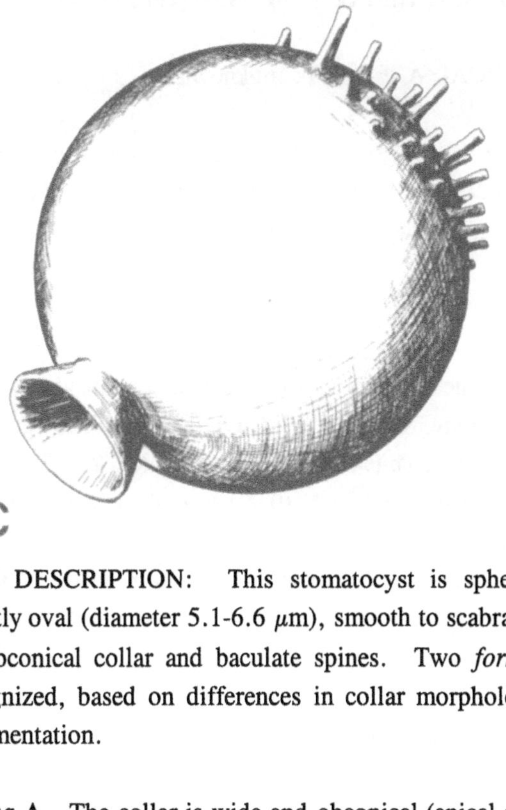

SEM DESCRIPTION: This stomatocyst is spherical to slightly oval (diameter 5.1-6.6 μm), smooth to scabrate, with an obconical collar and baculate spines. Two *formae* are recognized, based on differences in collar morphology and ornamentation.

***Forma* A** - The collar is wide and obconical (apical diameter 1.7-2.9 μm; height 1.0-1.1 μm). One SEM specimen had either a swollen annulus or a conical primary collar, but a clear view could not be obtained. The ornamentation consists of approximately three to >50 baculate spines (basal diameter 0.2-0.3 μm; maximum height ca. 0.9 μm) located in the posterior hemisphere only. The spines do not bifurcate, and may be laterally compressed and/or bent.

***Forma* B** - The collar is complex. The primary collar is cylindrical (diameter 0.9-1.1 μm). The secondary collar is obconical (apical diameter 1.7-2.6 μm; height 0.6-2.2 μm). The primary and secondary collars may be connected by siliceous struts or folds. Short echinate spines (basal diameter 0.3-0.5 μm; maximum height ca. 0.5 μm) with blunt apices are located randomly over the entire cyst surface. It is possible that *forma* B specimens are simply more mature than those of *forma* A.

LM DESCRIPTION: Stomatocyst 185 *forma* A is similar to stomatocysts 217 and 184, but can be recognized by its single obconical collar and non-bifurcating spines. When the spines are not visible (Fig. 82D), this *forma* can still be distinguished by its size and collar morphology.

Forma B is distinguished by the complex collar, and spines found over the entire cyst surface. It could be confused with stomatocyst 133 *forma* B if the spines cannot be distinguished from ridges.

REFERENCES: Stomatocyst 185 *forma* A has been found in British Columbia, Canada (Zeeb, 1994) and Greenland (Brown *et al.*, 1994). We have not found any other references to *forma* B.

ECOLOGY: We cannot yet estimate the ecological optima and tolerances of stomatocyst 185.

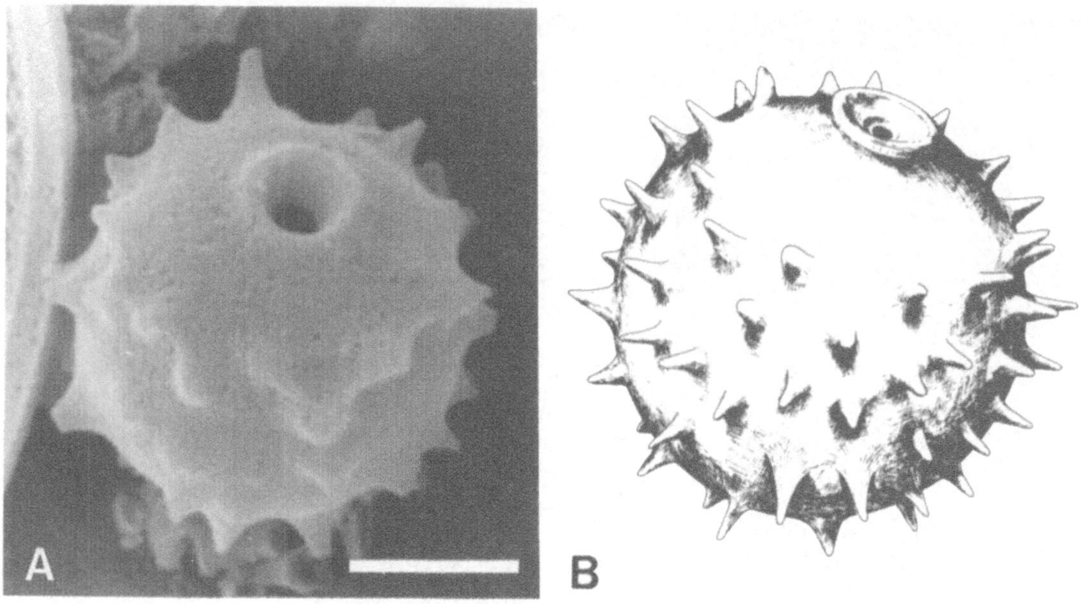

Fig. 83. Stomatocyst 215. A: SEM. B: Line drawing of SEM. Scale bar = 2 μm.

Stomatocyst 215, Duff & Smol, 1994 (4) Fig. 83

BIOLOGICAL AFFINITY: unknown

LOCALITY: Dahl Lake (53°50'N; 123°15'W), British Columbia, Canada. Ekman dredge, surface (1 cm) sediment sample.

SEM DESCRIPTION: This stomatocyst is small and spherical to widely depressed obovate (diameter 4.4-5.0 μm). The low collar varies from conical to widely obconical with a reflexed apex (collar diameter 1.1-1.4 μm; height 0.2-0.3 μm) and surrounds a conical pore (inner diameter 0.4-0.5 μm). The ornamentation consists of broad-based echinate spines to low ridges (basal diameter 0.4-1.2 μm; height 0.2-1.0 μm). The cyst surface nearest to the collar is devoid of ornamentation, after which a latitudinal ring of spines encircles the cyst (ring diameter ca. 3.6 μm); below this ring the spines are randomly distributed.

LM DESCRIPTION: We have not yet observed this morphotype with LM.

REFERENCES: This stomatocyst has been found in British Columbia, Canada (Duff & Smol, 1994), and the Northwest Territories, Canada (Webb & Smol, unpubl.).

ECOLOGY: No autecological data are yet available.

Stomatocyst 4, Duff & Smol, 1988 (19) Fig. 84

BIOLOGICAL AFFINITY: unknown

LOCALITY: "Rock Basin Lake" (unofficial name) (78°29.5'N; 76°44.8'W), Ellesmere Island, Northwest Territories, Canada. Piston core, core depth 15-16 cm.

SEM DESCRIPTION: This is a spherical to widely depressed obovate stomatocyst (diameter 3.4-4.9 μm) without a collar. The pore varies from regular to conical to slightly concave (pore diameter 0.3-0.5 μm). The posterior hemisphere is ornamented with numerous (ca. 20-45) echinate spines, of variable length (0.2-1.5 μm), which may have hooked apices.

Stomatocyst 66 (Chapter 5) has now been reassigned to stomatocyst 4.

LM DESCRIPTION: Stomatocyst 4 was counted as a collective category by Duff & Smol (1991), but we now believe that the cysts included in this category are distinguishable with LM. This morphotype can be recognized by the number, length, and posterior placement of the spines, and by the body shape.

REFERENCES: This cyst has been observed in the Northwest Territories, Canada (Duff & Smol, 1988, unpubl.), British Columbia, Canada (Duff & Smol, 1994; Zeeb, 1994), and New York, U.S.A. (Duff & Smol, 1991 - stomatocyst 66; Duff & Smol, unpubl.).

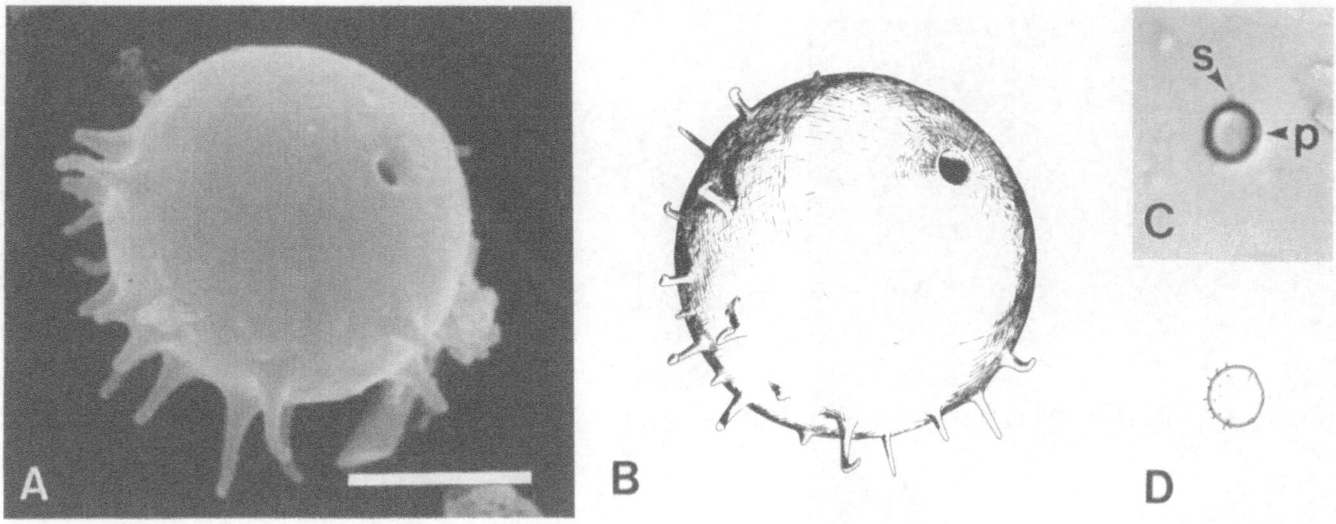

Fig. 84. Stomatocyst 4. A: SEM. B: Line drawing of SEM. C: LM. p = pore; s = spine. D: Line drawing of LM. Scale bar = 2 μm. LM x2000.

ECOLOGY: As a member of a collective category, stomatocyst 4 was tentatively classified as acidophilic by Duff & Smol (1991), although it demonstrated a more alkaliphilic distribution in British Columbia, Canada (Duff, 1994). Its presence in arctic (Duff & Smol, 1988, 1989) and alpine (Duff & Smol, 1991; Duff, 1994) lakes indicates that it is produced by a cold-tolerant species. It may also be more common in shallow sites (Duff, 1994).

Stomatocyst 83, Duff & Smol, 1991 (8) Fig. 85

BIOLOGICAL AFFINITY: unknown

LOCALITY: Upper Wallface Pond (44°08.8'N; 74°03.3'W), Adirondack Park, New York, U.S.A. Gravity core, core depth 40.0-40.5 cm.

SEM DESCRIPTION: This stomatocyst is spherical to oblate, with a fairly wide size range (diameter 3.9-6.1 μm). It has a regular pore (diameter 0.5-0.8 μm) and a conical primary collar (primary collar diameter 0.7-1.5 μm; height 0.1-1.4 μm); the primary collar has a slightly obconical apex when mature. Basal struts may be present on the primary collar. The collar may have a sloping inner margin, or a sloping planar annulus. A smooth planar interannulus separates the primary collar from the secondary collar, which is low and conical (diameter 2.5-4.9 μm). The surface below the secondary collar is ornamented with numerous echinate spines of variable length (to 2.0 μm).

LM DESCRIPTION: With LM, the complex collar (Fig. 85C) and ornamentation (Fig. 85D) of stomatocyst 83 are visible. Stomatocyst 83 can be distinguished from stomatocyst 84 because the latter morph lacks the secondary collar and has more pronounced collar struts on the primary collar. Stomatocyst 217 is similar to stomatocyst 83, but the secondary collar of stomatocyst 217 is narrower.

REFERENCES: This stomatocyst has been observed in Ontario, Canada (Rybak *et al.*, 1991 - cyst 63; Zeeb & Smol, unpubl.); British Columbia, Canada (Duff & Smol, 1994; Zeeb, 1994); New York, U.S.A. (Duff & Smol, 1991); California, U.S.A. (Adam & Mahood, 1979b - types 62, 63, 64, 65, 66, and 67); and Michigan, U.S.A. (Sandgren & Carney, 1983 - cyst 22).

ECOLOGY: This stomatocyst has been classified as pH-indifferent (Rybak *et al.*, 1991). It was most common in deep, oligotrophic, montane lakes in British Columbia, Canada (Duff, 1994).

Stomatocyst 84, Duff & Smol, 1991 (8) Fig. 86

BIOLOGICAL AFFINITY: unknown

LOCALITY: Upper Wallface Pond (44°08.8'N; 74°03.3'W), Adirondack Park, New York, U.S.A. Gravity core, core depth 10.0-10.5 cm.

SEM DESCRIPTION: This stomatocyst is obovate to

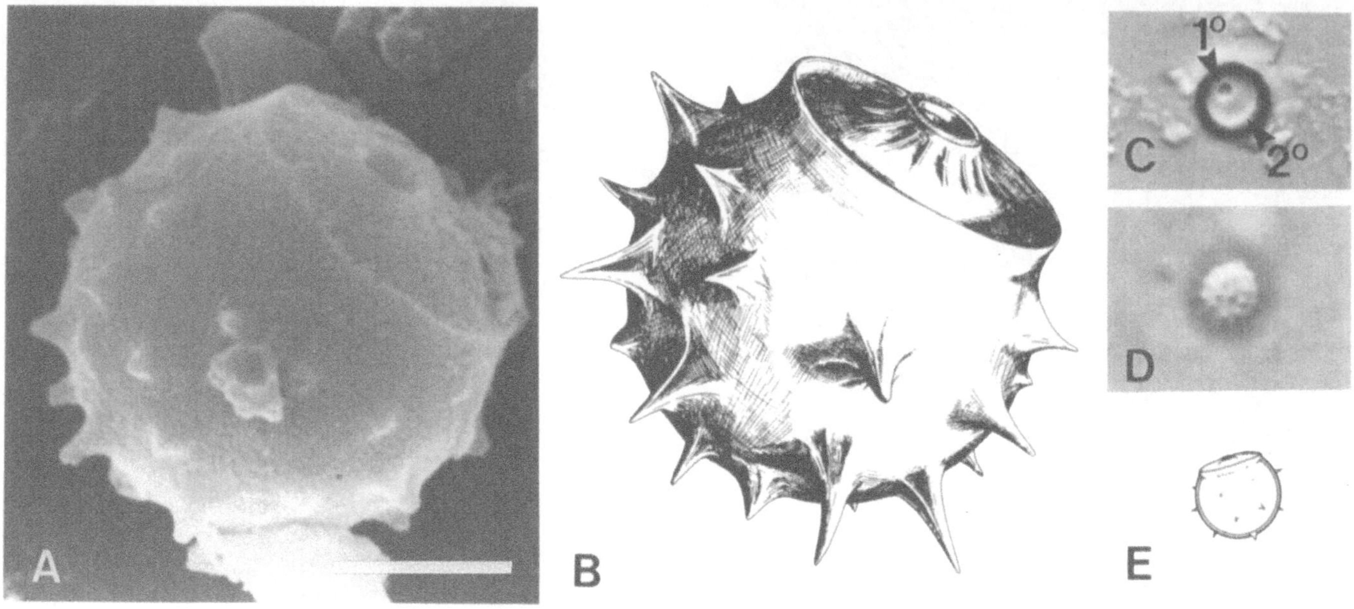

Fig. 85. Stomatocyst 83. A: SEM. B: Line drawing of SEM. C: LM, collar view. Primary (1°) and secondary (2°) collars indicated by arrows. D: LM, posterior view of spines. E: Line drawing of LM. Scale bar = 2 μm. LM x2000.

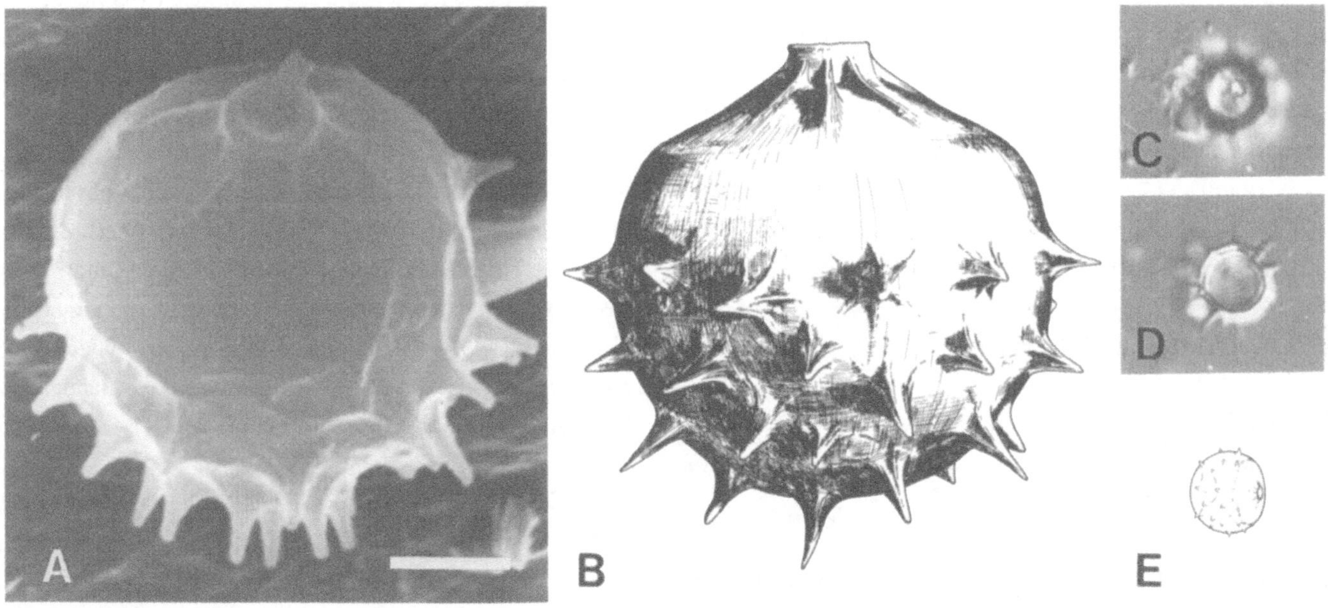

Fig. 86. Stomatocyst 84. A: SEM. B: Line drawing of SEM. C: LM, collar view showing struts. D: LM, cross-sectional view. E: Line drawing of LM. Scale bar = 2 μm. LM x2000.

spherical (4.4-5.5 x 4.3-5.0 μm), with a cylindrical to slightly obconical collar, a sloping planar annulus, and a conical pore (collar diameter 1.0-1.5 μm; height 0.3-0.8 μm; annulus diameter 0.8-1.0 μm; inner pore diameter 0.4-0.5 μm). The collar base possesses struts, which usually extend farther than those of stomatocyst 83, and may intertwine. Numerous echinate spines are located subequatorially; the spine bases may have struts or may be connected by low ridges (spine height <0.1-1.1 μm).

LM DESCRIPTION: Stomatocyst 84 is similar to stomatocyst 83, but is distinguished by the lack of a secondary collar and the more pronounced collar struts (Fig. 86C). It is differentiated from stomatocyst 4 by the presence of a collar. The obovate body shape of stomatocyst 84 is often very distinctive.

REFERENCES: This stomatocyst has been observed in Ontario, Canada (Rybak *et al.*, 1991 - cyst 61; Zeeb & Smol,

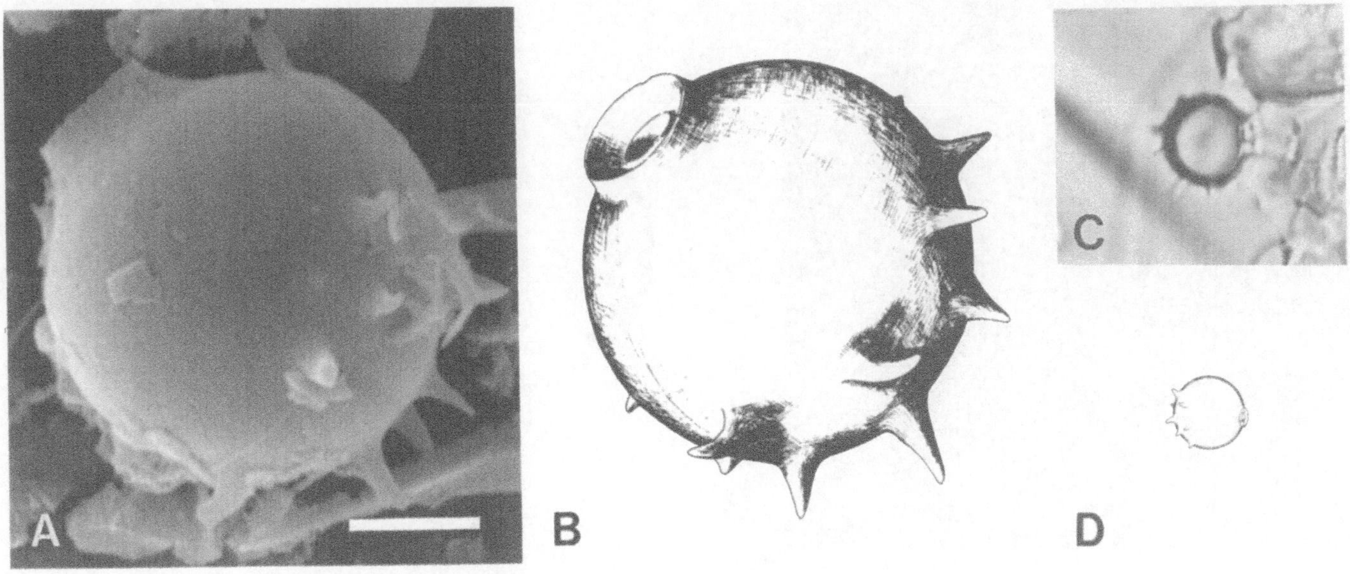

Fig. 87. Stomatocyst 217. A: SEM. B: Line drawing of SEM. C: LM. D: Line drawing of LM. Scale bar = 2 μm. LM x2000.

unpubl.), the Yukon Territory, Canada (Brown & Smol, unpubl.), and New York, U.S.A. (Duff & Smol, 1991, unpubl.).

ECOLOGY: Stomatocyst 84 has been classified as acidophilic (Rybak *et al.*, 1991; Duff & Smol, 1991).

Stomatocyst 217, Duff & Smol, 1994 (3) Fig. 87

BIOLOGICAL AFFINITY: unknown

LOCALITY: Durrance Lake (48°30'N; 123°25'W), British Columbia, Canada. Ekman dredge, surface (1 cm) sediment sample.

SEM DESCRIPTION: This small, spherical stomatocyst (diameter 5.4-6.1 μm) has a complex collar. The regular pore (diameter ca. 0.7 μm) is surrounded by a small conical primary collar with an acute apex (diameter 0.8-0.9 μm). A narrow planar interannulus (diameter ca. 1.5 μm) separates the primary collar from the conical secondary collar (diameter 1.7-2.0 μm; height 0.3-0.8 μm). A few to many echinate spines, which are variable in length and may be crooked (basal diameter 0.5-0.8 μm; maximum length 0.3-1.1 μm), are located in the posterior hemisphere.

LM DESCRIPTION: This morphotype can be recognized by the complex collar and the posterior spines. Stomatocyst 185 *forma* A is similar, but has a single obconical collar. Stomatocyst 184 is also similar, but its spines bifurcate, its primary and secondary collars differ in height, and its secondary collar is obconical. Stomatocyst 83 has a wider secondary collar than stomatocyst 217.

REFERENCES: This cyst has been observed in British Columbia, Canada (Duff & Smol, 1994; Zeeb, 1994), and Ontario, Canada (Rybak *et al.*, 1991 - cyst 74).

ECOLOGY: Rybak *et al.* (1991) cyst 74 was classified as acidophilic.

Stomatocysts with long, non-bifurcating spines

Stomatocyst 114, Zeeb *et al.*, 1990 (5) Fig. 88

BIOLOGICAL AFFINITY: unknown

LOCALITY: Little Round Lake (44°47.5'N; 76°41.5'W), Ontario, Canada. One metre piston core, core depth 15 cm.

SEM DESCRIPTION: This stomatocyst is spherical (diameter 5.0-6.3 μm) with a smooth to psilate surface. The conical pore (inner diameter 0.4-0.5 μm) is surrounded by a small conical primary collar and a flat planar interannulus located near the collar apex. The secondary collar is cylindrical (diameter 1.6-1.9 μm; height ca. 0.6 μm) with an acute apex. Up to four long, echinate spines (diameter 0.3-

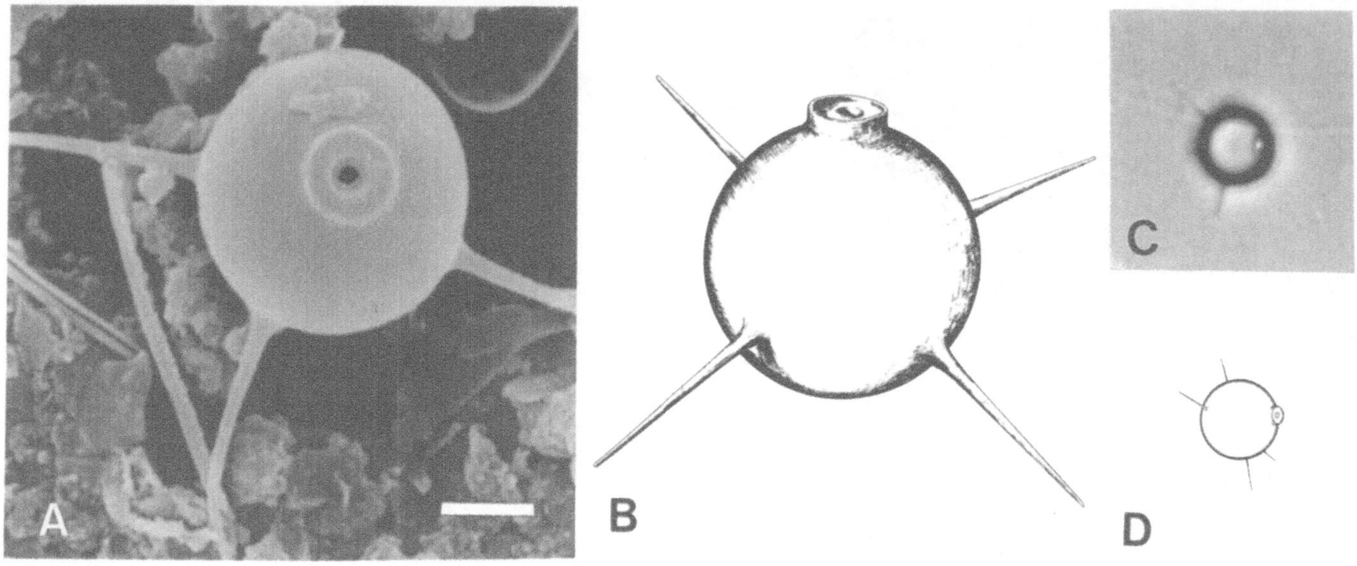

Fig. 88. Stomatocyst 114. A: SEM. B: Line drawing of SEM. C: LM. D: Line drawing of LM. Scale bar = 2 μm. LM x2000.

0.4 μm; maximum length to 6.7 μm) are located approximately equatorially and approximately equally spaced.

LM DESCRIPTION: Using LM, this stomatocyst is indistinguishable from stomatocysts 115 and 218.

REFERENCES: Stomatocyst 114 has been observed in Ontario, Canada (Zeeb *et al.*, 1990), and the Yukon Territory, Canada (Brown & Smol, unpubl.). It was probably included within *cysta subastroidea* Nygaard (Nygaard, 1956 - Denmark; Leventhal, 1970 - Italy; Carney & Sandgren, 1983 - Michigan, U.S.A.) and Sandgren & Carney (1983 - Michigan, U.S.A.) cyst 26.

ECOLOGY: In paleolimnological studies, this cyst appears to be most common during eutrophic, alkaline conditions (Nygaard, 1956; Carney & Sandgren, 1983; Zeeb *et al.*, 1990).

Stomatocyst 115, Zeeb *et al.*, 1990 (5) Fig. 89

BIOLOGICAL AFFINITY: unknown

LOCALITY: Little Round Lake (44°47.5'N; 76°41.5'W), Ontario, Canada. One metre piston core, core depth 20 cm.

SEM DESCRIPTION: This stomatocyst is spherical (diameter 5.4-6.5 μm) with a smooth to scabrate surface. The conical pore (inner diameter 0.3-0.8 μm) is surrounded by a flat planar to swollen annulus. The collar is cylindrical (diameter 1.7-2.3 μm; height 0.3-0.4 μm) with a slightly rounded apex. Up to four long, echinate spines (basal diameter 0.6-1.0 μm; maximum length to 2.0 μm, but broken on all SEM specimens; Fig. 89A) are located equatorially, approximately equidistantly.

LM DESCRIPTION: With LM, this stomatocyst cannot be distinguished from stomatocysts 114 (Figs. 88C, D) and 218.

REFERENCES: We have observed stomatocyst 115 in Ontario, Canada (Zeeb *et al.*, 1990), British Columbia, Canada (Duff & Smol, 1994), and Minnesota, U.S.A. (Zeeb & Smol, 1993a). It was probably included within Sandgren & Carney (1983 - Michigan, U.S.A.) cyst 26; their morphotype exhibited considerable variation and was thought to be a collective category. Similarly, stomatocyst 115 was probably included within *cysta subastroidea* Nygaard (Nygaard, 1956 - Denmark; Leventhal, 1970 - Italy; Carney & Sandgren, 1983 - Michigan, U.S.A.).

ECOLOGY: In paleolimnological studies, this cyst has been most common during periods of inferred eutrophy (Nygaard, 1956; Carney & Sandgren, 1983; Zeeb *et al.*, 1990; Zeeb & Smol, 1993b). As a member of a collective category with stomatocyst 218, it was most abundant in large, slightly acidic to circumneutral, productive lakes (Duff, 1994).

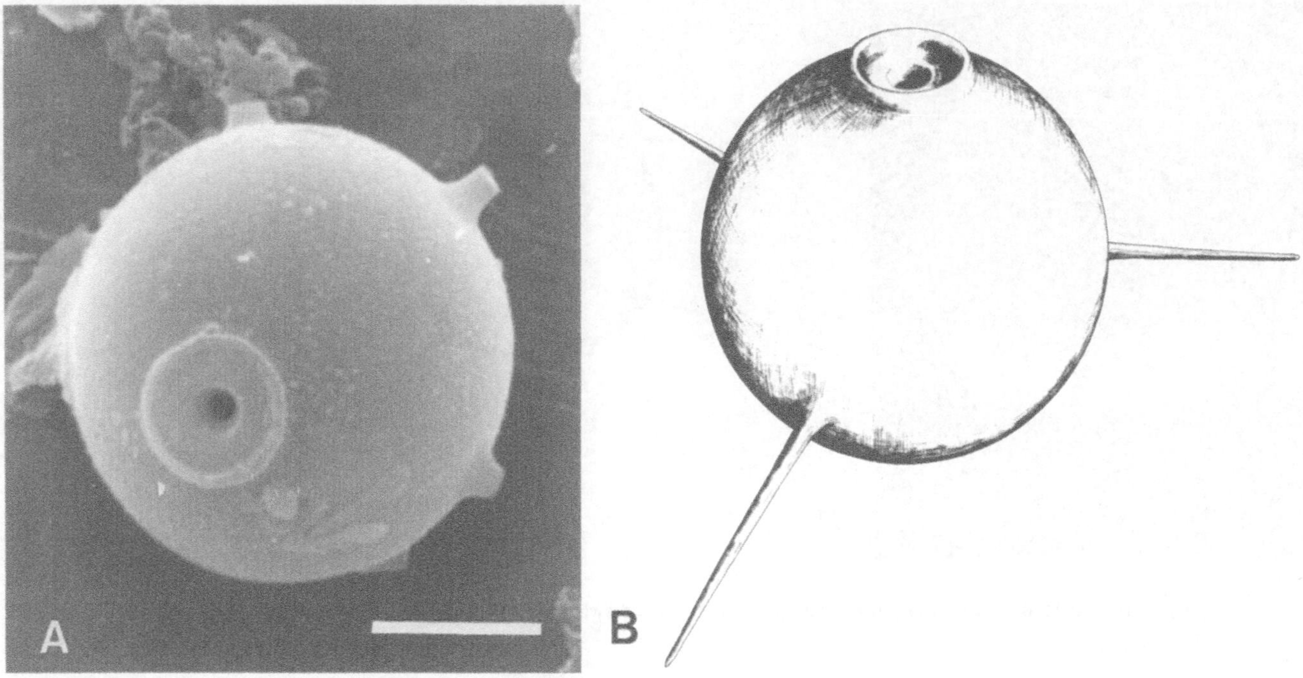

Fig. 89. Stomatocyst 115. A: SEM. B: Line drawing of SEM. Scale bar = 2 μm. See Fig. 88C, D for LM.

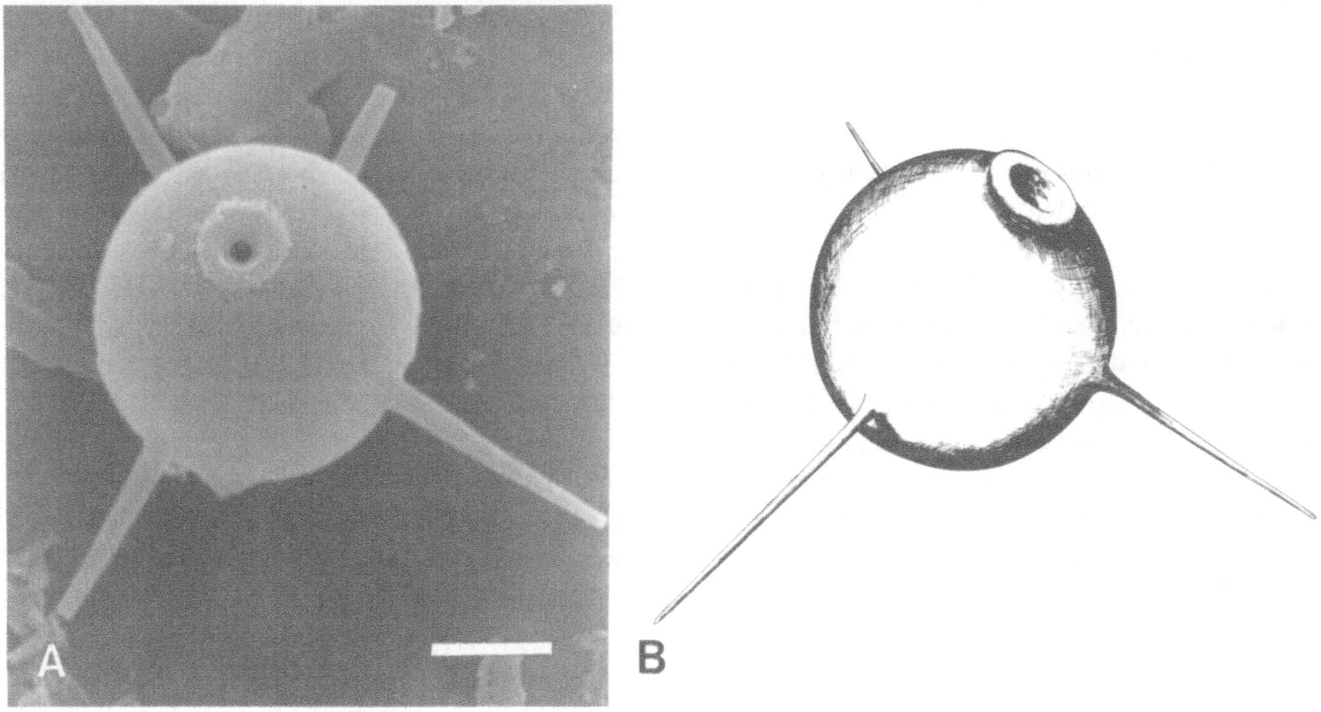

Fig. 90. Stomatocyst 218. A: SEM. B: Line drawing of SEM. Scale bar = 2 μm. See Fig. 88C, D for LM.

Stomatocyst 218, Duff & Smol, 1994 (6) Fig. 90

BIOLOGICAL AFFINITY: unknown

LOCALITY: Lang Lake (52°00'N; 121°00'W), British Columbia, Canada. Ekman dredge, surface (1 cm) sediment sample.

SEM DESCRIPTION: This spherical stomatocyst (diameter 4.9-5.6 μm) has a very low, obconical collar (diameter 1.2-2.0 μm; height 0.2-0.4 μm) with a thickened, rounded apex and a conical pore (inner diameter 0.3-0.5 μm). The surface

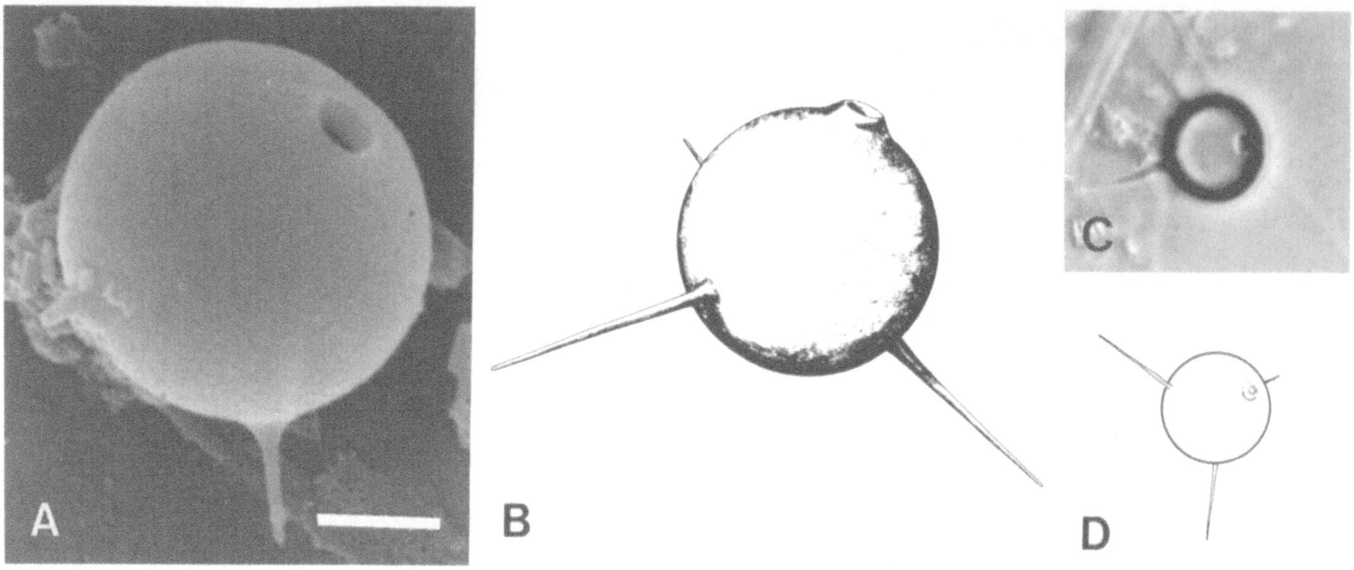

Fig. 91. Stomatocyst 219. A: SEM. B: Line drawing of SEM. C: LM. D: Line drawing of LM. Scale bar = 2 μm. LM x2000.

is smooth and ornamented with three to four approximately equatorial, long, thin, echinate spines (maximum diameter 0.7-1.0 μm; length to 5.5 μm).

LM DESCRIPTION: Using LM, this cyst cannot be distinguished from stomatocysts 114 (Figs. 88C, D) and 115.

REFERENCES: Stomatocyst 218 has been observed in British Columbia, Canada (Duff & Smol, 1994), Ontario, Canada (Rybak *et al.*, 1987 - cyst 12), the Yukon Territory, Canada (Pienitz *et al.*, 1992 - misidentified as stomatocyst 115; Brown & Smol, unpubl.), and Minnesota, U.S.A. (Zeeb & Smol, 1993a - misidentified as stomatocyst 115). Rybak *et al.* (1991; Ontario, Canada) cyst 68 is similar. Stomatocyst 218 was probably included within *cysta subastroidea* Nygaard (Nygaard, 1956 - Denmark; Leventhal, 1970 - Italy; Carney & Sandgren, 1983 - Michigan, U.S.A.), and Sandgren & Carney (1983 - Michigan, U.S.A.) cyst 26.

ECOLOGY: This cyst may be produced by a eutrophic taxon (Nygaard, 1956; Carney & Sandgren, 1983; Rybak *et al.*, 1987) that is intolerant of high (> 10mg/L) salinities (Pienitz *et al.*, 1992). As a member of a collective category with stomatocyst 115, it was most abundant in large, slightly acidic to circumneutral, productive lakes (Duff, 1994).

Stomatocyst 219, Duff & Smol, 1994 (6) Fig. 91

BIOLOGICAL AFFINITY: unknown

LOCALITY: Glen Lake (48°30'N; 123°30'W), British Columbia, Canada. Ekman dredge, surface (1 cm) sediment sample.

SEM DESCRIPTION: This is a smooth to psilate, spherical stomatocyst (diameter 5.9-7.0 μm). The collar is low and conical (diameter 0.9-1.5 μm; height 0.2-0.5 μm) with a very gradual basal margin and an acute, and possibly slightly obconical, apex. We observed two or three subequatorial spines on all SEM specimens, but we estimate a total of four spines are present. The spines are straight and echinate (basal diameter 0.7-1.0 μm; maximum length to 5.0 μm); they were non-bifurcating but broken (Fig. 91A) in all but one of the SEM specimens that we examined.

LM DESCRIPTION: Careful attention must be paid to differentiate small specimens (≤6.5 μm) of stomatocyst 219 from stomatocysts 114/115/218. The collar of stomatocyst 219 is lower and less distinct and the spines tend to be more posteriorly placed than stomatocysts 114/115/218.

REFERENCES: This stomatocyst has been observed in British Columbia, Canada (Duff & Smol, 1994; Zeeb, 1994), and the Yukon Territory, Canada (Brown & Smol, unpubl.). A similar morph has been found in California, U.S.A. (Adam & Mahood, 1980c - type 349).

ECOLOGY: No autecological details are yet available.

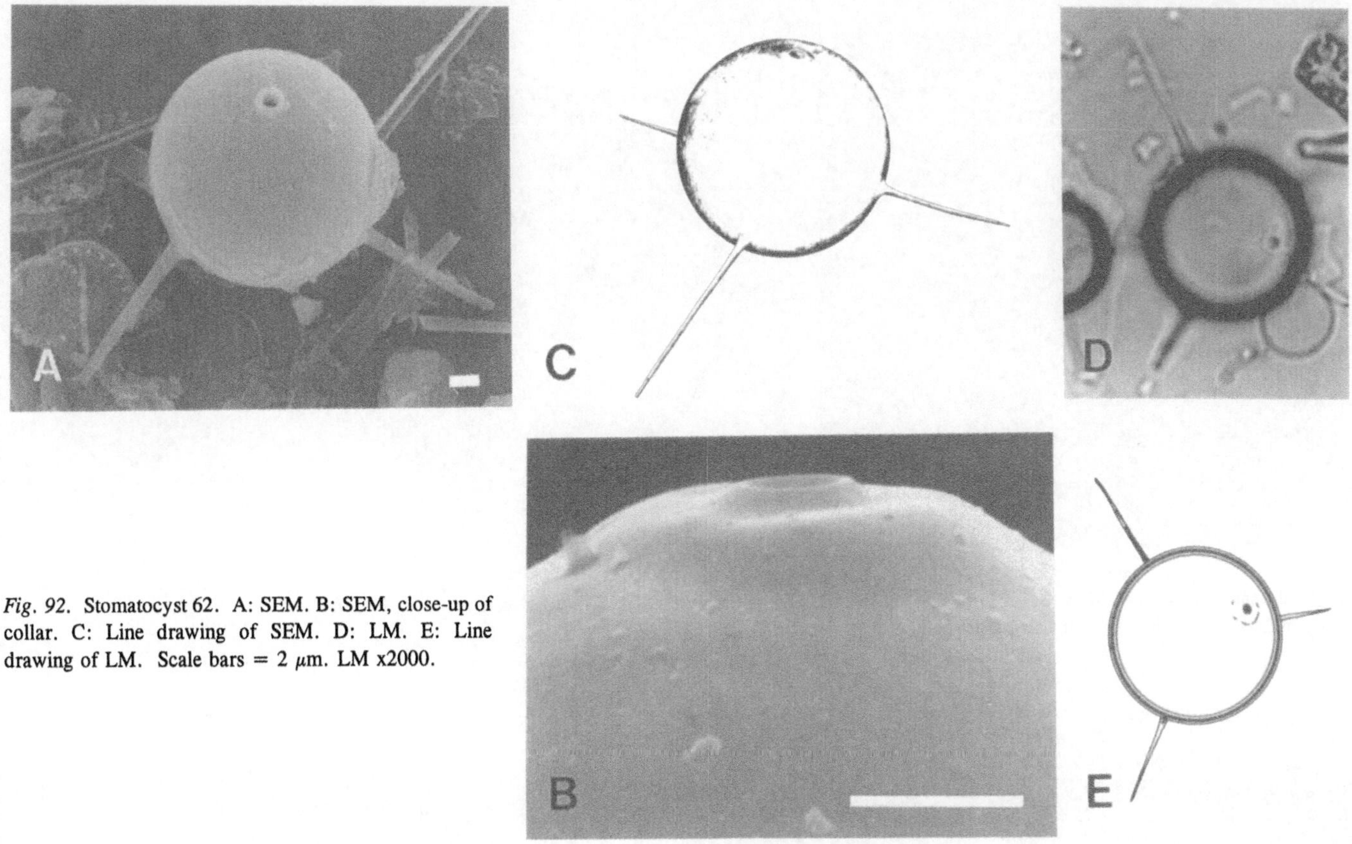

Fig. 92. Stomatocyst 62. A: SEM. B: SEM, close-up of collar. C: Line drawing of SEM. D: LM. E: Line drawing of LM. Scale bars = 2 μm. LM x2000.

Stomatocyst 62, Duff & Smol, 1991 (9) Fig. 92

BIOLOGICAL AFFINITY: unknown

LOCALITY: Upper Wallface Pond (44°08.8'N; 74°03.3'W), Adirondack Park, New York, U.S.A. Gravity core, core depth 20.0-20.5 cm.

SEM DESCRIPTION: This is a very large, spherical stomatocyst (diameter 13.2-17.6 μm). The collar is quite variable. There is usually a narrow, low, cylindrical to conical collar (Fig. 92A) with a thickened apex and no annulus (diameter 1.2-2.2 μm; height ca. 0.4 μm; collar diameter : cyst diameter ratio 0.06-0.13) surrounding a regular pore (diameter 0.8-1.1 μm). The collar may also consist of a low, very broadly rounded conical collar (basal diameter 5.4-5.7 μm), either with a sloping inner margin to the pore, or with a narrow, very low, conical distal portion (diameter 0.9-1.2 μm) immediately around the pore (Fig. 92B). The cyst surface is smooth, with two to three long, straight, echinate posterior spines, which are usually broken (basal diameter 1.5-2.0 μm; length to 23.8 μm).

Stomatocyst 62 now includes stomatocyst 63 (Chapter 5). It can be distinguished from stomatocyst 128 *forma* B by the following features: the collar of stomatocyst 62 is generally narrower; its spines are always straight, never bifurcating; and no more than three spines were observed on stomatocyst 62.

LM DESCRIPTION: This cyst is easily identified by its size, low collar, and straight, non-bifurcating posterior spines.

REFERENCES: Stomatocyst 62 has been observed in Ontario, Canada (Rybak *et al.*, 1991 - cysts 69 and 73; Zeeb *et al.*, 1994), Minnesota, U.S.A. (Zeeb & Smol, 1993a), and New York, U.S.A. (Smol, 1988a - Plate 1D; Duff & Smol, 1991, unpubl.). The SEM specimen shown in Zeeb & Smol (1993a, Fig. 69) has now been reassigned to stomatocyst 128 *forma* B.

ECOLOGY: This stomatocyst is typically found in acidic, unproductive sites (Duff & Smol, 1991; Rybak *et al.*, 1991; Zeeb *et al.*, 1994). Rybak *et al.* (1991) cyst 69 was classified as an acidobiont.

Stomatocysts with bifurcating spines

Stomatocyst 169, Zeeb & Smol, 1993a (3) Fig. 93

BIOLOGICAL AFFINITY: unknown

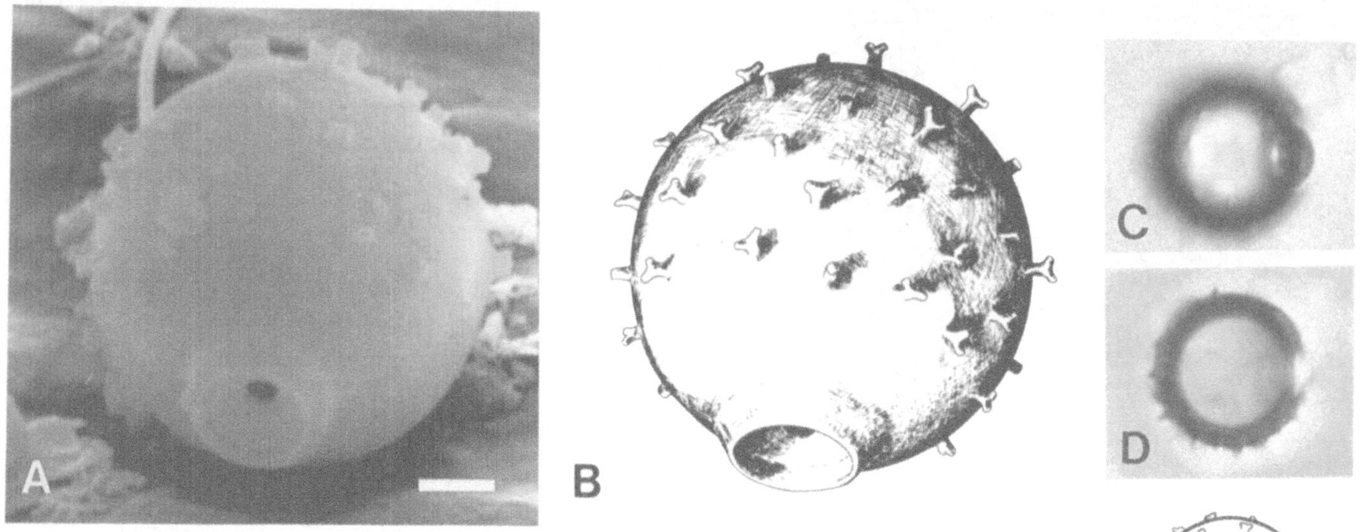

Fig. 93. Stomatocyst 169. A: SEM. B: Line drawing of SEM. C: LM, surface view. D: LM, cross-sectional view. E: Line drawing of LM. Scale bar = 2 μm. LM x2000.

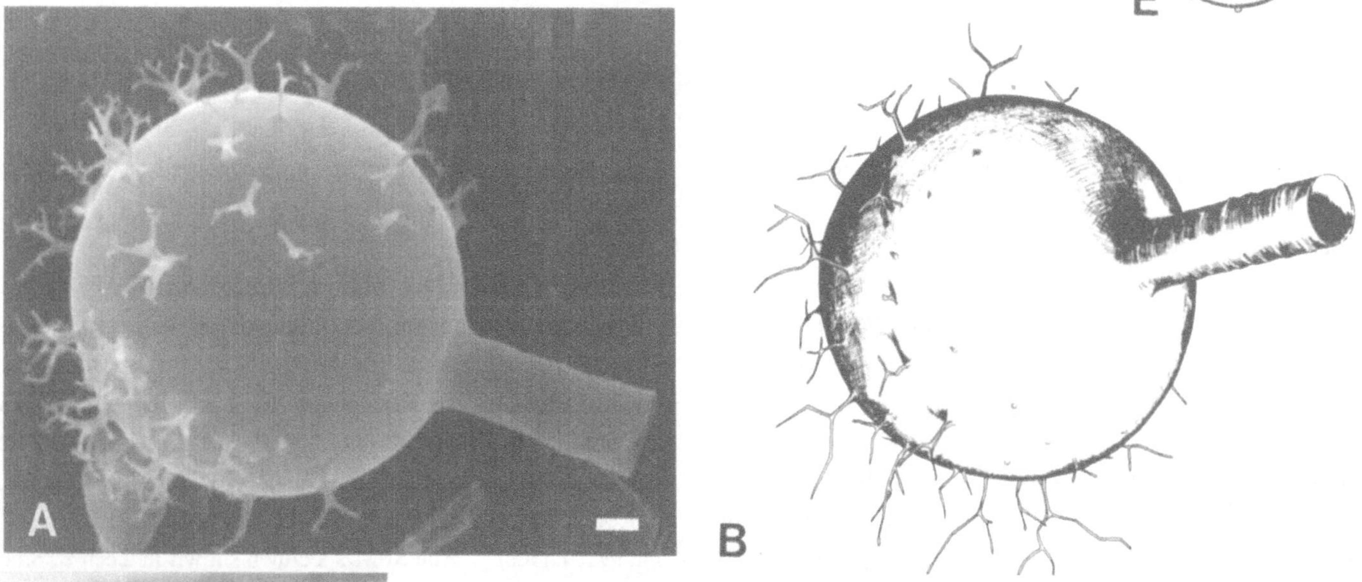

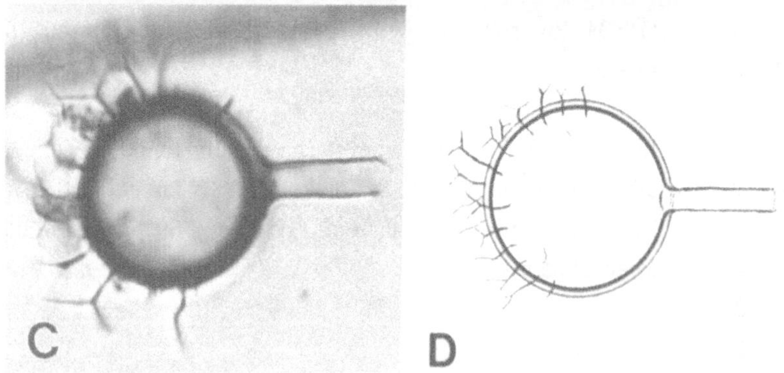

Fig. 94. Stomatocyst 237. A: SEM. B: Line drawing of SEM. C: LM. D: Line drawing of LM. Scale bar = 2 μm. LM x1500.

LOCALITY: Elk Lake (47°11'N; 95°13'W), Clearwater County, Minnesota, U.S.A. Piston core, varve year 10 159 B.P.

SEM DESCRIPTION: This stomatocyst is spherical (diameter 9.8-10.9 μm). The collar is short, conical to cylindrical (basal diameter 3.2-4.1 μm; apical diameter 2.6-2.9 μm; height ca. 1.2 μm) with a broadly rounded outer margin, a straight inner margin, and a sloping planar annulus that begins near the apex of the inner collar margin. The

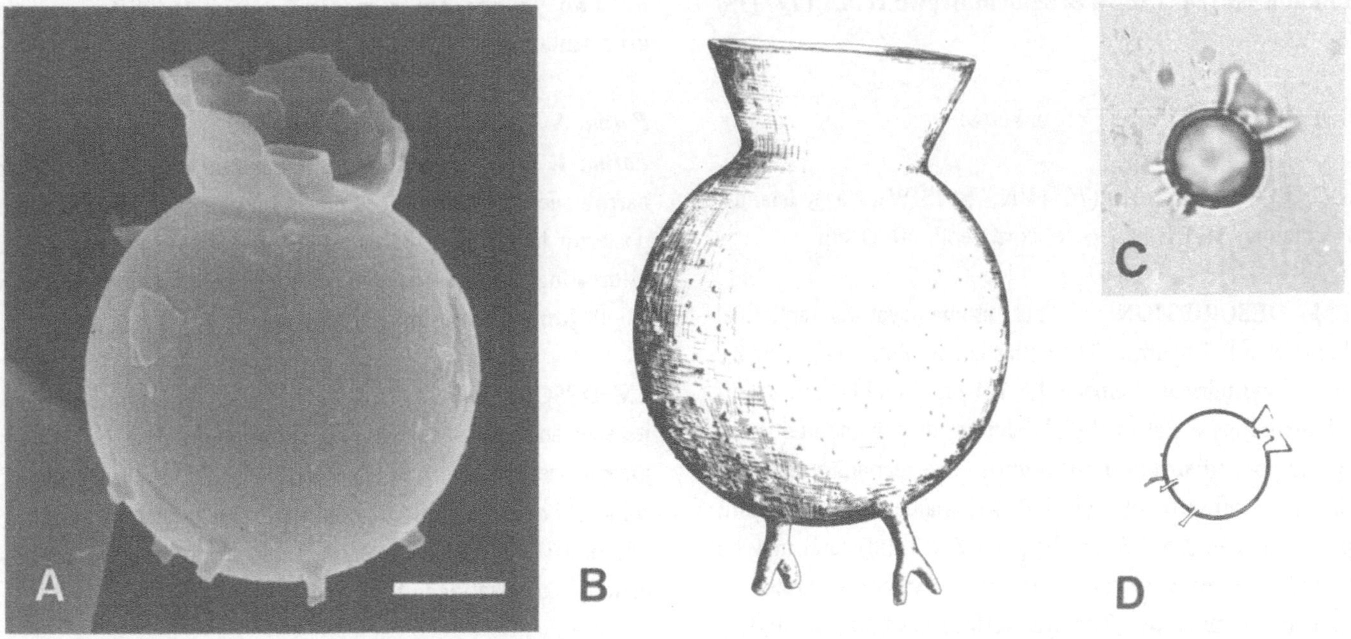

Fig. 95. Stomatocyst 184. A: SEM. B: Line drawing of SEM. C: LM. D: Line drawing of LM. Scale bar = 2 μm. LM x2000.

pore is regular (diameter ca. 0.8 μm). The surface is ornamented with short, baculate spines with flattened or bifurcated apices (diameter 0.4-1.2 μm; height 0.4-0.7 μm).

Stomatocyst 169 may be equivalent to stomatocyst 106 (Chapter 5), but this cannot be determined without more SEM specimens of the latter morph.

LM DESCRIPTION: With LM, this cyst can be recognized by its size, rounded collar margin, and baculate spines. The bifurcated apices of the spines are not always visible.

REFERENCES: We have observed stomatocyst 169 in Ontario, Canada (Zeeb & Smol, unpubl.), British Columbia, Canada (Duff & Smol, 1994; Zeeb, 1994), and Minnesota, U.S.A. (Zeeb & Smol, 1993a).

ECOLOGY: This cyst has been observed in alkaline lakes (Zeeb & Smol, 1993a, unpubl.; Duff & Smol, 1994).

Stomatocyst 237, Zeeb & Smol in Duff *et al.* (this volume) (3) Fig. 94

BIOLOGICAL AFFINITY: unknown

NEGATIVE NUMBER: J.P. Smol 584

LOCALITY: Lake 227 (49°42'N; 93°42'W), Experimental Lakes Area, Ontario, Canada. Freeze core, varve year A.D. 1989.

SEM DESCRIPTION: This is a large, spherical stomatocyst (diameter 16.6-18.1 μm) with a very long, cylindrical to gently obconical collar (diameter ca. 3.4 μm; height 9.1-12.1 μm). The collar wall is ornamented with fine spiralling striations. Very narrow, bifurcating (primary to tertiary bifurcations), echinate spines are located primarily in the posterior hemisphere; spine density is variable (spine length to primary bifurcation 1.1-3.4 μm; total height to 7.2 μm).

LM DESCRIPTION: This stomatocyst is easily identified with LM by its large size, its long, narrow collar, and its spines.

REFERENCES: We have not observed this morphotype elsewhere.

ECOLOGY: Autecological details are as yet unknown.

Stomatocyst 184, Brown & Smol in Brown *et al.*, 1994 (15) Fig. 95

BIOLOGICAL AFFINITY: unknown

LOCALITY: Nordvestø (76°44'N; 73°13'W), Carey Islands, Greenland. SIPRE peat core, core depth 50-51 cm.

SEM DESCRIPTION: This stomatocyst is spherical (diameter 5.9-7.4 μm). The collar is complex. The primary collar is cylindrical (diameter 0.8-1.4 μm; height 0.3-0.8 μm) and surrounds a flat to slightly sloping planar annulus and a regular pore (diameter 0.4-0.6 μm). The secondary collar is wide and abruptly obconical (basal diameter 3.1-4.2 μm; apical diameter 3.5-5.3 μm; height 1.8-2.3 μm), and may be striated. The primary and secondary collars are separated by a planar interannulus. The cyst surface is smooth to slightly scabrate, with three to nine baculate spines (basal diameter 0.4-0.8 μm; maximum length ca. 1.5 μm) located in the posterior hemisphere. The spines bifurcate when mature, but are often broken (Fig. 95A).

LM DESCRIPTION: Stomatocyst 184 resembles both stomatocysts 185 and 217, but neither of these have bifurcating spines. Stomatocyst 185 has a single collar, and the primary and secondary collars of stomatocyst 217 are conical and equal in height.

REFERENCES: This stomatocyst has not yet been observed outside of Greenland (Brown *et al.*, 1994).

ECOLOGY: We do not yet know the ecological optima and tolerances of this cyst.

Stomatocyst 128, Duff & Smol in Duff *et al.*, 1992 *emend.* Duff & Smol, 1994 (20) Fig. 96

BIOLOGICAL AFFINITY: unknown

LOCALITY: "Willow Pond" (unofficial name), Cape Herschel (78°37'N; 74°42'W), Ellesmere Island, Northwest Territories, Canada. Moss sample, collected August 1, 1986.

SEM DESCRIPTION: This stomatocyst is spherical with a wide size range (diameter 10.0-18.6 μm). It has an abruptly cylindrical, narrow, high collar (diameter 2.0-2.3 μm; height to 4.8 μm; collar diameter : cyst diameter ratio 0.12-0.20).

Two *formae* are recognized based on differences in ornamentation:

***Forma* A** - is completely unornamented.
***Forma* B** - is ornamented with approximately three to four narrow, echinate spines, often with blunt apices, located in the posterior hemisphere. The spines may be very long and/or bifurcating; on most specimens they are broken (maximum length from 1.6 μm to >32.0 μm).

LM DESCRIPTION: This stomatocyst can be identified by its size and the very narrow, cylindrical collar. The two *formae* are differentiated based on the presence or absence of spines. *Forma* A could be confused with stomatocysts 116B, 153 or 161, but its collar base is abrupt, not gradual, and its collar lacks striations.

REFERENCES: Stomatocyst 128 *forma* A has been observed in the Northwest Territories, Canada (Duff *et al.*, 1992; Webb & Smol, unpubl.), and in Denmark (*cysta siphonata* Nygaard (1956)). *Forma* B has been observed in British Columbia, Canada (Duff & Smol, 1994; Zeeb, 1994); Ontario, Canada (Rybak *et al.*, 1991 - cysts 66 and 67); the Northwest Territories, Canada (Webb & Smol, unpubl.); Michigan, U.S.A. (Sandgren & Carney, 1983 - cyst 25); Minnesota, U.S.A. (Zeeb & Smol, 1993a - misidentified as stomatocyst 62); Denmark (*cysta longispinosa* Nygaard (1956)); Poland (Rybak, 1986 - *cysta longispinosa*); and Sweden (Cronberg & Kristiansen, 1980 - *Chrysastrella paradoxa* Chodat).

ECOLOGY: Cronberg & Kristiansen (1980) observed this cyst across a wide range of trophic status and pH (5.47-7.29), but noted that it was most numerous in oligotrophic, humic lakes. Rybak *et al.* (1991) classified it as pH-indifferent. In his paleolimnological analysis, Rybak (1986) observed it during periods of inferred eutrophy.

Stomatocyst 64, Duff & Smol, 1991 (15) Fig. 97

BIOLOGICAL AFFINITY: unknown

LOCALITY: Upper Wallface Pond (44°08.8'N; 74°03.3'W), Adirondack Park, New York, U.S.A. Gravity core, core depth 10.0-10.5 cm.

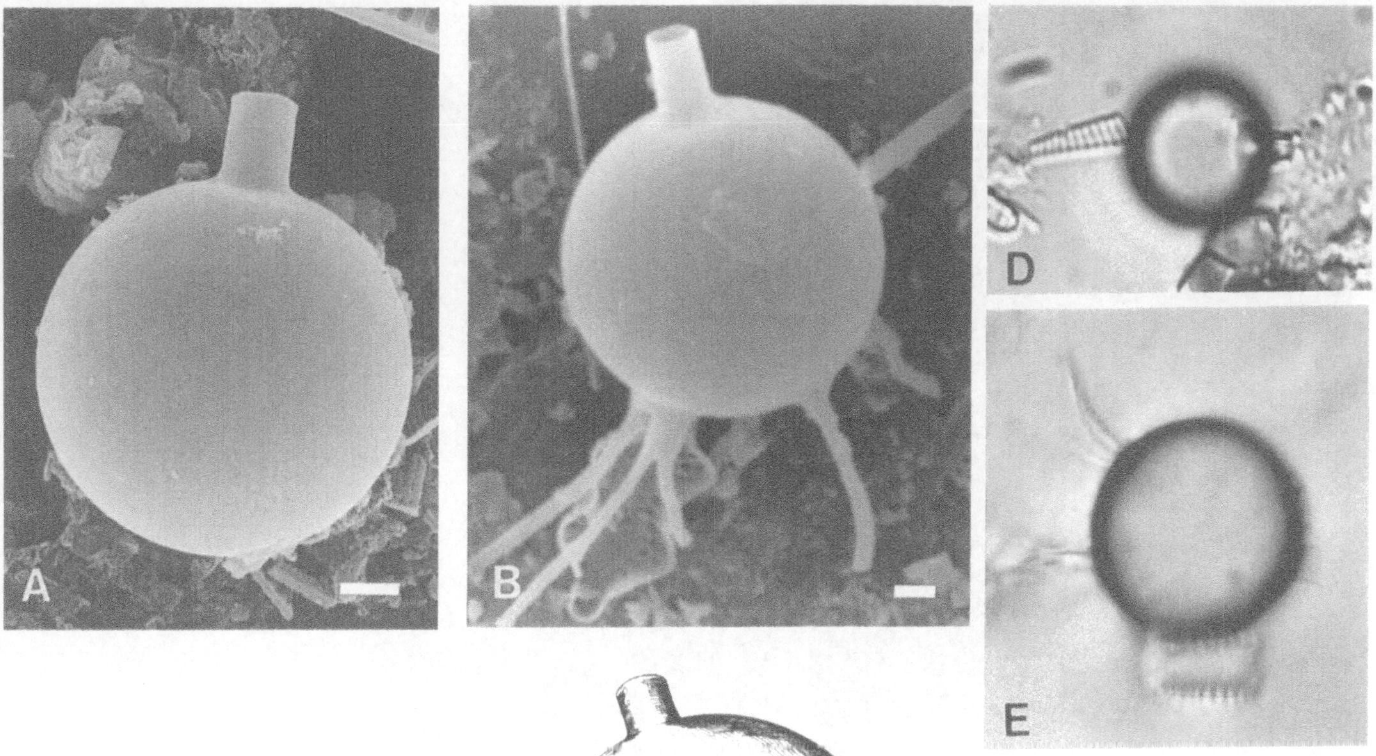

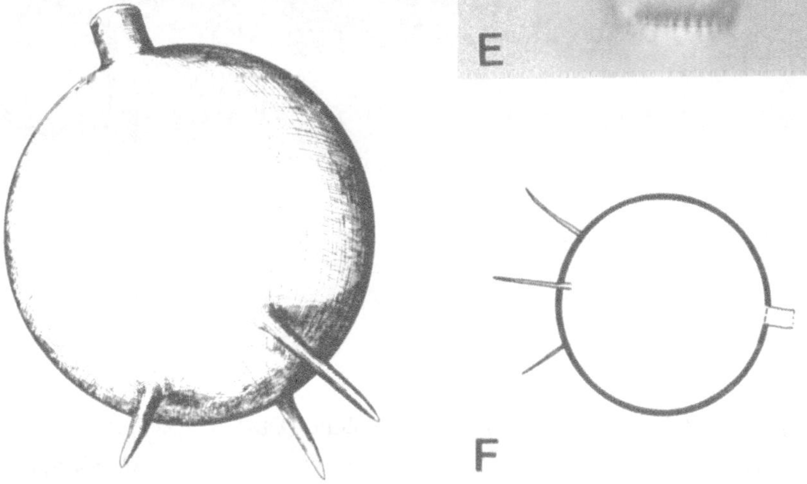

Fig. 96. Stomatocyst 128 *formae* A and B. A: *Forma* A, SEM. B: *Forma* B, SEM. C: *Forma* B, line drawing of SEM. D: *Forma* A, LM. E: *Forma* B, LM. F: *Forma* B, line drawing of LM. Scale bars = 2 μm. LM x2000.

SEM DESCRIPTION: This stomatocyst is spherical (diameter 5.1-8.8 μm) or slightly irregular in shape. The collar is slightly to broadly obconical (apical diameter 1.7-3.4 μm; height 0.7-2.0 μm) with an acute apex and a sloping inner margin. A slightly sloping planar annulus may surround the regular pore (diameter 0.4-0.8 μm). The collar and/or the annulus may be striated. The ornamentation is quite variable. The surface may be smooth or scabrate. There are two to four long, echinate spines (basal diameter 0.4-2.2 μm; length to 23.0 μm) located approximately equidistantly in the equatorial region or the posterior hemisphere. The spines may be straight or bent, non-bifurcating or with a primary bifurcation, and may be twisted or rope-like in appearance (not shown in Fig. 97).

Stomatocyst 65 (Chapter 5) is now included within stomatocyst 64.

LM DESCRIPTION: Stomatocyst 64 is distinguished from stomatocysts 114/115/218 by its generally larger size; the longer, obconical collar; and the more posterior spines. It differs from stomatocyst 130 (Duff et al., 1992) in that it is usually smaller, with more than one posterior spine.

REFERENCES: This stomatocyst has been observed in Ontario, Canada (Rybak *et al.*, 1991 - cyst 68), California, U.S.A. (Adam & Mahood, 1979b - type 94), and New York, U.S.A. (Duff & Smol, 1991, unpubl.). The SEM specimen depicted as stomatocyst 130 by Zeeb & Smol (1993a - Minnesota, U.S.A.) has now been reassigned to stomatocyst 64. This morphotype is probably identical to *cysta paucispinosa* Nygaard (Nygaard, 1956 - Denmark; Rybak, 1986 - Poland), but SEM confirmation is needed.

ECOLOGY: This stomatocyst has been classified as

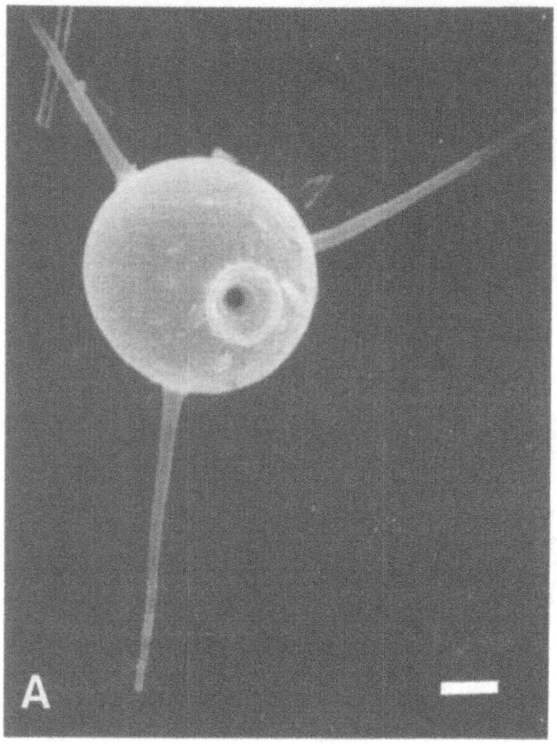

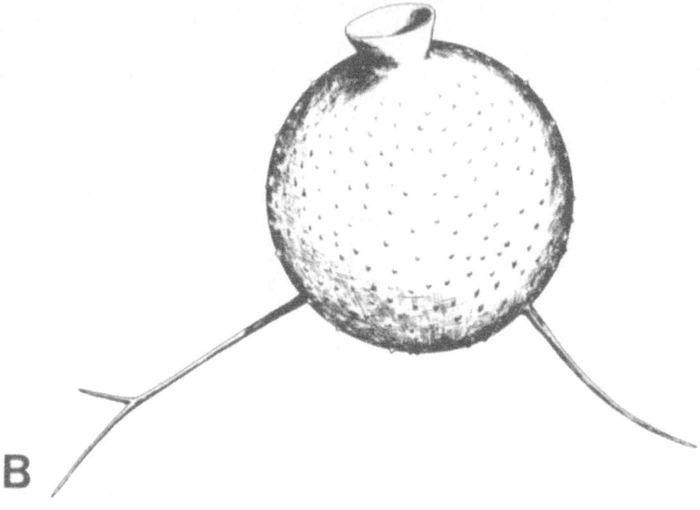

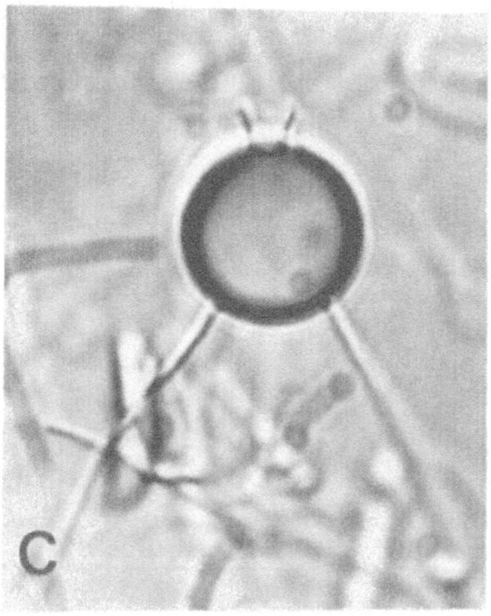

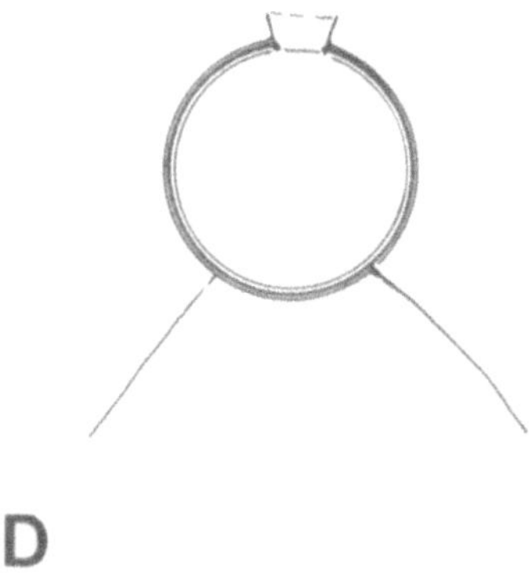

Fig. 97. Stomatocyst 64. A: SEM. B: Line drawing of SEM. C: LM. D: Line drawing of LM. Scale bar = 2 μm. LM x2000.

acidobiontic (Rybak *et al.*, 1991) or acidophilic (Duff & Smol, 1991).

Stomatocyst 130, Duff & Smol in Duff *et al.*, 1992 *emend.* Duff & Smol, 1994 (34) Fig. 98

BIOLOGICAL AFFINITY: This cyst was originally assigned to the artificial genus *Chrysastrella* (*C. furcata* (Dolgoff) Deflandre, 1934), and Cronberg & Kristiansen (1980) speculated that it was produced by a *Chromulina* or *Ochromonas* species. However, Nicholls (1981) identified it as the stomatocyst of *Chrysococcus furcatus* (Dolgoff) Nicholls. He demonstrated that this species also has a silicified vegetative cell, which we had identified previously as "stomatocyst 17" (Duff & Smol, 1988).

LOCALITY: "Elison Pond" (unofficial name), Cape Herschel (78°37'N; 74°42'W), Ellesmere Island, Northwest Territories, Canada. Epilithon, collected August 3, 1986.

SEM DESCRIPTION: This stomatocyst is spherical with a considerable range in diameter (8.3-12.3 μm). The collar is cylindrical (Fig. 98A) to widely obconical (Fig. 98B) with an abrupt or gradual base; it is usually long but varies considerably in length (apical diameter 1.7-5.4 μm; height 1.0-9.6 μm), and it surrounds a regular pore (diameter 1.0-1.1 μm). The outer basal margin of the collar is ornamented with vertical to spiralling striae or struts (Fig. 98C) that may extend quite far onto the cyst body, on approximately half of our specimens. The ornamentation is also variable. On all specimens, a single, echinate spine, bifurcating when mature, is located at or very close to the posterior pole. It usually has a twisted or rope-like appearance (Figs. 98B, C), and appears to be hollow; its length and degree of bifurcation (primary to secondary) are variable (spine length to primary bifurcation 1.0-8.4 μm; length after bifurcation to 22.2 μm; maximum spine diameter to 2.1 μm). The rest of the cyst body varies from nearly smooth to scabrate (Fig. 98B) to spinose (spine diameter 0.2-1.8 μm; height to 1.1 μm; Fig. 98A). The scabrae or spines are irregularly distributed, and may be

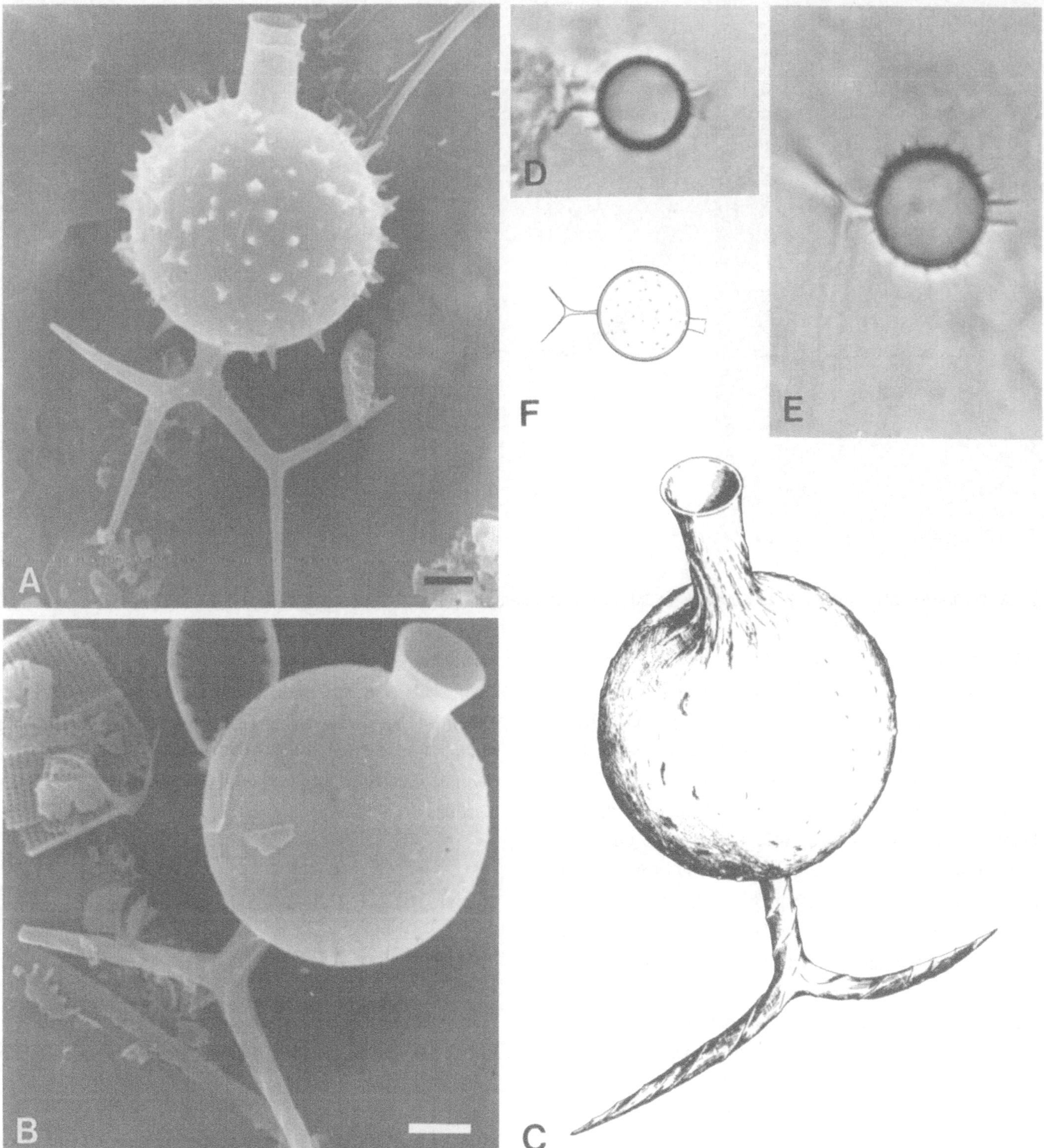

Fig. 98. Stomatocyst 130. A: SEM. Highly ornamented specimen. B: SEM, showing rope-like spine. C: Line drawing of SEM, showing spiralling struts. D: LM, relatively unornamented specimen. E: LM, highly ornamented specimen. F: Line drawing of LM. Scale bars = 2 μm. LM x2000.

joined by low ridges.

LM DESCRIPTION: This stomatocyst can be identified by its single, bifurcating posterior spine and its collar, which is either cylindrical (but wider than that of stomatocyst 128 *forma* B) or obconical. The surface ornamentation (scabrae or small spines) is visible with LM.

REFERENCES: This taxon is very widespread. The stomatocyst has been found in Ontario, Canada (Nicholls,

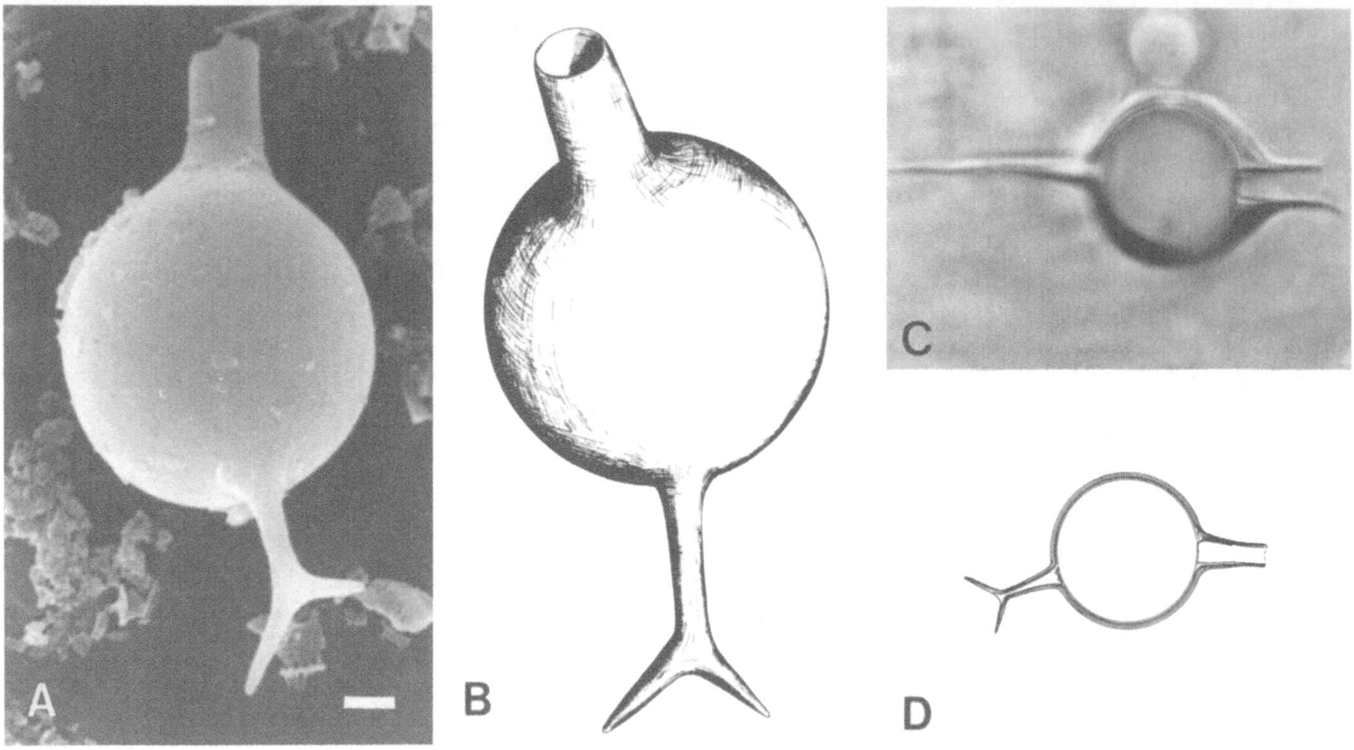

Fig. 99. Stomatocyst 221. A: SEM. B: Line drawing of SEM. C: LM. D: Line drawing of LM. Scale bar = 2 μm. LM x2000.

1981; Zeeb & Smol, unpubl.); the Northwest Territories, Canada (Duff *et al.*, 1992; Webb & Smol, unpubl.); the Yukon Territory, Canada (Brown & Smol, unpubl.); British Columbia, Canada (Duff & Smol, 1994; Zeeb, 1994); New York, U.S.A. (Duff & Smol, unpubl.); Minnesota, U.S.A. (Zeeb & Smol, 1993a); California, U.S.A. (Adam & Mahood, 1980a, b, c - types 143, 144 & 212); Greenland (Brown *et al.*, 1994); France (Bourrelly, 1957 - *Chrysastrella furcata*); and Sweden (Cronberg & Kristiansen, 1980 - *C. furcata*). The SEM specimen depicted in Zeeb & Smol (1993a - Fig. 64) as stomatocyst 130 has now been reassigned to stomatocyst 64, but the other specimen observed by these authors was correctly identified as stomatocyst 130.

The silicified vegetative cell (our "stomatocyst 17") has been found in Ontario, Canada (Nicholls, 1981; Smol, 1985 - Fig. 1C), Québec, Canada (Cumming & Smol, unpubl.), the Northwest Territories, Canada (Smol, 1983; Duff & Smol, 1988, 1989; A. Wolfe, pers. commun.), and New York, U.S.A. (Duff & Smol, unpubl.).

ECOLOGY: This taxon appears to have an affinity for cold water, as it is frequently found in arctic sites (Smol, 1983; Duff & Smol, 1988, 1989; Duff *et al.*, 1992; A. Wolfe, pers. commun.) and in the early post-glacial sediments of temperate lakes (Cumming & Smol, unpubl.). Cronberg & Kristiansen (1980) collected it at a water temperature of 5.8°C. It is most common in oligotrophic sites (Cronberg & Kristiansen, 1980; Duff & Smol, 1988, 1989; Duff *et al.*, 1992; Smol, 1983), but has a wide tolerance for pH, being found in both acidic (Duff & Smol, 1988, 1989; Smol, 1983) and alkaline (Duff *et al.*, 1992; Duff & Smol, 1994; Zeeb & Smol, 1993a) locations.

Stomatocyst 221, Duff & Smol, 1994 (3) Fig. 99

BIOLOGICAL AFFINITY: unknown

LOCALITY: Champion Lake #3 (49°15'N; 117°30'W), British Columbia, Canada. Ekman dredge, surface (1 cm) sediment sample.

SEM DESCRIPTION: This is a large, spherical stomatocyst (diameter 10.4-12.0 μm). It has a long, cylindrical collar with a very wide, conical base, but without striae (collar base diameter 3.4-5.7 μm; apical diameter 2.5-2.6 μm; height to 5.5 μm but broken on all our SEM specimens). A single spine is situated near, but not on, the posterior pole; the spine is long, wide (especially at the base), echinate, and bifurcating (spine base diameter 2.2-3.6 μm; length to bifurcation ca. 4.6 μm; length after bifurcation ca. 3.8 μm).

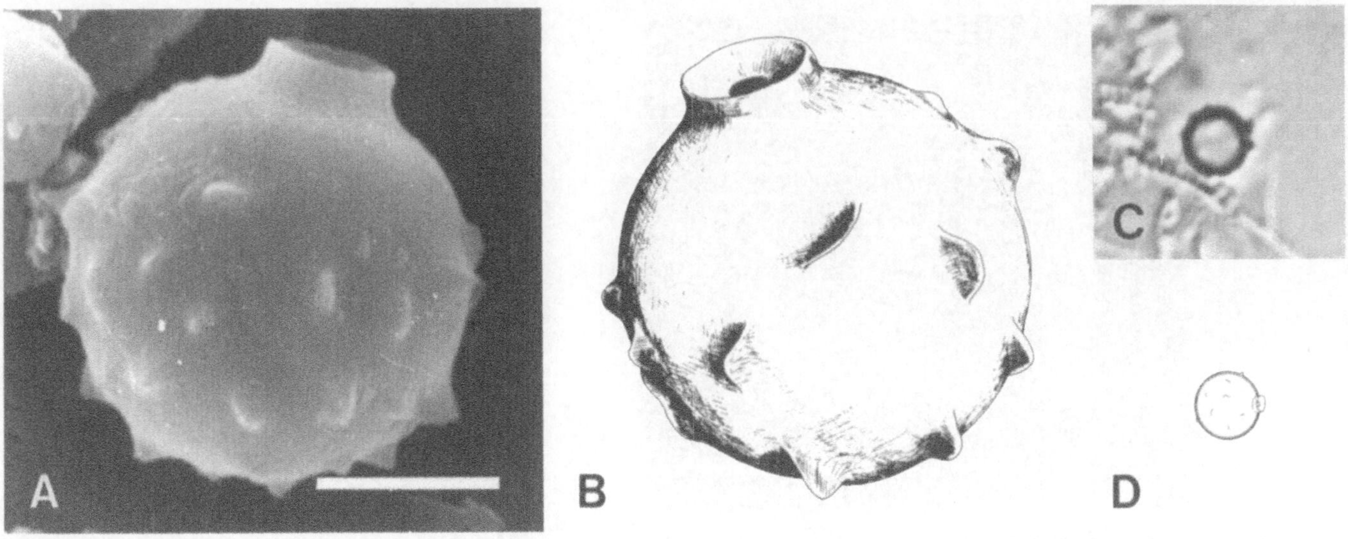

Fig. 100. Stomatocyst 33. A: SEM. B: Line drawing of SEM. C: LM. D: Line drawing of LM. Scale bar = 2 μm. LM x2500.

LM DESCRIPTION: This stomatocyst can be distinguished from similar cysts (e.g. stomatocysts 130 and 128 *forma* B) by the conical base and very straight inner margin of the collar, and by the single, wide spine.

REFERENCES: We have only found this cyst in British Columbia, Canada (Duff & Smol, 1994; Zeeb, 1994).

ECOLOGY: We can not yet estimate the ecological optima and tolerances of this morph.

Stomatocysts with short ridges; length of ridges is less than half of the cyst circumference

Stomatocyst 33, Duff & Smol, 1989 (11) Fig. 100

BIOLOGICAL AFFINITY: unknown

LOCALITY: Tasikutaaq Lake (66°23'N; 65°57'W), Baffin Island, Northwest Territories, Canada. Percussion core, core depth 75 cm.

SEM DESCRIPTION: This is a very small, spherical stomatocyst (diameter 3.0-4.5 μm). It has a low, wide collar, varying in shape from conical to obconical (diameter 1.0-1.9 μm; height 0.3-0.6 μm), surrounding a regular pore (diameter ca. 0.4 μm). Short, low, randomly oriented lunate ridges are fairly widely scattered on the cyst surface (ridge height 0.2-0.5 μm; maximum ridge length 0.5-0.9 μm). Collar and/or ridge striae may or may not be present.

The only characteristic separating stomatocyst 33 from stomatocyst 113 is maximum ridge length. This is a decision made after careful observation of natural groupings of many SEM specimens.

LM DESCRIPTION: Stomatocyst 33 can be distinguished with LM from small cysts with spines (i.e. stomatocyst 170) or with longer ridges (i.e. stomatocysts 32, 91, 113) with high resolution optics. It cannot be distinguished from stomatocyst 222.

REFERENCES: Stomatocyst 33 has been found in the Northwest Territories, Canada (Duff & Smol, 1989; Webb & Smol, unpubl.), Ontario, Canada (Rybak *et al.*, 1987 - cyst 19; Zeeb *et al.*, 1994), British Columbia, Canada (Duff & Smol, 1994), and Minnesota, U.S.A. (Zeeb & Smol, 1993a). Rybak *et al.* (1991 - Ontario, Canada) cyst 85 may be identical. *Cysta stellata* Nygaard (Nygaard, 1956 - Denmark; Rybak, 1986 - Poland) probably includes both stomatocysts 33 and 113.

ECOLOGY: This cyst may be produced by a cold-water taxon (Rybak, 1986; Duff & Smol, 1989; Duff, 1994), which may have a wide pH tolerance (Duff, 1994). It may be produced by an oligotrophic species; in Zeeb *et al.'s* (1994) study, it decreased in relative abundance after high nutrient loading, and in Duff's (1994) survey, it displayed a significant negative correlation with total phosphorus. It was particularly common in deep lakes (Duff, 1994).

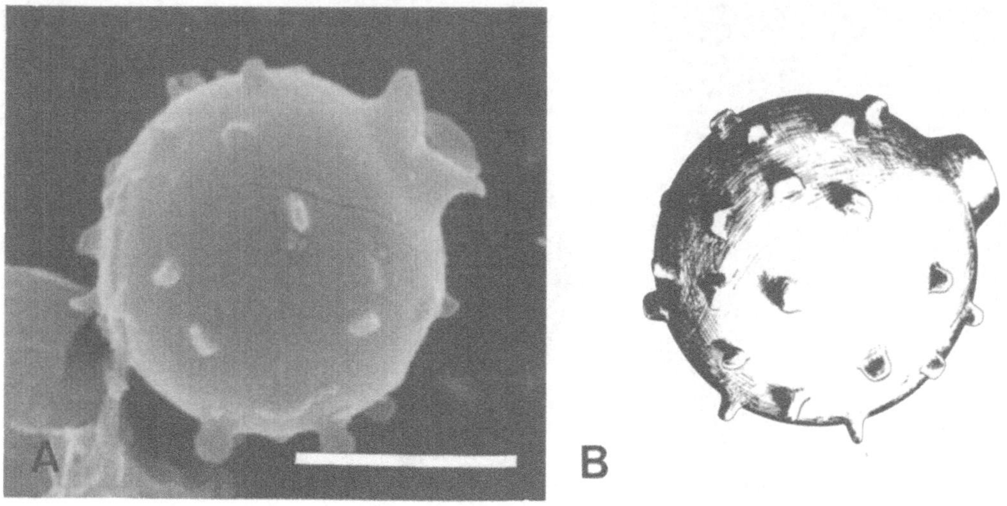

Fig. 101. Stomatocyst 222. A: SEM. B: Line drawing of SEM. Scale bar = 2 μm. See Figs. 100C, D for LM.

Stomatocyst 222, Duff & Smol, 1994 (6) Fig. 101

BIOLOGICAL AFFINITY: unknown

LOCALITY: Lac de Roche (51°45'N; 120°45'W), British Columbia, Canada. Ekman dredge, surface (1 cm) sediment sample.

SEM DESCRIPTION: This stomatocyst is very small and spherical (diameter 3.1-3.7 μm). The collar is cylindrical to obconical, often with a very irregular apex (diameter 0.8-1.3 μm; height 0.3-0.5 μm); pore morphology is unknown. The ornamentation consists of short ridges to spines (base diameter or length 0.2-0.4 μm; height 0.3-0.5 μm), which are distinguished by their very rounded apices, and often basal constrictions. The ridge/spine morphology is the only feature that separates stomatocyst 222 from stomatocyst 33.

LM DESCRIPTION: With LM, stomatocyst 222 cannot be distinguished from stomatocyst 33 (Figs. 100C, D). See stomatocyst 33 for the features used to distinguish these cysts from other morphs.

REFERENCES: We have observed this stomatocyst in British Columbia, Canada (Duff & Smol, 1994), the Northwest Territories, Canada (Webb & Smol, unpubl.), and Minnesota, U.S.A. (Zeeb & Smol, 1993a - misidentified as stomatocyst 33).

ECOLOGY: We cannot yet estimate the ecological optima and tolerances of this cyst. As a member of a collective category with stomatocyst 33, it was most common in deep, oligotrophic, montane lakes (Duff, 1994).

Stomatocyst 113, Zeeb *et al.*, 1990 (8) Fig. 102

BIOLOGICAL AFFINITY: unknown

LOCALITY: Little Round Lake (44°47.5'N; 76°41.5'W), Ontario, Canada. One metre piston core, core depth 15 cm.

SEM DESCRIPTION: This is a small, spherical stomatocyst (diameter 3.1-4.2 μm), with a low, conical to obconical collar and a regular pore (collar diameter 0.8-1.5 μm; height 0.2-0.4 μm; pore diameter 0.3-0.4 μm). The collar diameter : cyst diameter ratio is always <0.41. The outer collar never has struts, but may be striated. Lunate ridges, which may be striated, are randomly oriented on the cyst surface; they may be concentrated in the posterior hemisphere. The maximum ridge length is ≥1.0 μm. This is the single feature that we use to separate stomatocyst 113 from stomatocyst 33, and is based on natural groupings of our SEM specimens.

LM DESCRIPTION: This stomatocyst can be distinguished from stomatocyst 33/222 if the microscope is adequate to determine the ridge length. Usually, it can be distinguished from stomatocyst 91 because its ridges are straight and often shorter than those of stomatocyst 91, and tend to be found in the posterior hemisphere.

REFERENCES: We have observed this stomatocyst in Ontario, Canada (Zeeb *et al.*, 1990); the Northwest Territories, Canada (Duff *et al.*, 1992 - misidentified as

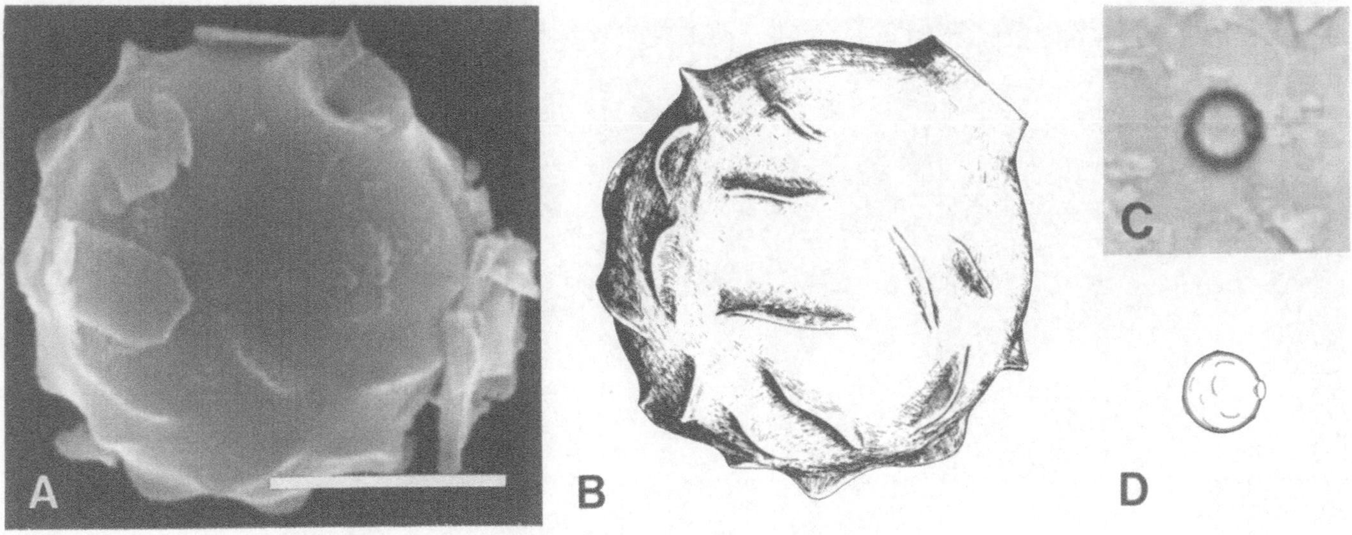

Fig. 102. Stomatocyst 113. A: SEM. B: Line drawing of SEM. C: LM. D: Line drawing of LM. Scale bar = 2 μm. LM x2500.

stomatocyst 91); British Columbia, Canada (Duff & Smol, 1994; Zeeb, 1994); Minnesota, U.S.A. (Zeeb & Smol, 1993a - misidentified as stomatocyst 33); and New York, U.S.A. (Duff & Smol, unpubl.). Adam & Mahood (1979b - California, U.S.A.) type 114 is similar. Both stomatocysts 113 and 33 are probably included within *cysta stellata* Nygaard (Nygaard, 1956 - Denmark; Rybak, 1986 - Poland).

ECOLOGY: This stomatocyst is likely produced by an oligotrophic taxon (Zeeb *et al.*, 1990; Duff, 1994). It has been found across a wide range of pH (Rybak, 1986; Zeeb *et al.*, 1990; Duff *et al.*, 1992; Zeeb & Smol, 1993a; Duff, 1994). It is possibly a cool-water form (Rybak, 1986; Duff *et al.*, 1992; Duff, 1994).

Stomatocyst 223, Duff & Smol, 1994 (8) Fig. 103

BIOLOGICAL AFFINITY: unknown

LOCALITY: Spout Lake (52°00'N; 121°20'W), British Columbia, Canada. Ekman dredge, surface (1 cm) sediment sample.

SEM DESCRIPTION: This is a very small, spherical stomatocyst (diameter 3.0-3.8 μm), with a relatively high, conical to cylindrical collar (diameter 0.9-1.2 μm; height 0.3-1.0 μm). On more mature specimens, the outer collar margin possesses struts which blur the distinction between the collar and the cyst body, and the collar may have a rounded, thickened apex (Fig. 103A). The pore was observed on only one of our SEM specimens, and appeared to be conical. The ornamentation consists of lunate ridges (maximum length 0.8-2.5 μm; height 0.2-0.7 μm), randomly oriented, which may have reflexing apices when mature.

With SEM, this stomatocyst can be distinguished from similar cysts by the presence of collar struts on mature specimens, the collar height, the ridge height on mature specimens, and the collar shape, which is never obconical.

LM DESCRIPTION: This stomatocyst can be distinguished from similar cysts (i.e. stomatocysts 33/222, 91, and 113) by its very small size, and the collar shape and height. The ridges are most similar to those of stomatocyst 113.

REFERENCES: We have observed this stomatocyst in British Columbia, Canada (Duff & Smol, 1994; Zeeb, 1994), the Northwest Territories (Webb & Smol, unpubl.), and Minnesota, U.S.A. (Zeeb & Smol, 1993a - misidentified as stomatocyst 33).

ECOLOGY: This stomatocyst was most common in small, deep, slightly acidic to circumneutral, oligotrophic lakes in British Columbia, Canada (Duff, 1994). It may produced by an aluminum-tolerant taxon.

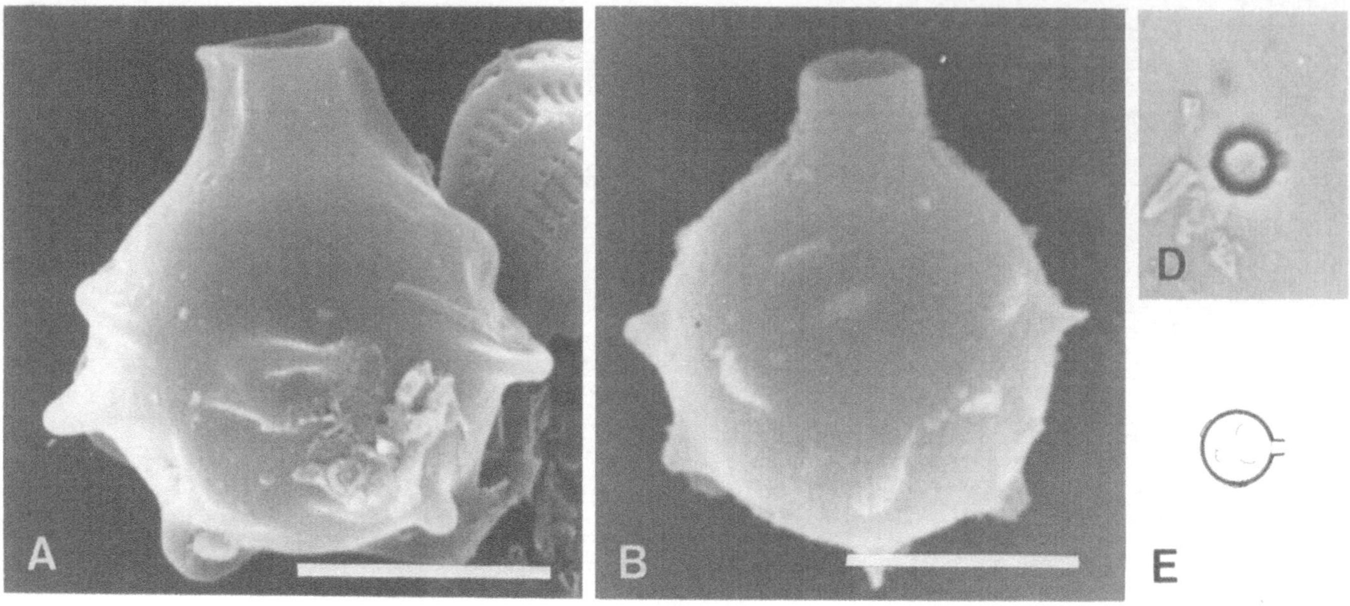

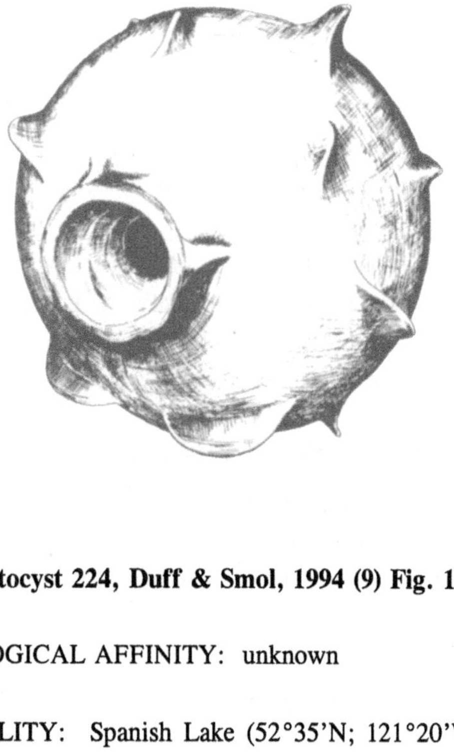

Fig. 103. Stomatocyst 223. A: SEM, mature specimen. B: SEM, immature specimen. C: Line drawing of SEM. D: LM. E: Line drawing of LM. Scale bars = 2 μm. LM x2500.

(diameter 0.3-0.5 μm). The cyst surface is ornamented with short lunate to longer ridges, which are usually straight and randomly oriented (maximum ridge length 0.6-1.8 μm).

LM DESCRIPTION: This stomatocyst can be differentiated from similar morphs (i.e. stomatocysts 33/222, 91, 113 and 223) by the collar width.

REFERENCES: This stomatocyst has been observed in British Columbia, Canada (Duff & Smol, 1994; Zeeb, 1994), the Northwest Territories, Canada (Webb & Smol, unpubl.), and New York, U.S.A. (Duff & Smol, 1991 - misidentified as stomatocyst 91; Duff & Smol, unpubl.).

ECOLOGY: This cyst was most abundant in small, deep, oligotrophic, montane lakes in Duff's (1994) survey.

Stomatocyst 224, Duff & Smol, 1994 (9) Fig. 104

BIOLOGICAL AFFINITY: unknown

LOCALITY: Spanish Lake (52°35'N; 121°20'W), British Columbia, Canada. Ekman dredge, surface (1 cm) sediment sample.

SEM DESCRIPTION: This stomatocyst is small and spherical (diameter 2.6-4.0 μm) with a low, wide, cylindrical to obconical collar (diameter 1.4-2.5 μm; height 0.1-0.5 μm). The collar diameter : cyst diameter is always ≥0.41; this feature is used to distinguish this stomatocyst from similar morphs (i.e. stomatocysts 33, 91, 92 (Chapter 5), 113, 222 and 223). The inner collar margin is gradually sloping to almost flat, and the pore is regular to slightly conical

Stomatocyst 32, Duff & Smol, 1989 (10) Fig. 105

BIOLOGICAL AFFINITY: unknown

LOCALITY: Tasikutaaq Lake (66°23'N; 65°57'W), Baffin Island, Northwest Territories, Canada. Percussion core, core depth 75 cm.

SEM DESCRIPTION: This stomatocyst is spherical (diameter 5.0-6.7 μm), with a low, conical to cylindrical

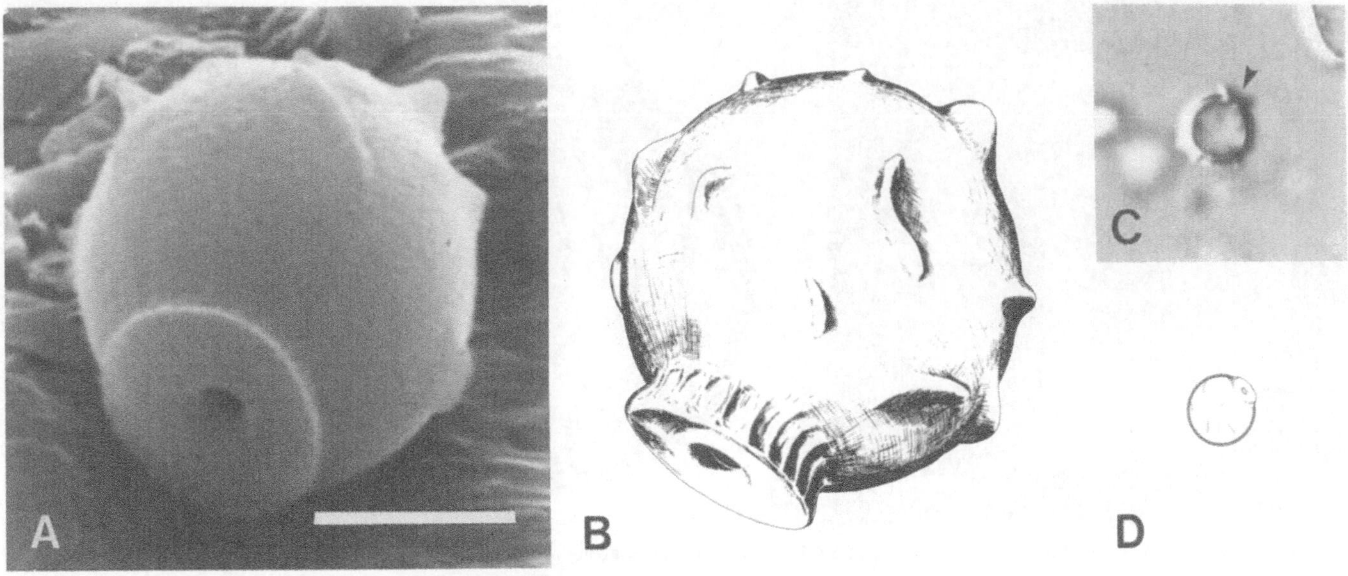

Fig. 104. Stomatocyst 224. A: SEM. B: Line drawing of SEM. C: LM. Collar indicated by arrow. D: Line drawing of LM. Scale bar = 2 μm. LM x2500.

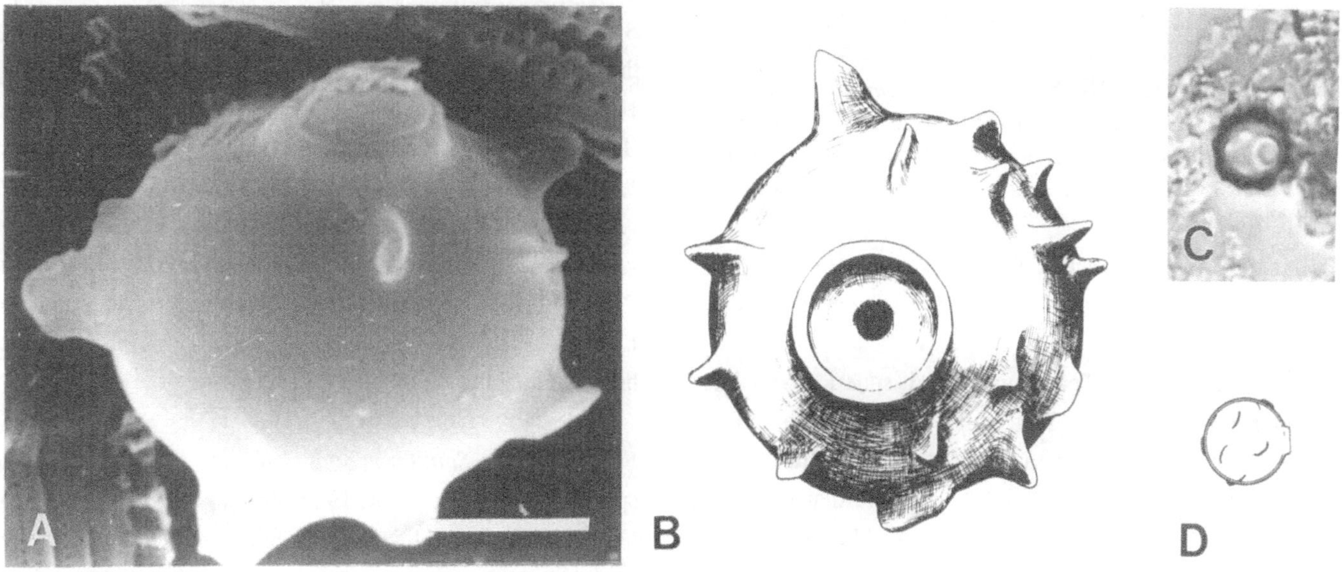

Fig. 105. Stomatocyst 32. A: SEM. Pore covered by cap. B: Line drawing of SEM. Pore visible. C: LM. Pore covered by cap. D: Line drawing of LM. Scale bar = 2 μm. LM x2000.

collar, often with a flattened or rounded apex (collar diameter 1.8-2.6 μm; height ca. 0.6 μm). A planar annulus was observed on one SEM specimen, surrounding a regular or conical pore (diameter 0.5 μm). On the remaining specimens, a thin, convex siliceous cap (diameter 1.4-2.1 μm) obscures the pore (Fig. 105A). Irregular ridges are randomly oriented over the cyst body; these vary considerably in height, even on the same specimen (length 0.7-1.1 μm; height 0.3-1.1 μm).

LM DESCRIPTION: This stomatocyst was grouped with stomatocyst 33 for LM analysis by Duff & Smol (1989). However, it can be distinguished from stomatocyst 33 by its larger size, wider collar, and thicker and higher ridges.

REFERENCES: This stomatocyst has been found in Ontario, Canada (Rybak *et al.*, 1991 - cyst 77), and the Northwest Territories, Canada (Duff & Smol, 1989; Webb & Smol, unpubl.).

ECOLOGY: Stomatocyst 32 has been found in oligotrophic, slightly acidic to circumneutral lakes (Duff & Smol, 1989; Rybak *et al.*, 1991; Webb & Smol, unpubl.), and was

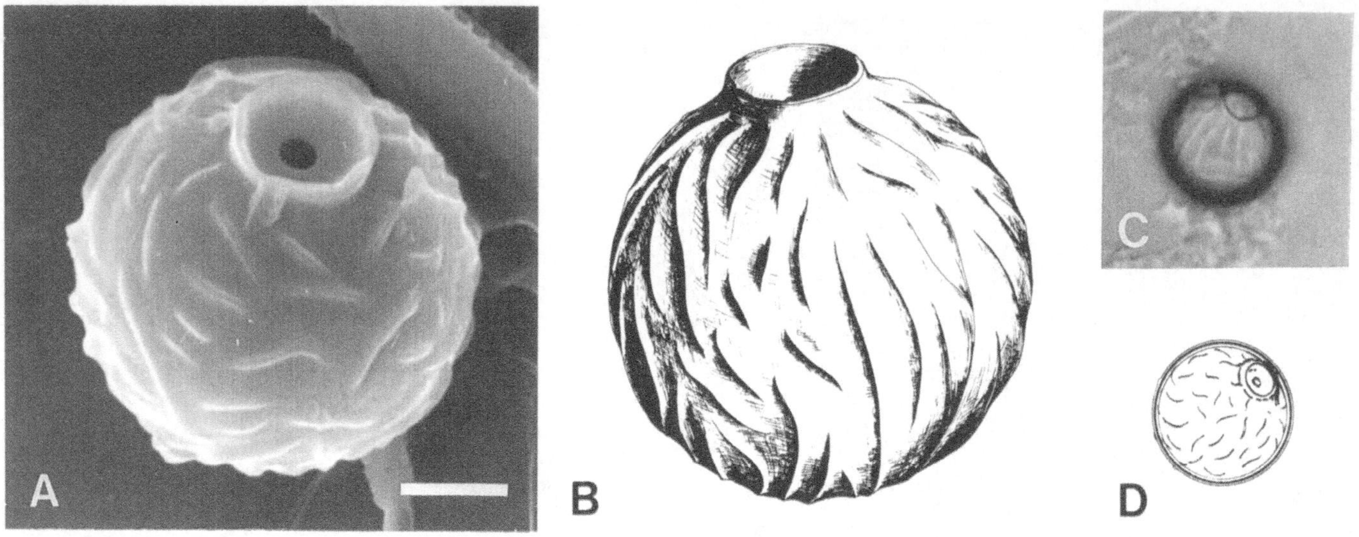

Fig. 106. Stomatocyst 111. A: SEM. B: Line drawing of SEM. C: LM. D: Line drawing of LM. Scale bar = 2 μm. LM x2000.

classified as acidophilic by Rybak *et al.* (1991).

Stomatocyst 111, Zeeb *et al.,* 1990 (15) Fig. 106

BIOLOGICAL AFFINITY: *Spiniferomonas trioralis* (Takahashi) Preisig *et* Hibberd (Skogstad, 1986 - *Chromophysomonas trioralis* (Takahashi) Preisig & Hibberd; McKenzie & Kling, 1989).

LOCALITY: Little Round Lake (44°47.5'N; 76°41.5'W), Ontario, Canada. One metre piston core, core depth 20 cm.

SEM DESCRIPTION: This stomatocyst is spherical to oblate (diameter 6.4-9.7 μm). The collar is cylindrical with a sloping inner margin and a regular pore (collar apical diameter 1.3-3.0 μm; height 0.4-1.1 μm; pore diameter 0.5-0.7 μm). The collar apex is often irregular. Skogstad (1986) observed spiralling striations on the inner collar margin. The ornamentation consists of low, rounded ridges (height 0.1-0.3 μm). These vary greatly with respect to orientation and length, from short (length ca. 0.7 μm) and randomly oriented, to longer (>7.0 μm) and usually longitudinally oriented.

Zeeb *et al.* (1990) unidentified stomatocyst 1 (Chapter 5), which was described based on LM only, is now reassigned to stomatocyst 111.

LM DESCRIPTION: This stomatocyst can be easily identified by its shape, wide collar, and surface pattern. The light micrograph given by Zeeb *et al.* (1990 - Fig. 8), however, is a misidentification and is properly attributed to stomatocysts 146/156.

REFERENCES: This stomatocyst is widespread. It has been observed in Ontario, Canada (Rybak *et al.*, 1987 - cyst 27; Zeeb *et al.*, 1990, 1994; Rybak *et al.*, 1991 - cyst 105); British Columbia, Canada (Duff & Smol, 1994; Zeeb, 1994); the Northwest Territories, Canada (Duff & Smol, 1988; Fig. 16 - stomatocyst 7 (invalidly described)); New York, U.S.A. (Duff & Smol, 1991 - Fig. 152); Michigan, U.S.A. (Sandgren & Carney, 1983 - cyst 34; Carney & Sandgren, 1983 - *cysta vermicularis* Nygaard); Denmark (Nygaard, 1956 - *cysta vermicularis*); and Poland (Rybak, 1986 - *cysta vermicularis*; Rybak, 1987 - cyst 22). The stomatocysts identified by Leventhal (1970) and Gritten (1977) as *cysta vermicularis* may be questionable.

ECOLOGY: *Spiniferomonas trioralis* is a cosmopolitan species (e.g. Dürrschmidt, 1980; Roijackers & Kessels, 1981; Croome & Tyler, 1985; Siver, 1988b; and many more) found across a wide range of environmental conditions. Siver (1989a) reported an estimated pH optimum of 6.7. This is a spring blooming species (Dürrschmidt, 1980; Kristiansen, 1988), with cyst formation often observed during, or prior to, spring ice break-up (Skogstad, 1986). Paleolimnological reports frequently link the stomatocyst with eutrophic conditions (Nygaard, 1956; Rybak, 1986; Carney & Sandgren, 1983; Zeeb *et al.*, 1990, 1994).

Stomatocyst 133, Duff & Smol in Duff *et al.*, 1992 *emend.* Zeeb *et al.*, subm. *emend.* Duff & Smol, 1994 (28) Fig. 107

BIOLOGICAL AFFINITY: This stomatocyst is possibly

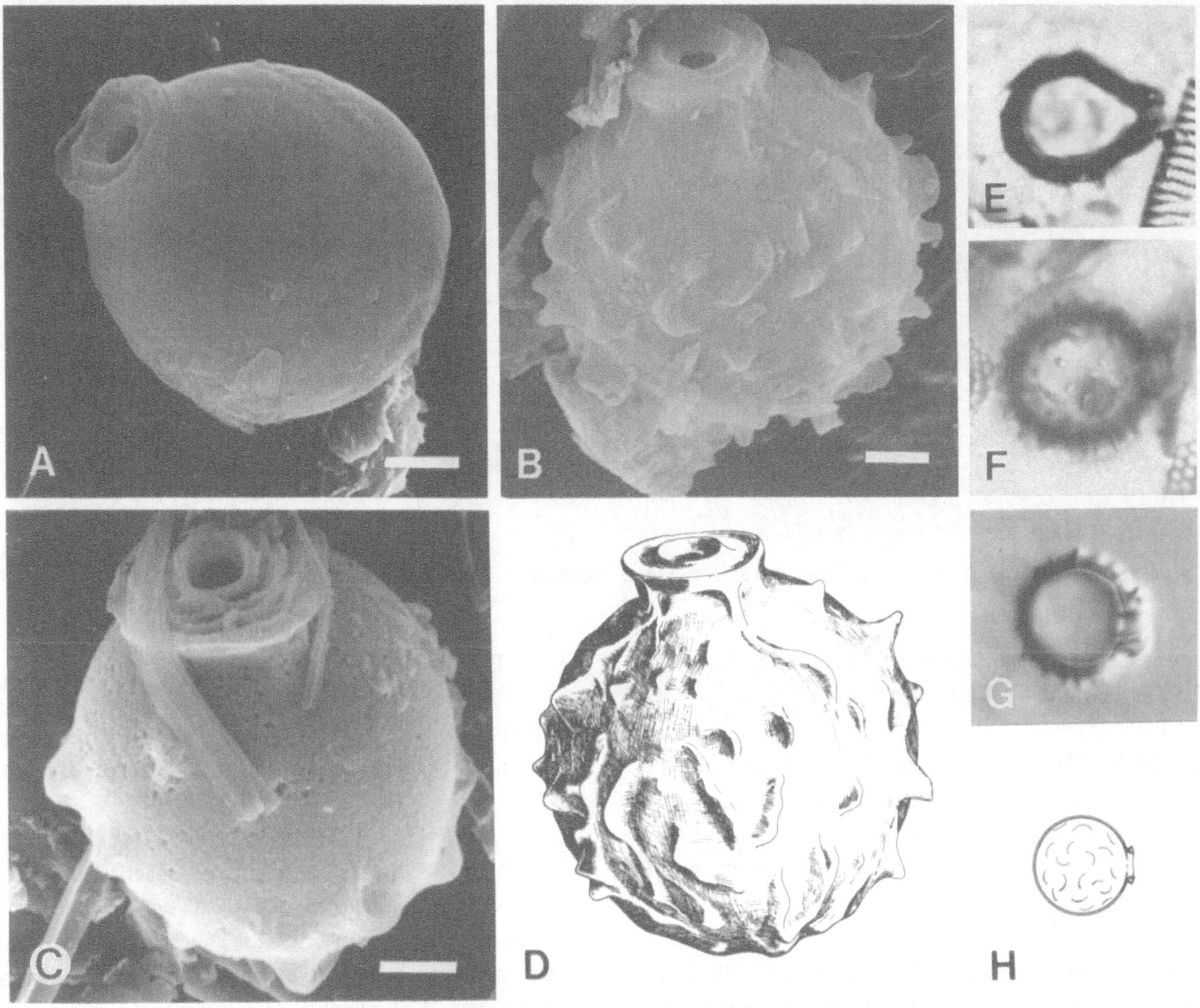

Fig. 107. Stomatocyst 133 *formae* A, B, and C. A: *Forma* A, SEM. B: *Forma* B, SEM. C: *Forma* C, SEM. D: *Forma* B, line drawing of SEM. E: *Forma* A, LM. F: *Forma* B, LM. G: *Forma* C, LM. H: *Forma* B, line drawing of LM. Scale bars = 2 μm. LM x2000.

produced by an *Ochromonas* species, as it resembles the cyst of *O. sphaerocystis* Matvienko (Andersen, 1982), but is much smaller (see the description of stomatocyst 186).

LOCALITY: "Col Pond" (unofficial name), Cape Herschel (78°37'N; 74°42'W), Ellesmere Island, Northwest Territories, Canada. Moss sample, collected July 29, 1986.

SEM DESCRIPTION: This is a spherical to very widely ovate stomatocyst (diameter 8.2-12.2 μm) with a complex collar. The regular pore (diameter ca. 0.9 μm) is surrounded by a conical primary collar (diameter 1.2-2.4 μm) with a sloping inner margin, and sometimes with an inwardly reflexing apex. The secondary collar (diameter 2.7-6.4 μm) is conical to obconical, and may have basal collar struts; there is no interannulus. Three *formae* were recognized based on differences in the degree and type of collar and ornamentation development.

***Forma* A** - This *forma* (Figs. 107A, E) has a complex collar as described above, and is smooth or ornamented with very wide, very low verrucae (diameter 1.3-1.5 μm; height 0.1-0.2 μm).

***Forma* B** - The collar complex is identical to that of *forma* A, but the cyst (Figs. 107B, D, F, and H) is ornamented with short echinate spines to randomly oriented, lunate ridges of varying length (ridge length 1.1-2.4 μm; height 0.5-1.5 μm).

***Forma* C** - Spiralling ridges are found on the inner margin of the secondary collar, and the secondary collar apex reflexes strongly towards the cyst surface (Figs. 107C, G). The

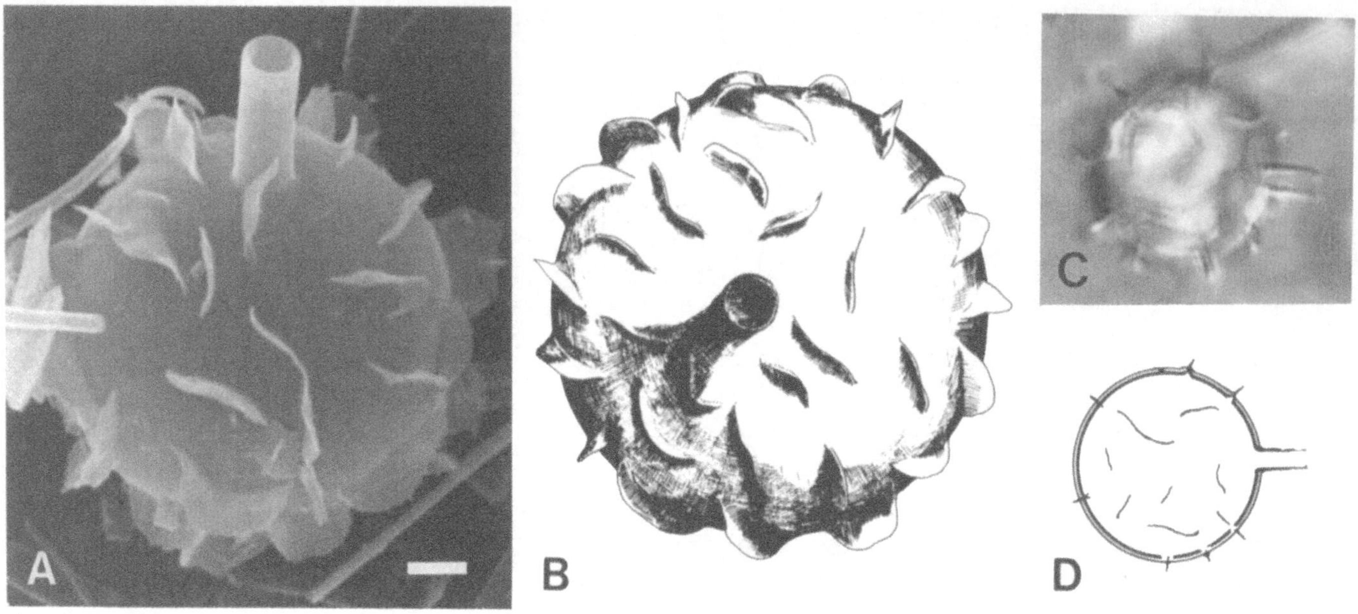

Fig. 108. Stomatocyst 238. A: SEM. B: Line drawing of SEM. C: LM. D: Line drawing of LM. Scale bar = 2 μm. LM x2000.

ornamentation is identical to that of *forma* B (ridge length 0.6-2.5 μm; height 0.4-1.4 μm).

LM DESCRIPTION: This stomatocyst can be recognized by its fairly wide, complex collar and by its ornamentation. If the ridges cannot be distinguished from spines, stomatocyst 133 *forma* B could be confused with stomatocyst 185 *forma* B.

REFERENCES: Stomatocyst 133 *forma* A has been observed in the Northwest Territories, Canada (Duff *et al.*, 1992), and British Columbia, Canada (Zeeb, 1994). *Forma* B has been observed in California, U.S.A. (Zeeb *et al.*, subm.), and Poland (Rybak, 1987 - cyst 9). *Forma* C has been found in British Columbia, Canada (Duff & Smol, 1994; Zeeb, 1994), and Ontario, Canada (Rybak *et al.*, 1991 - cyst 39). Sandgren & Carney (1983 - Michigan, U.S.A.) cyst 20 resembles *forma* B but has a false, rather than a true, complex collar. Stomatocyst 133 is probably not identical to *cysta infundibularis* Nygaard (1956), as indicated by Rybak (1986) and Sandgren & Carney (1983).

ECOLOGY: Autecological details are not yet available for any of these *formae*.

Stomatocyst 238, Duff *et al.* (this volume) (4) Fig. 108

BIOLOGICAL AFFINITY: This stomatocyst is produced by an unidentified *Ochromonas* species (Sandgren, 1983b).

NEGATIVE NUMBER: J.P. Smol 558

LOCALITY: Lake 227 (49°42'N; 93°42'W), Experimental Lakes Area, Ontario, Canada. Freeze core, varve year A.D. 1978.

SEM DESCRIPTION: This is a large, spherical to widely oval stomatocyst (diameter 10.8-12.5 μm) with a long cylindrical collar (diameter 1.7-2.0 μm; height 2.3-3.1 μm) with an abrupt base. It is ornamented with randomly oriented, very thinly silicified, high, wavy ridges, which may have reflexed apices (ridge height 0.2-1.8 μm; length 0.8-6.0 μm).

LM DESCRIPTION: Stomatocyst 238 can be distinguished using LM from stomatocyst 173 because the latter cyst is much larger, with a wider collar and fewer, longer ridges that are longitudinally, not randomly, oriented.

REFERENCES: This cyst is newly described based on specimens found in Ontario, Canada and New York, U.S.A. It has also been observed elsewhere in Ontario, Canada (Rybak *et al.*, 1987 - cyst 28); New York, U.S.A. (Duff & Smol, 1991 - Fig. 154); California, U.S.A. (Carney *et al.*, 1992 - cyst 20 in Fig. 21 but not Fig. 22); Denmark (Nygaard, 1956 - *cysta hieroglyphica* Nygaard); and Poland (Rybak, 1987 - cysts 27 and 28).

ECOLOGY: This cyst may be produced by an acidophilic

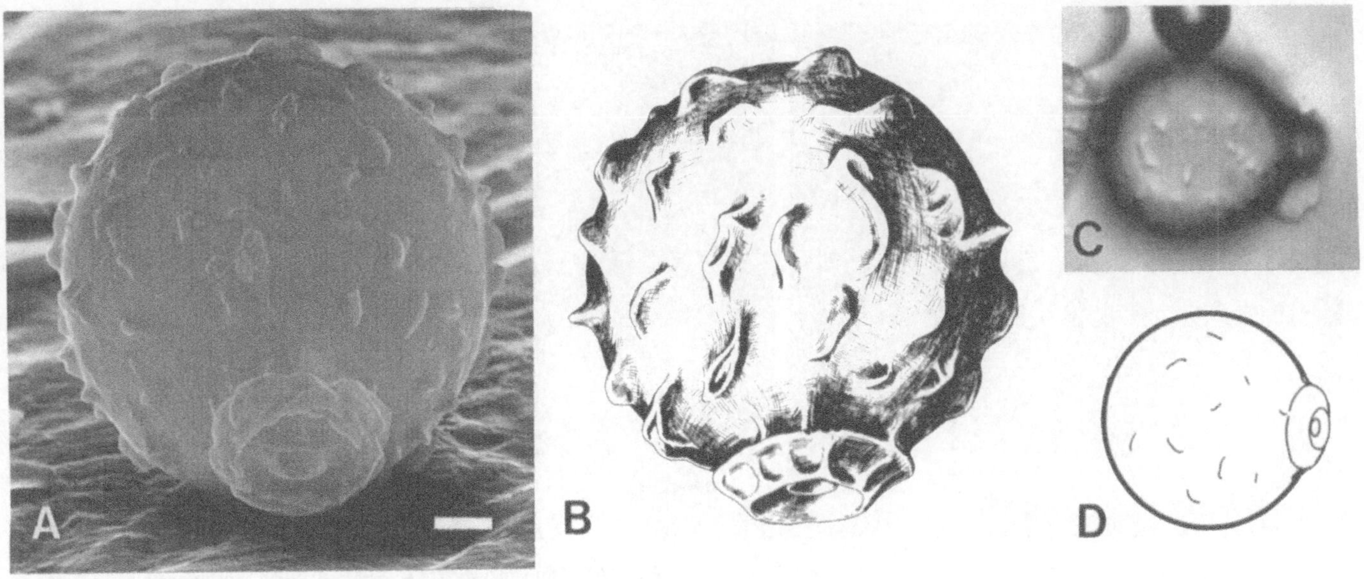

Fig. 109. Stomatocyst 186. A: SEM. B: Line drawing of SEM. C: LM. D: Line drawing of LM. Scale bar = 2 μm. LM x2000.

(Nygaard, 1956; Duff & Smol, 1991) or pH-indifferent species (Carney *et al.*, 1992).

Stomatocyst 186, Brown & Smol in Brown *et al.*, 1994 (9) Fig. 109

BIOLOGICAL AFFINITY: Stomatocyst 186 resembles the cyst of *Ochromonas sphaerocystis* Matvienko (Andersen, 1982), but differs in the collar details and is slightly smaller.

LOCALITY: Nordvestø (76°44'N; 73°13'W), Carey Islands, Greenland. SIPRE peat core, core depth 80-81 cm.

SEM DESCRIPTION: This stomatocyst is large, spherical to ovate (diameter 10.0-15.5 μm) with a complex collar. The primary collar is conical (diameter 1.6-2.3 μm) and surrounds a regular pore (diameter ca. 1.3 μm). The secondary collar is cylindrical with a medial outward distension (apical diameter 3.0-4.3 μm; medial diameter 4.2-7.4 μm). A planar interannulus appears to separate the two collars. Some specimens possess a distinct convoluted pattern on the outer margin of the secondary collar. The basal margin of the secondary collar may be strutted. The cyst surface is densely ornamented with short lunate to wavy ridges (length 0.9-5.3 μm; height to 0.5-1.7 μm).

LM DESCRIPTION: This cyst is recognized by its size and collar complex.

REFERENCES: This stomatocyst has been found in British Columbia, Canada (Zeeb, 1994), and Greenland (Brown *et al.*, 1994). *Trachelomonas yberiensis* var. *membranosa* Frenguelli (1932 - Argentina) and *Outesia yberiensis* var. *reticulato-spinosa* Conrad (1940 - Connecticut, U.S.A.) both resemble stomatocyst 186.

ECOLOGY: Autecological details are as yet unknown.

Stomatocyst 57, Duff & Smol, 1991 (13) Fig. 110

BIOLOGICAL AFFINITY: unknown

LOCALITY: Upper Wallface Pond (44°08.8'N; 74°03.3'W), Adirondack Park, New York, U.S.A. Gravity core, core depth 20.0-20.5 cm.

SEM DESCRIPTION: This stomatocyst is usually obovate, but is occasionally oval (4.9-6.0 x 5.4-6.4 μm). It has a high, cylindrical collar, sometimes with a thickened apex, rising abruptly from the cyst wall (collar diameter 0.6-0.9 μm; height 0.8-1.4 μm). Internal collar details are unknown. The posterior hemisphere is usually ornamented with a network of curved ridges, radiating from the posterior pole, and extending approximately half-way to the equatorial region; this network was misidentified as debris in our original description (Duff & Smol, 1991).

LM DESCRIPTION: This cyst can be recognized by its size,

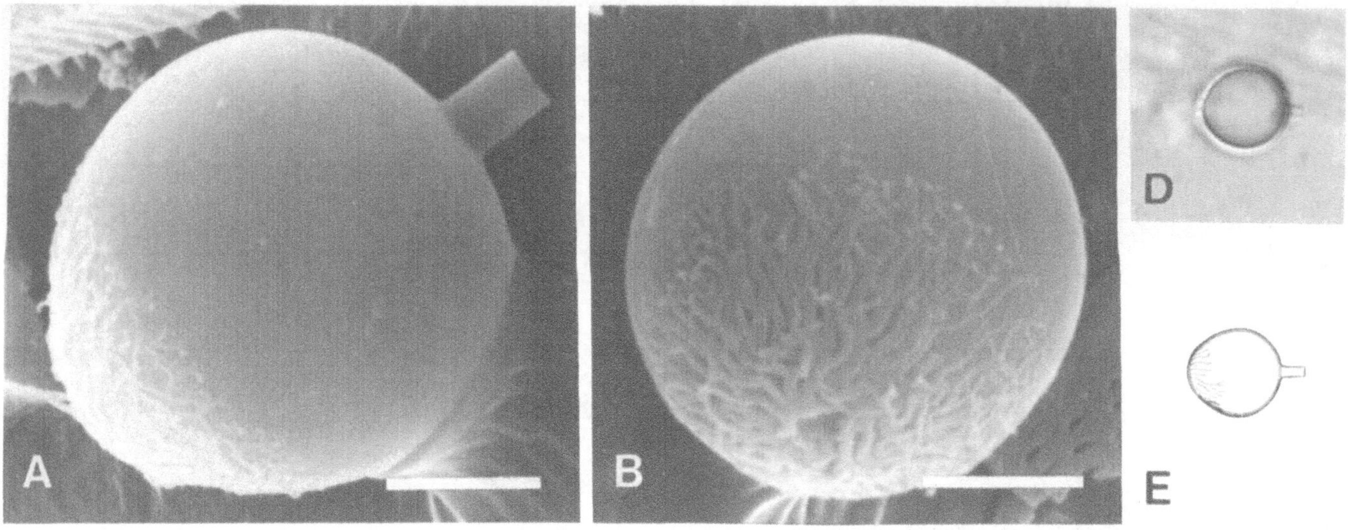

Fig. 110. Stomatocyst 57. A: SEM. B: SEM, posterior pole. C: Line drawing of SEM. D: LM. E: Line drawing of LM. Scale bars = 2 μm. LM x2000.

narrow, cylindrical collar, and obovate body shape. Both the cyst wall, and particularly the collar, are thinly silicified. The posterior ornamentation is sometimes, but not always, visible with LM. The LM specimen depicted in Duff & Smol (1991 - Fig. 31) is misidentified, and is most likely a specimen of stomatocyst 128 *forma* A, although SEM specimens of the latter were not observed in the Upper Wallface core.

REFERENCES: This cyst has been observed in Ontario, Canada (Rybak *et al.*, 1991 - cyst 180; Zeeb & Smol, unpubl.), British Columbia, Canada (Duff, 1994), New York, U.S.A. (Duff & Smol, 1991), and California, U.S.A. (Adam & Mahood, 1979b - type 135). It was probably included within Carney *et al.* (1992 - California, U.S.A.) cyst 3.

ECOLOGY: This stomatocyst is most likely produced by an acidophilic (Duff & Smol, 1991; Rybak *et al.*, 1991) or pH-indifferent (Carney *et al.*, 1992) species. It is found in oligotrophic locations (Duff & Smol, 1991; Rybak *et al.*, 1991; Carney *et al.*, 1992; Duff, 1994). It was associated with deep lakes in Duff's (1994) survey.

Stomatocysts with long ridges; ridge length is equal to or greater than half of the cyst circumference

Stomatocyst 239, Duff *et al.* (this volume) (8) Fig. 111

BIOLOGICAL AFFINITY: unknown

NEGATIVE NUMBER: J.P. Smol 425

LOCALITY: Dry Channel Pond (44°22'N; 74°28'W), Adirondack Park, New York, U.S.A. Gravity core, surface (0-1 cm) sediment sample.

SEM DESCRIPTION: This stomatocyst is small and spherical (diameter 3.1-4.5 μm) with a regular pore (diameter 0.3-0.5 μm). The ornamentation is highly variable, consisting of ridges (basal width ca. 0.3 μm; height 0.1-0.4 μm), randomly oriented and varying in length (to ≥4.5 μm). There is no true collar, but two or more ridges are usually located near the pore and may fuse to form an incomplete or false collar. The ridges in the pore region may be higher than elsewhere on the cyst body.

This stomatocyst appears to be identical to stomatocyst 94, except that it lacks depressions; nevertheless, they may be produced by the same, or closely related, species.

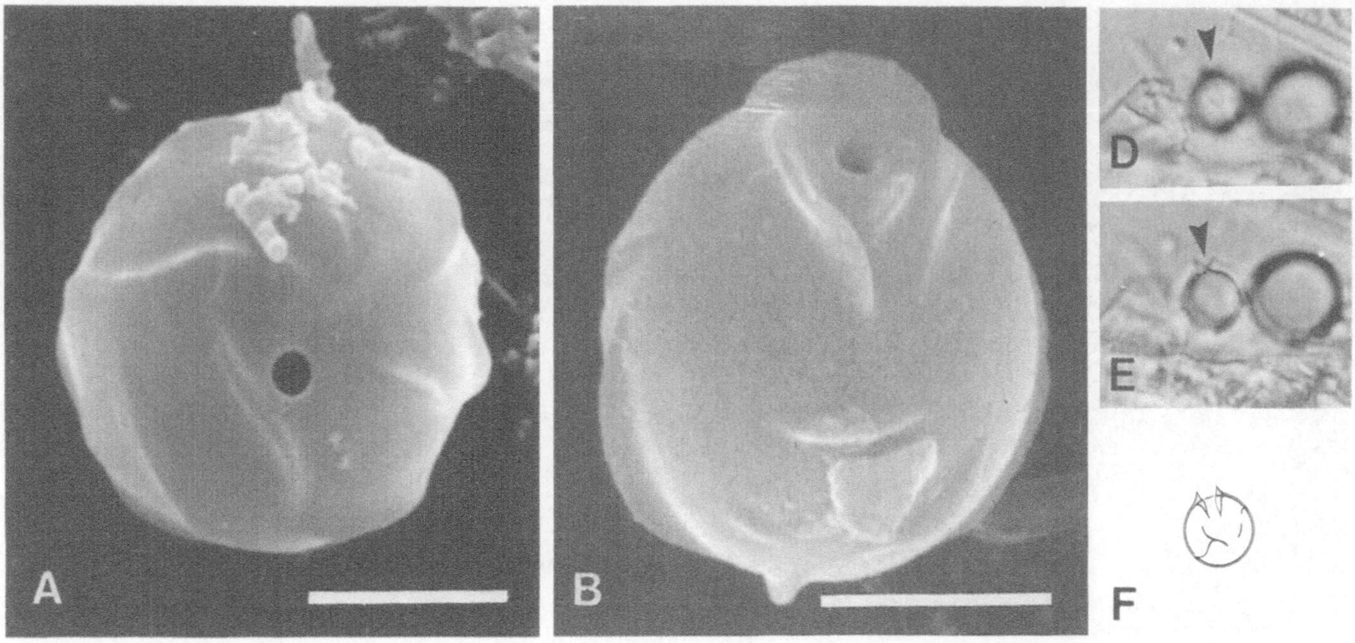

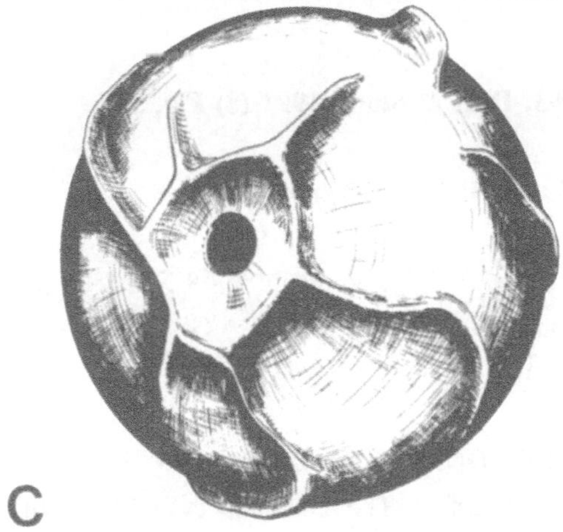

Fig. 111. Stomatocyst 239. A: SEM, apical view. B: SEM. C: Line drawing of SEM. D: LM, surface view. Pore indicated by arrow. E: LM, cross-sectional view. Pore indicated by arrow. F: Line drawing of LM. Scale bars = 2 μm. LM x2500.

LM DESCRIPTION: With LM, this stomatocyst cannot be distinguished from stomatocyst 94. Both morphs are very small with visible ridges; they could be confused with small specimens of stomatocyst 91. However, stomatocyst 91 tends to have wavier ridges and the true collar is usually more distinct.

REFERENCES: Stomatocyst 239 has been found in Ontario, Canada (Zeeb & Smol, unpubl.), Washington, U.S.A. (Smith & White - Fig. 14), California, U.S.A. (Adam & Mahood, 1979b - types 113 and 117), and New York, U.S.A. (Duff *et al.*, this volume).

ECOLOGY: Smith & White (1985) found stomatocyst 239 in a hardwater lake, newly formed as a result of volcanic damming.

Stomatocyst 91, Duff & Smol, 1991 (14) Fig. 112

BIOLOGICAL AFFINITY: unknown

LOCALITY: Upper Wallface Pond (44°08.8'N; 74°03.3'W), Adirondack Park, New York, U.S.A. Gravity core, core depth 10.0-10.5 cm.

SEM DESCRIPTION: This is a small, spherical stomatocyst (diameter 2.9-5.2 μm) with quite variable collar morphology and ornamentation. There is a low, wide, obconical collar, usually with a thickened apex (collar diameter 0.7-1.6 μm; height 0.1-1.0 μm; average height 0.4 μm). The cyst surface is covered with low ridges which vary from short and lunate to long and wavy, frequently varying even on the same specimen (ridge height 0.1-0.6 μm; maximum ridge length 0.7-3.4 μm; average maximum ridge length 1.9 μm). Branched ridges may be present. Struts were always present on the collar base, but collar or ridge striations were not observed.

Stomatocyst 91 can be distinguished from similar stomatocysts by the presence of collar struts, and the length and variable nature of the ridges. It was originally described based on five SEM specimens; of these, only the characteristic specimen (Duff & Smol, 1991 - Fig. 111) has been retained. Two other specimens, including Fig. 112

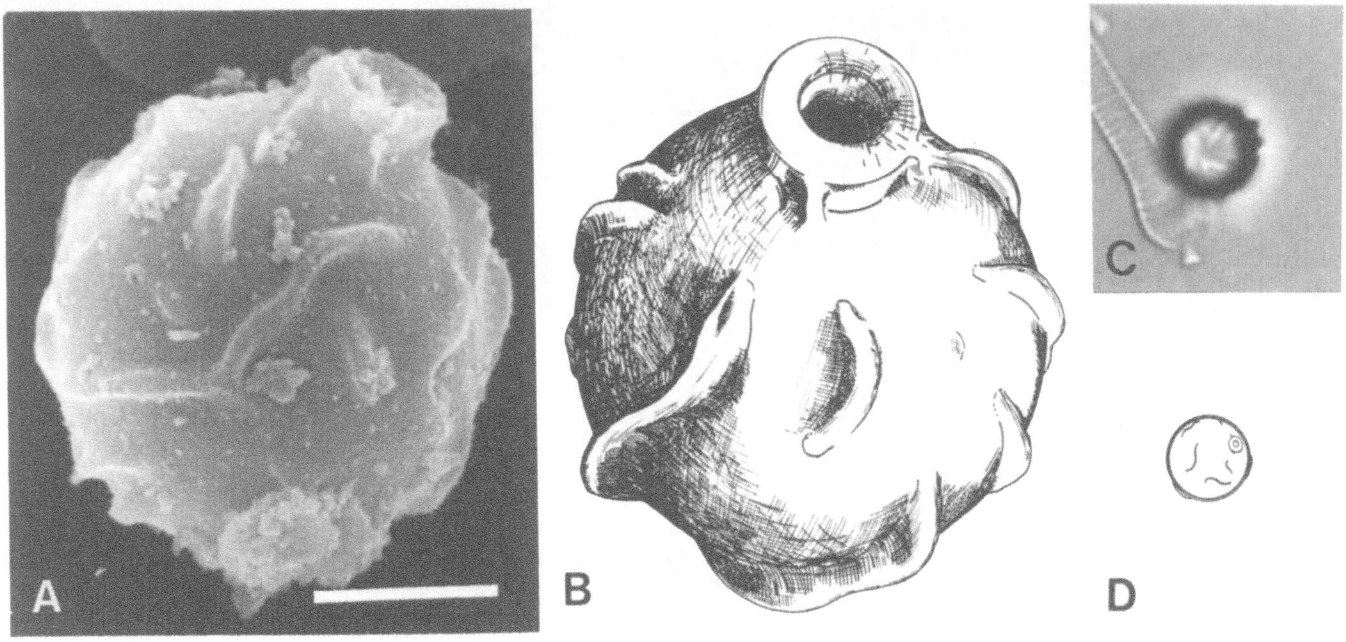

Fig. 112. Stomatocyst 91. A: SEM. B: Line drawing of SEM. C: LM. D: Line drawing of LM. Scale bar = 2 μm. LM x2500.

(Duff & Smol, 1991), have been reassigned to stomatocyst 224; one specimen has been reassigned to stomatocyst 34; and the final specimen cannot yet be assigned definitely to any morphotype. The cyst identified as stomatocyst 91 by Duff *et al.* (1992) has been reassigned to stomatocyst 113.

LM DESCRIPTION: This cyst apparently cannot be distinguished with LM from stomatocyst 92 (Chapter 5), but can be distinguished from stomatocysts 33/222 based on ridge length, and from stomatocyst 113 based on ridge morphology.

REFERENCES: This is a common morphotype. It has been found in British Columbia, Canada (Duff & Smol, 1994; Zeeb, 1994); Ontario, Canada (Rybak *et al.*, 1991 - cyst 81); Minnesota, U.S.A. (Zeeb & Smol, 1993a); California, U.S.A. (Carney *et al.*, 1992 - cyst 9); and New York, U.S.A. (Duff & Smol, 1991, unpubl.).

ECOLOGY: The pH preference of this cyst has not yet been ascertained; it has been classified variously as acidophilic (Duff & Smol, 1991), pH-indifferent (Carney *et al.*, 1992), and alkaliphilic (Rybak *et al.*, 1991). In Duff & Smol's (1991) paleolimnological study, high abundance of this morph appeared to be linked to a watershed disturbance involving increased erosion, and possibly higher productivity. Stomatocyst 91 has been reported from alpine lakes (Duff & Smol, 1991; Carney *et al.*, 1992); this suggests that it is produced by a cold-water tolerant form.

Stomatocyst 93, Duff & Smol, 1991 (8) Fig. 113

BIOLOGICAL AFFINITY: unknown

LOCALITY: Upper Wallface Pond (44°08.8'N; 74°03.3'W), Adirondack Park, New York, U.S.A. Gravity core, core depth 0.0-0.5 cm.

SEM DESCRIPTION: This stomatocyst is spherical (diameter 5.2-6.9 μm). The collar is cylindrical with a flattened, obconical or reflexed apex, a sloping inner margin, and a regular pore (collar base diameter 0.8-1.5 μm; apical diameter 1.5-2.9 μm; height 0.3-2.2 μm; pore diameter ca. 0.4 μm). The ornamentation consists of narrow ridges with overturned, irregular apices, usually but not always longitudinally oriented (base width 0.1-0.2 μm; height 0.3-2.1 μm). The ridges are straight or slightly curved.

LM DESCRIPTION: Stomatocysts 93 and 34 are very similar, but can be distinguished by the following features: stomatocyst 93 is usually larger; stomatocyst 34 has shorter, wavier ridges and occasional spines; and the collar of stomatocyst 34 is usually shorter with a more reflexed apex.

REFERENCES: To date, this stomatocyst has only been observed in New York, U.S.A. (Duff & Smol, 1991, unpubl.).

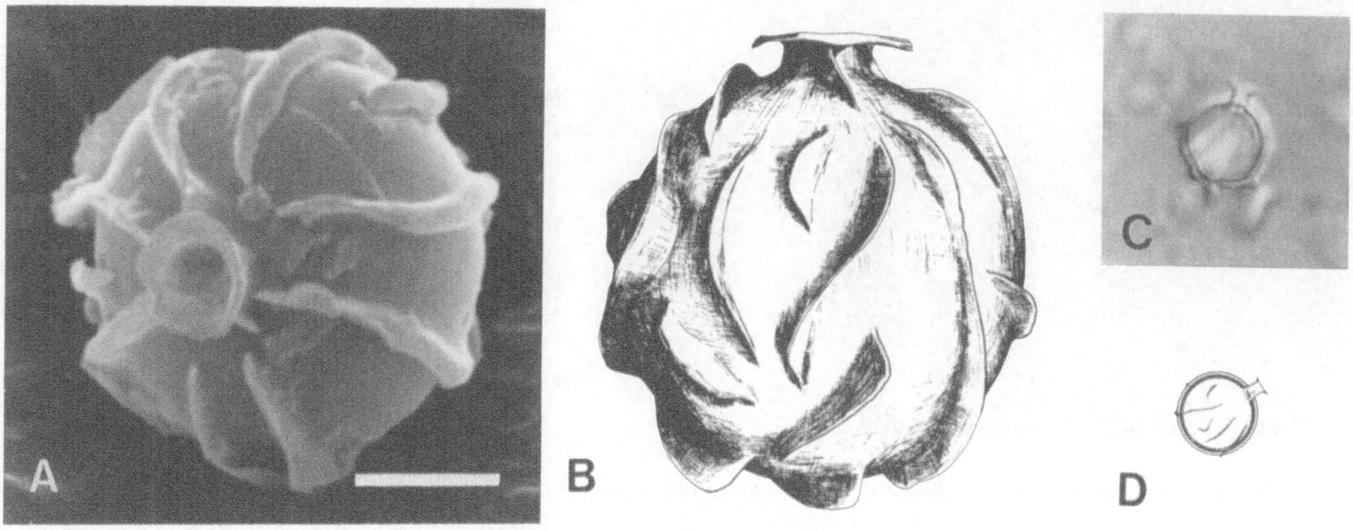

Fig. 113. Stomatocyst 93. A: SEM. B: Line drawing of SEM. C: LM. D: Line drawing of LM. Scale bar = 2 μm. LM x2000.

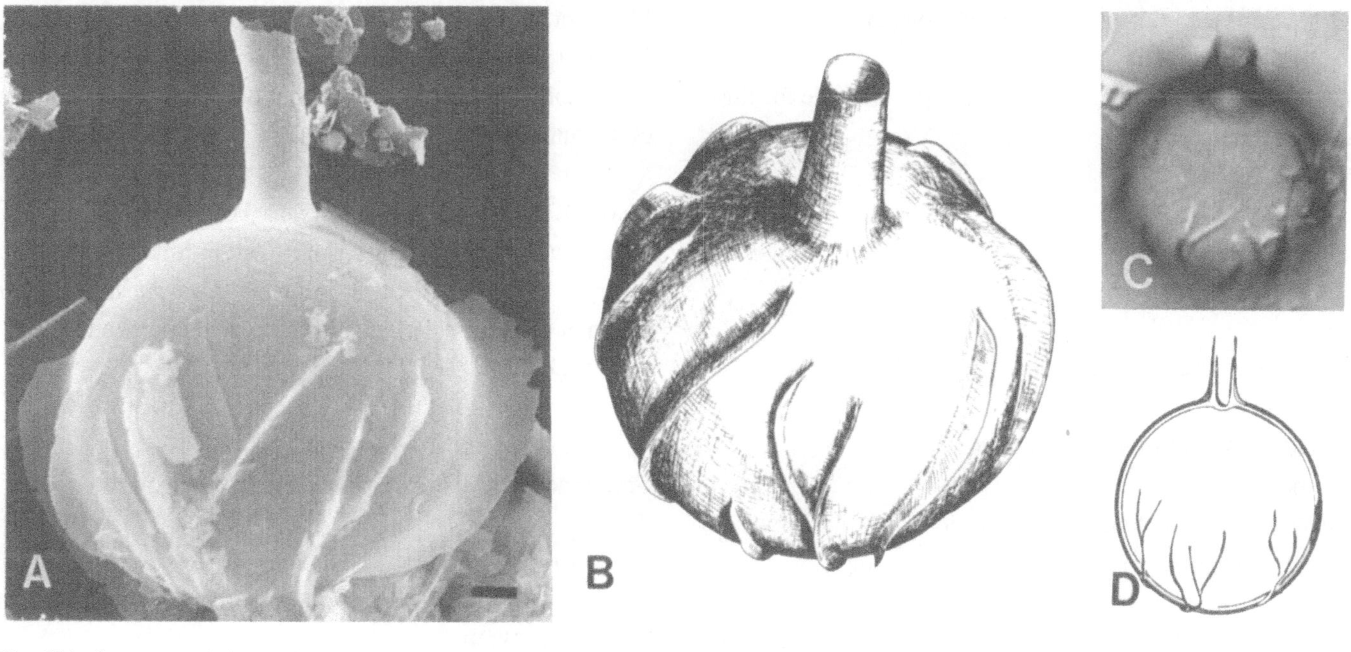

Fig. 114. Stomatocyst 173. A: SEM. B: Line drawing of SEM. C: LM. D: Line drawing of LM. Scale bar = 2 μm. LM x2000.

ECOLOGY: This cyst may be produced by an acidobiontic, and possibly aluminum-tolerant species (Duff & Smol, 1991).

Stomatocyst 173, Zeeb & Smol, 1993a (5) Fig. 114

BIOLOGICAL AFFINITY: unknown

LOCALITY: Elk Lake (47°11'N; 95°13'W), Clearwater County, Minnesota, U.S.A. Piston core, varve year 7972 B.P.

SEM DESCRIPTION: This is a large, spherical stomatocyst (diameter 16.3-18.1 μm). The collar is long and cylindrical with a conical base (collar diameter 2.2-3.3 μm; height to 7.2 μm or greater); inner collar details are unknown. Long ridges run more-or-less longitudinally, usually starting some distance from the collar base, but extending to the posterior pole. The ridges are thinly silicified and quite high when mature (height to 2.5 μm; length 6.9 to >23.2 μm).

LM DESCRIPTION: This cyst is easily recognized by its size, collar, and ridge morphology.

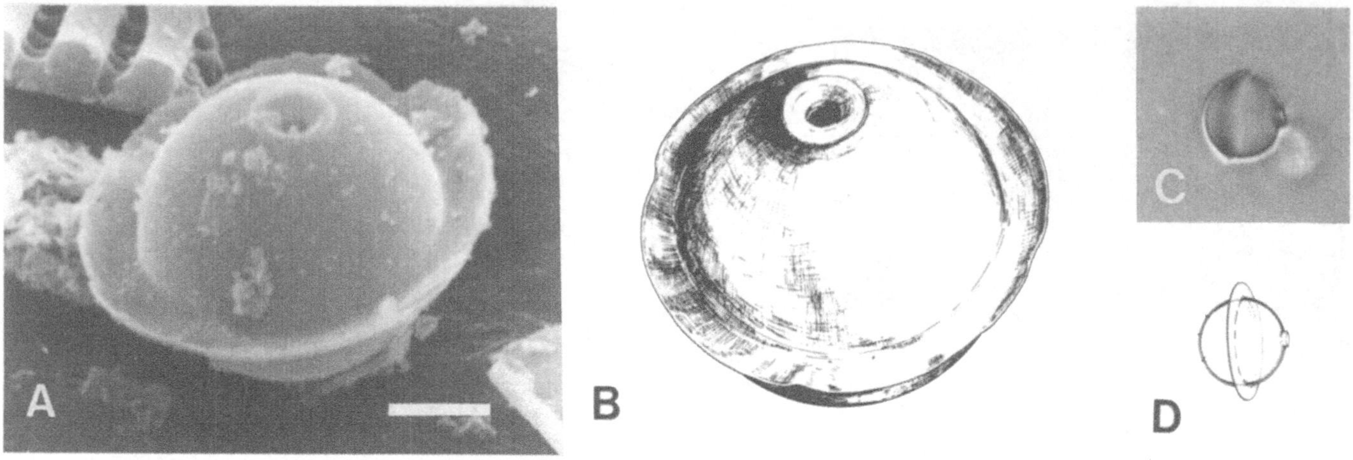

Fig. 115. Stomatocyst 174. A: SEM. B: Line drawing of SEM. C: LM. D: Line drawing of LM. Scale bar = 2 μm. LM x2000.

REFERENCES: This stomatocyst has been observed in British Columbia, Canada (Duff & Smol, 1994; Zeeb, 1994), and Minnesota, U.S.A. (Zeeb & Smol, 1993a).

ECOLOGY: Autecological data are not yet available for this rare morphotype.

Stomatocysts with one or more circuli

Stomatocyst 174, Zeeb & Smol, 1993a (8) Fig. 115

BIOLOGICAL AFFINITY: unknown

LOCALITY: Elk Lake (47°11'N; 95°13'W), Clearwater County, Minnesota, U.S.A. Sediment trap sample.

SEM DESCRIPTION: This is a small, spherical stomatocyst (diameter 4.8-6.3 μm). The collar is short and abruptly obconical with a sloping planar annulus, and a regular pore (collar diameter 1.2-2.1 μm; height 0.3-0.7 μm; pore diameter ca. 0.6 μm). This cyst possesses a single equatorial circulus with an anteriorly reflexed apex (circulus height 0.2-1.4 μm); on several specimens the circulus base has indistinct struts. The posterior hemisphere on some, but not all, SEM specimens has several small conula (diameter 0.7-1.0 μm; height 0.3-0.4 μm).

LM DESCRIPTION: This stomatocyst is recognized by its short, thick collar and equatorial circulus. If conula are present, they are visible with LM.

REFERENCES: This cyst has been observed in British Columbia, Canada (Duff & Smol, 1994; Zeeb, 1994), Minnesota, U.S.A. (Zeeb & Smol, 1993a), and Michigan, U.S.A. (Sandgren & Carney, 1983 - cyst 37). Stomatocyst 21 (Chapter 5) may be identical, but the collar location of this cyst is unknown.

ECOLOGY: Stomatocyst 174 has been found in alkaline lakes (Sandgren & Carney, 1983; Zeeb & Smol, 1993a). In a survey of lakes in British Columbia, Canada, it was most common in deep, clear, alkaline lakes with high conductivity values (Duff, 1994).

Stomatocyst 97, Duff & Smol, 1991 (8) Fig. 116

BIOLOGICAL AFFINITY: *Conradiella iserina* Ettl & Permann apparently produces an identical stomatocyst (Bourrelly, 1968).

LOCALITY: Upper Wallface Pond (44°08.8'N; 74°03.3'W), Adirondack Park, New York, U.S.A. Gravity core, core depth 20.0-20.5 cm.

SEM DESCRIPTION: This stomatocyst is spherical to slightly oval (diameter 5.1-8.9 μm) with a regular pore (diameter 0.4-0.6 μm). The ornamentation consists of five to eight parallel latitudinal circuli. The circuli are thinly silicified (height 0.1-0.5 μm; distance between circuli 0.9-1.3 μm) with curved apices; those in the anterior hemisphere curving towards the pore, and those in the posterior hemisphere curving towards the posterior pole. The ornamentation at the posterior pole is variable; either a small

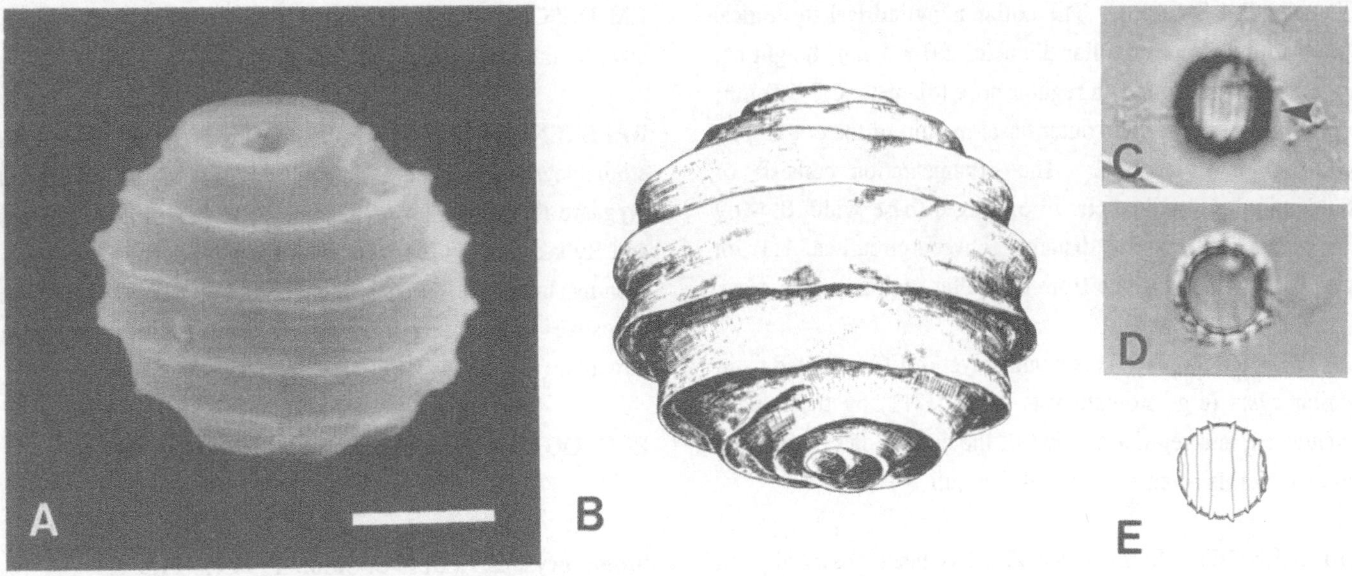

Fig. 116. Stomatocyst 97. A: SEM. Pore visible. B: Line drawing of SEM. Posterior pole visible. C: LM, surface view.Collar indicated by arrow. D: LM, cross-sectional view. E: Line drawing of LM. Scale bar = 2 μm. LM x2000.

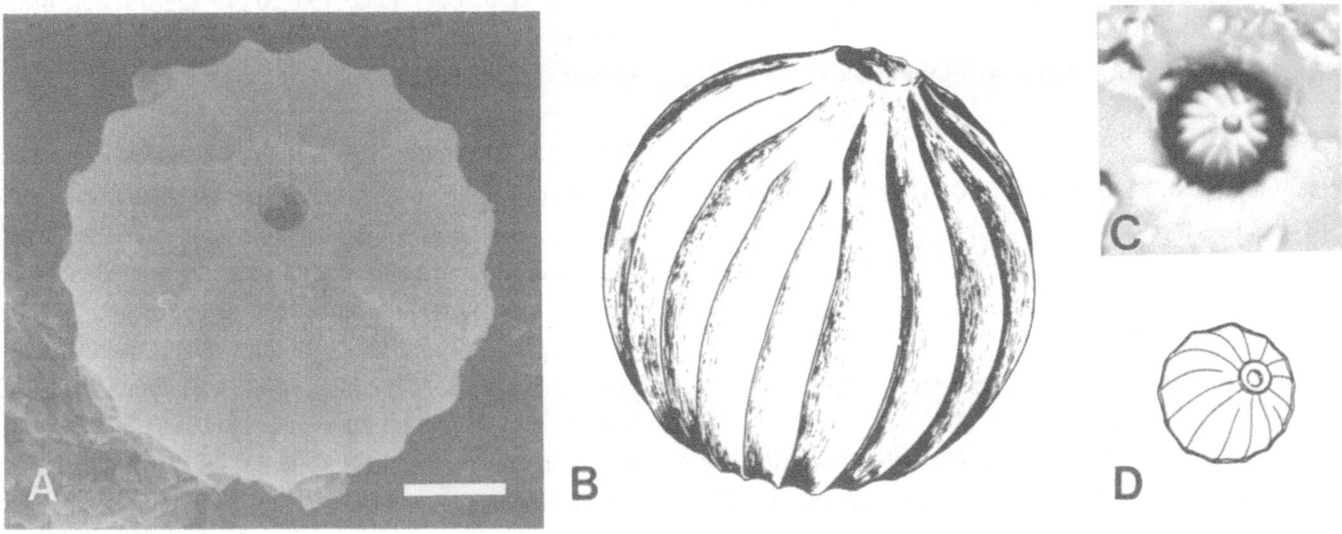

Fig. 117. Stomatocyst 228. A: SEM. B: Line drawing of SEM. C: LM. D: Line drawing of LM. Scale bar = 2 μm. LM x2000.

circulus is present, or a spiralling ridge.

LM DESCRIPTION: Stomatocyst 97 is easily recognizable by its surface ornamentation.

REFERENCES: This stomatocyst has been observed in the Yukon Territory, Canada (Brown & Smol, unpubl.), California, U.S.A. (Mahood & Adam, 1979 - type 13), and New York, U.S.A. (Duff & Smol, 1991, unpubl.).

ECOLOGY: Stomatocyst 97 has been reported from alpine or low arctic lakes (Mahood & Adam, 1979; Duff & Smol, 1991, unpubl.; Brown & Smol, unpubl.). In Duff & Smol's (1991) analysis, it appeared to be acidophilic but not acidobiontic.

Stomatocyst 228, Duff & Smol, 1994 (3) Fig. 117

BIOLOGICAL AFFINITY: unknown

LOCALITY: Wilgress Lake (49°10'N; 118°40'W), British Columbia, Canada. Ekman dredge, surface (1 cm) sediment sample.

SEM DESCRIPTION: This stomatocyst is spherical to oblate

(diameter 7.4-8.4 μm). The collar is cylindrical to conical with a rounded apex (collar diameter 2.0-3.4 μm; height ca. 0.5 μm), and surrounds a regular pore (diameter 0.7-1.0 μm; depth 0.2-0.3 μm). The outer basal margin of the collar may be irregular in outline. The ornamentation consists of equidistant longitudinal circuli or ridges (base width 0.4-0.9 μm; height 0.2-0.5 μm; distance between circuli ca. 1.1 μm at equator) which radiate from the collar region.

LM DESCRIPTION: This stomatocyst is differentiated from similar cysts (e.g. stomatocysts 93 and 111) by the length, orientation, and regular spacing of the circuli, and by the low collar which is often joined to the circuli.

REFERENCES: Stomatocyst 228 has been observed with SEM in British Columbia, Canada (Duff & Smol, 1994; Zeeb, 1994), and with LM in Minnesota, U.S.A. (Zeeb & Smol, 1993a - unidentified stomatocyst 8 (Chapter 5)), and the Northwest Territories, Canada (Duff & Smol, unpubl.).

ECOLOGY: Ecological details are as yet unknown.

Stomatocyst 240, Zeeb & Smol in Duff *et al.* (this volume) (3) Fig. 118

BIOLOGICAL AFFINITY: unknown

NEGATIVE NUMBER: J.P. Smol 645

LOCALITY: Lake 227 (49°42'N; 93°42'W), Experimental Lakes Area, Ontario, Canada. Freeze core, varve year A.D. 1989.

SEM DESCRIPTION: This stomatocyst is very large and spherical (diameter 21.4-25.8 μm). It has a long, cylindrical collar (diameter 3.3-4.0 μm; height 5.4-8.1 μm), the very base of which is gradual. The surface is ornamented with low, wavy ridges (length 1.7-4.5 μm; height 0.5-1.2 μm) with irregular apices, which may project into echinate spines (height ca. 1.2 μm). The ridges are densely scattered and randomly oriented. A high, latitudinal or tangential circulus is located at the posterior pole (diameter 13.6-20.5 μm; height 6.1-7.5 μm). The circulus is radially striated, and has an irregular, wavy apex. The cyst surface within the circulus is very smooth.

LM DESCRIPTION: This cyst is easily recognizable by its size, cylindrical collar, and ornamentation.

REFERENCES: No morphotypes exactly matching this stomatocyst have been found elsewhere. *Cysta excentrica* Nygaard (Nygaard, 1956 - Denmark; Rybak, 1986 - Poland) and Rybak *et al.* (1991 - Ontario, Canada) cyst 54 are nearly identical but possess spines, not ridges, and have regular rows of psilae ornamenting the cyst surface within the posterior circulus.

ECOLOGY: Autecological data are not yet available.

Stomatocyst 227, Duff & Smol, 1994 (7) Fig. 119

BIOLOGICAL AFFINITY: unknown

LOCALITY: Dahl Lake (53°50'N; 123°15'W), British Columbia, Canada. Ekman dredge, surface (1 cm) sediment sample.

SEM DESCRIPTION: This is a small, spherical stomatocyst (diameter 4.0-5.6 μm) with a narrow cylindrical collar (diameter 0.7 μm; height 0.2-0.6 μm); pore details are unknown. It is ornamented with several scattered tangential circuli, which may be somewhat asymmetric in outline (diameter 0.7-1.9 μm; height 0.1-0.5 μm; distance between circuli 1.7-2.5 μm). The interior of each circulus is plateau-like and may be raised above the cyst surface. On more mature specimens, the circuli walls curl inward to form nearly closed chambers (Fig. 119A).

LM DESCRIPTION: With LM, this stomatocyst is identified by the circuli. The collar is often not visible.

REFERENCES: We have observed this stomatocyst in British Columbia, Canada (Duff & Smol, 1994; Zeeb, 1994), the Yukon Territory, Canada (Brown & Smol, unpubl.), the Northwest Territories, Canada (Webb & Smol, unpubl.), and Minnesota, U.S.A. (Zeeb & Smol, unpubl.).

ECOLOGY: Autecological details are as yet unknown.

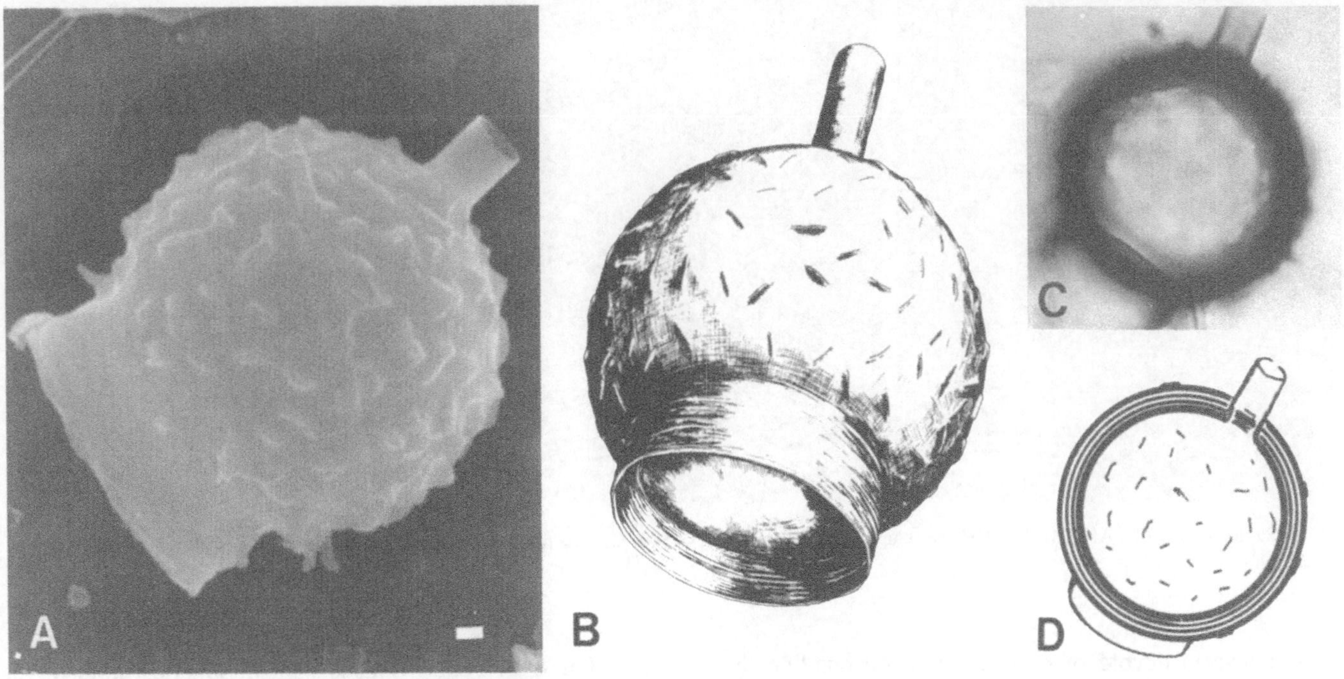

Fig. 118. Stomatocyst 240. A: SEM. B: Line drawing of SEM. C: LM. D: Line drawing of LM. Scale bar = 2 μm. LM x1500.

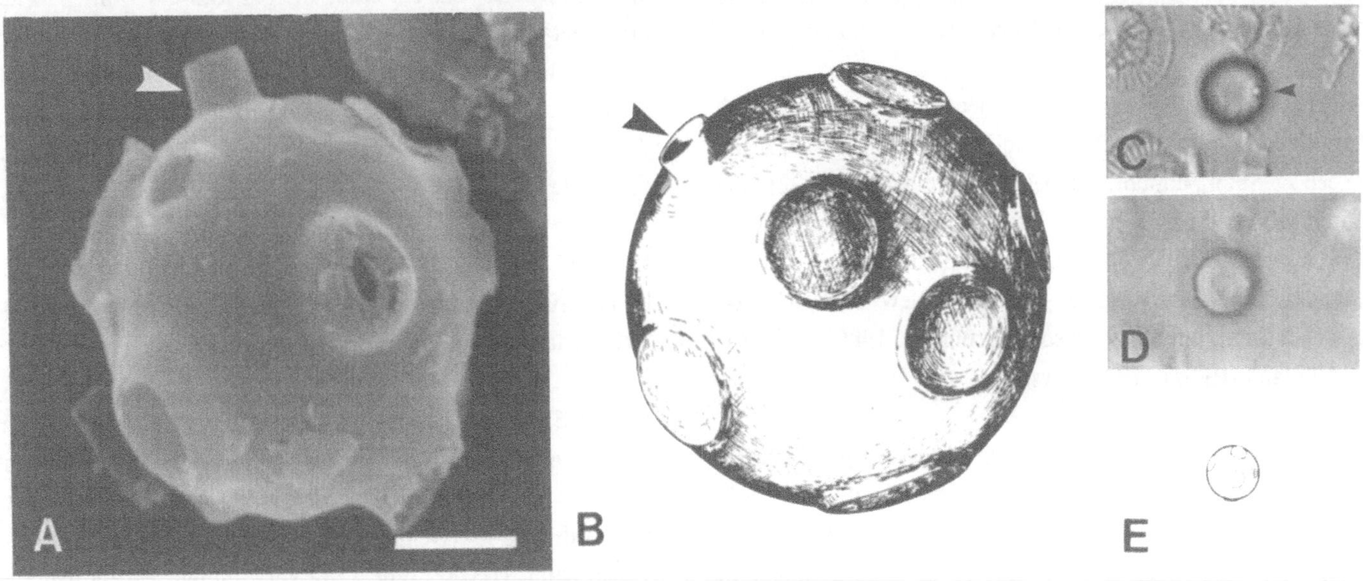

Fig. 119. Stomatocyst 227. A: SEM. Collar indicated by arrow. B: Line drawing of SEM. Collar indicated by arrow. C: LM, surface view. Collar indicated by arrow. D: LM, cross-sectional view. Circulus is visible. E: Line drawing of LM. Scale bar = 2 μm. LM x2000.

Stomatocysts with a reticulum

Stomatocyst 178, Zeeb & Smol, 1993a (33) Fig. 120

BIOLOGICAL AFFINITY: unknown

LOCALITY: Elk Lake (47°11'N; 95°13'W), Clearwater County, Minnesota, U.S.A. Sediment trap sample.

SEM DESCRIPTION: This is a spherical stomatocyst (diameter 4.5-6.3 μm). The collar is cylindrical with a slightly obconical apex and a conical pore (apical collar diameter 0.9-1.6 μm; height 0.3-1.0 μm; pore diameter 0.3-0.6 μm). The cyst surface is covered with a regular reticulum, consisting of rounded scabrae arranged in groups of six to form hexagonal lacunae (lacunae diameter 0.1-0.3 μm). The reticulum may extend approximately half-way up the collar, or alternatively, in less developed specimens, there

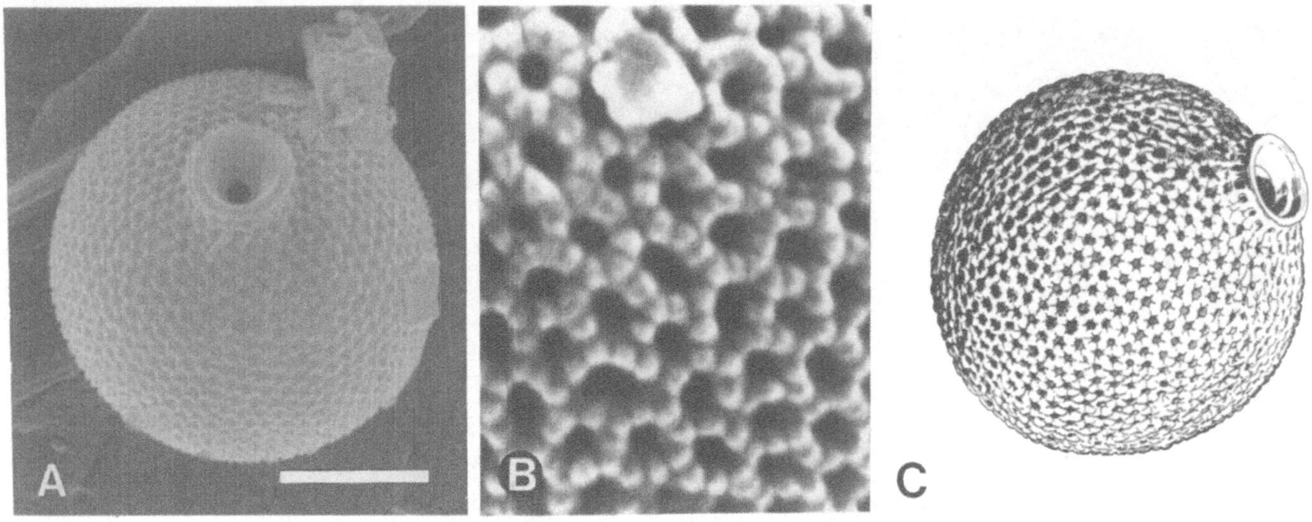

Fig. 120. Stomatocyst 178. A: SEM. B: SEM, close-up of ornamentation. x39,500. C: Line drawing of SEM. Scale bar = 2 μm. See Figs. 52C, D, and E for LM.

may be a margin devoid of ornamentation surrounding the collar (margin diameter 1.5-3.1 μm).

LM DESCRIPTION: This cyst is indistinguishable from cysts 117 (Figs. 52C, D, and E) and 210.

REFERENCES: This cyst has been observed in British Columbia, Canada (Duff & Smol, 1994); Ontario, Canada (Zeeb & Smol, unpubl.; Rybak *et al.*, 1991 - cyst 32); the Yukon Territory, Canada (Brown & Smol, unpubl.); Minnesota, U.S.A. (Zeeb & Smol, 1993a); Michigan, U.S.A. (Sandgren & Carney, 1983 - cyst 30); New York, U.S.A. (Duff & Smol, unpubl.); Sweden (Cronberg, 1986 - Fig. 24); and Poland (Rybak, 1987 - cyst 6).

ECOLOGY: This morph has been described as either alkaliphilic (Sandgren & Carney, 1983; Rybak, 1987; Zeeb & Smol, 1993a) or pH-indifferent (Rybak *et al.*, 1991). As a member of a collective category, it was most common in slightly acidic to circumneutral, productive lakes in Duff's (1994) survey.

Stomatocyst 231, Duff & Smol, 1994 (4) Fig. 121

BIOLOGICAL AFFINITY: unknown

LOCALITY: Champion Lake #3 (49°15'N; 117°30'W), British Columbia, Canada. Ekman dredge, surface (1 cm) sediment sample.

SEM DESCRIPTION: This is a small, spherical stomatocyst (diameter 4.6-4.8 μm), with a regular pore (diameter 0.3-0.6 μm) surrounded by a low, cylindrical collar with an obconical apex (diameter 1.1-1.3 μm; height 0.3-0.4 μm). The ornamentation consists of a regular reticulum with circular lacunae; the reticulum extends up the collar to the apical flare (lacuna diameter 0.4-0.5 μm; distance between lacunae ca. 0.5 μm). One SEM specimen appeared to have two echinate spines (length to 0.6 μm) in the posterior hemisphere.

LM DESCRIPTION: This stomatocyst is identified with LM by its reticulum and small collar. The lacunae of stomatocyst 231 are larger than those of stomatocyst 178.

REFERENCES: This cyst has been observed from British Columbia, Canada (Duff & Smol, 1994), and the Yukon Territory, Canada (Brown & Smol, unpubl.). *Cysta reticulata* Nygaard (1956 - Denmark) and VanLandingham (1964 - Washington, U.S.A.) type 3 are very similar, except that the reticulum lacunae are apparently polygonal rather than circular.

ECOLOGY: Stomatocyst 231 was most abundant in slightly acidic to circumneutral lakes with low conductivity values in British Columbia, Canada (Duff, 1994).

Stomatocyst 180, Zeeb & Smol, 1993a (8) Fig. 122

BIOLOGICAL AFFINITY: *Spiniferomonas bourrellyi* Takahashi (Skogstad & Reymond, 1989).

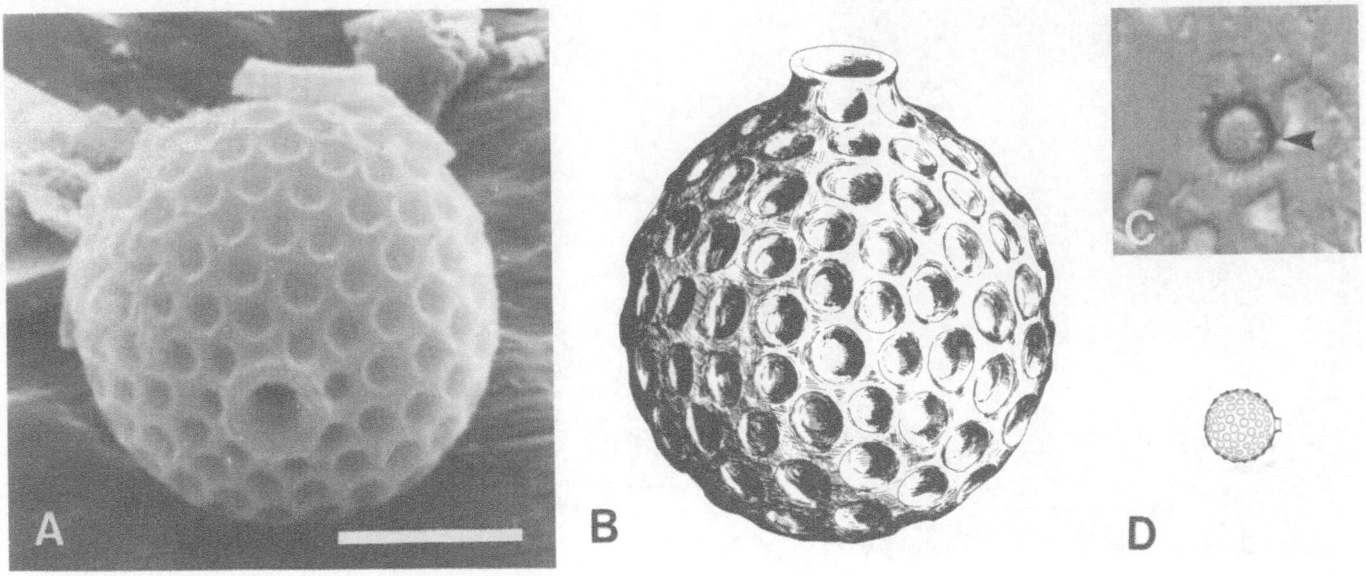

Fig. 121. Stomatocyst 231. A: SEM. B: Line drawing of SEM. C: LM. Collar indicated by arrow. D: Line drawing of LM. Scale bar = 2 μm. LM x2000.

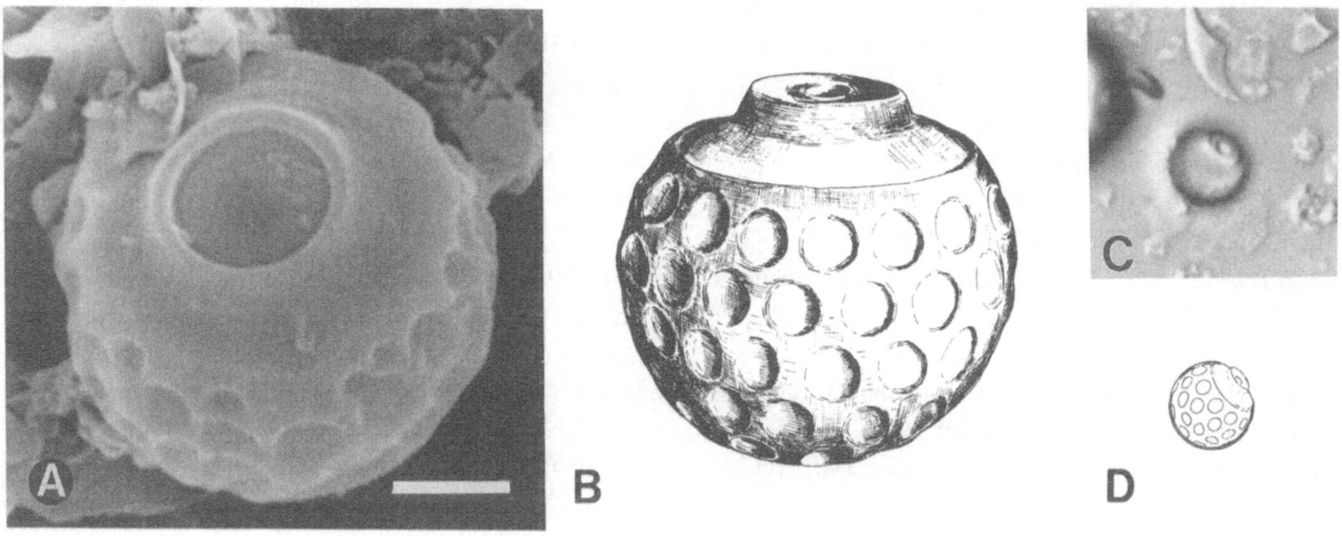

Fig. 122. Stomatocyst 180. A: SEM. Pore and primary collar covered by cap. B: Line drawing of SEM. Primary collar visible. C: LM. D: Line drawing of LM. Scale bar = 2 μm. LM x2000.

LOCALITY: Elk Lake (47°11'N; 95°13'W), Clearwater County, Minnesota, U.S.A. Sediment trap sample.

SEM DESCRIPTION: This is an obovate stomatocyst (6.2-6.9 x 5.4-6.4 μm). The complex collar consists of a conical pore (inner diameter 0.5-0.7 μm; depth 0.3-0.4 μm) set in a planar annulus (diameter 2.4-2.5 μm) located closer to the primary collar apex than the base. The primary collar is conical with an acute to rounded apex (diameter 2.6-3.5 μm; height 0.3-0.9 μm). An unornamented interannulus separates the primary collar from the low secondary collar (diameter 5.0-6.0 μm). A very low, regular reticulum with equidistant circular lacunae ornaments the wall below the secondary collar (lacuna diameter 0.5-1.7 μm).

LM DESCRIPTION: This cyst is easily identified with LM. Specimens with an underdeveloped reticulum can be recognized by the wide collar, deep pore and obovate body shape. Immature specimens also frequently appear thin-walled and faint with LM.

REFERENCES: This stomatocyst has been observed in Ontario, Canada (Rybak *et al.*, 1987 - cyst 29; Rybak *et al.*, 1991 - cyst 19; Zeeb & Smol, unpubl.); British Columbia,

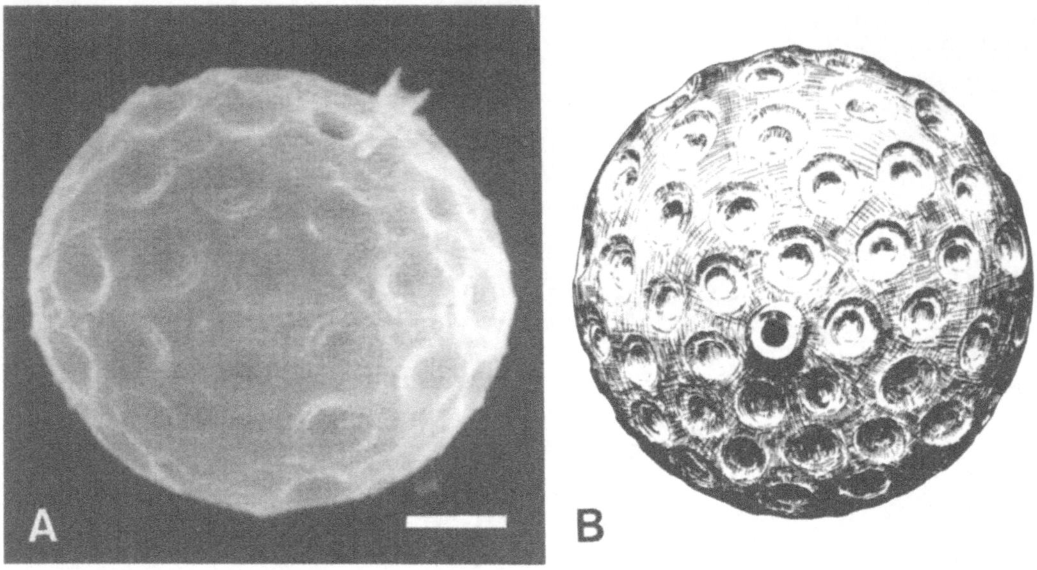

Fig. 123. Stomatocyst 99. A: SEM. B: Line drawing of SEM. Scale bar = 2 μm.

Canada (Duff & Smol, 1994; Zeeb, 1994); the Yukon Territory, Canada (Brown & Smol, unpubl.); New York, U.S.A. (Duff & Smol, unpubl.); Minnesota, U.S.A. (Zeeb & Smol, 1993a); Michigan, U.S.A. (Sandgren & Carney, 1983 - cyst 29); and California, U.S.A. (Adam & Mahood, 1980b - type 177).

ECOLOGY: *Spiniferomonas bourrellyi* is considered a cool-water, acidophilic species (Hällfors & Hällfors, 1988; Kristiansen, 1988; Kristiansen & Tong, 1989; Siver, 1989a; Siver & Hamer, 1990). The cyst was classified as acidophilic by Rybak *et al.* (1991); however, it has also been collected from alkaline lakes (Sandgren & Carney, 1983; Zeeb & Smol, 1993a; Duff & Smol, 1994), and its distribution was uncorrelated with pH in Duff's (1994) study. It was most common in deep, oligotrophic, montane lakes (Duff, 1994).

Stomatocyst 99, Duff & Smol, 1991 (2) Fig. 123

BIOLOGICAL AFFINITY: unknown

LOCALITY: Upper Wallface Pond (44°08.8'N; 74°03.3'W), Adirondack Park, New York, U.S.A. Gravity core, core depth 20.0-20.5 cm.

SEM DESCRIPTION: This stomatocyst is spherical (diameter ca. 9.3 μm). The collar is small and cylindrical with a rounded apex and a regular pore (collar diameter ca. 1.3 μm; pore diameter ca. 0.5 μm). The cyst surface is ornamented with a low, wide, regular reticulum with circular lacunae (lacuna diameter 0.8-1.9 μm; interlacunal distance 2.0-2.8 μm). Within each lacuna is a spherical depression (diameter 0.6-0.8 μm). Small verrucae, possibly undeveloped spines, were observed on the reticulum interstices on one specimen.

LM DESCRIPTION: Stomatocyst 99 has not been observed definitely with LM, and was grouped with stomatocyst 100 for counting purposes by Duff & Smol (1991). However, we believe this cyst could be identified with LM by the regular reticulum.

REFERENCES: This stomatocyst has been observed in Ontario, Canada (Rybak *et al.*, 1991 - cyst 87), New York, U.S.A. (Duff & Smol, 1991, unpubl.), and California, U.S.A. (Adam & Mahood, 1980a - cyst 140).

ECOLOGY: This stomatocyst may be produced by an oligotrophic species (Duff & Smol, 1991; Rybak *et al.*, 1991). It has been observed over a fairly wide range of pH (4.6 to ca. 7.0), and was classified as pH-indifferent by Rybak *et al.* (1991).

Stomatocyst 177, Zeeb & Smol, 1993a (6) Fig. 124

BIOLOGICAL AFFINITY: *Mallomonas caudata* Ivanoff *emend.* Krieger (Cronberg, 1988).

LOCALITY: Elk Lake (47°11'N; 95°13'W), Clearwater County, Minnesota, U.S.A. Sediment trap sample.

SEM DESCRIPTION: This stomatocyst is very large and spherical (diameter 24.3-28.9 μm; $\bar{x}$ = 27.0). It has a conical pore (inner diameter 1.3-1.6 μm; outer diameter 1.6-2.3 μm)

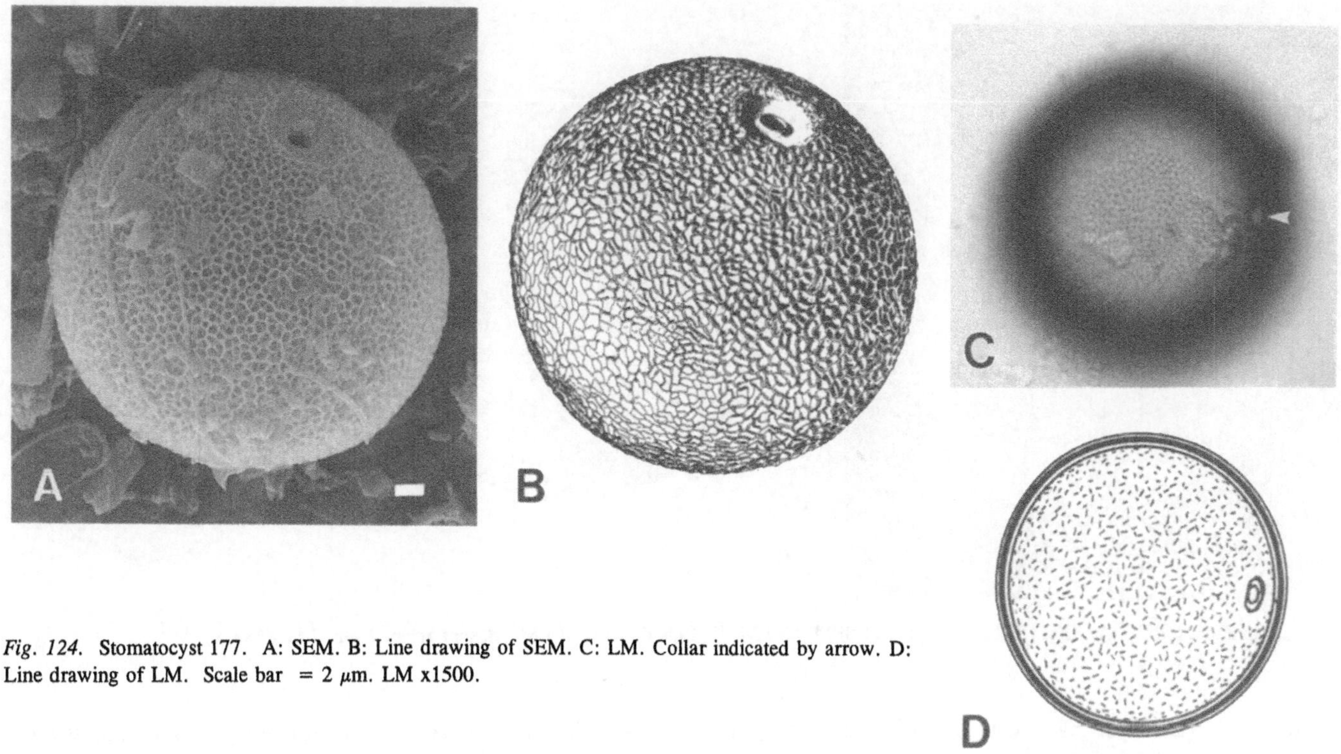

Fig. 124. Stomatocyst 177. A: SEM. B: Line drawing of SEM. C: LM. Collar indicated by arrow. D: Line drawing of LM. Scale bar = 2 μm. LM x1500.

which may or may not be surrounded by a circular, unornamented portion of the cyst wall (diameter 4.2-7.7 μm). The remainder of the cyst body is ornamented with a variable reticulum with small, very roughly circular lacunae.

LM DESCRIPTION: This stomatocyst can be recognized by its large size and multiple cyst walls, and by its surface ornamentation. It can be distinguished from stomatocyst 229 on the basis of size.

REFERENCES: This stomatocyst has been observed in British Columbia, Canada (Duff & Smol, 1994), Ontario, Canada (Rybak *et al.*, 1991 - cyst 132), Minnesota, U.S.A. (Zeeb & Smol, 1993a), and Sweden (Cronberg, 1988).

ECOLOGY: *Mallomonas caudata* is a cosmopolitan taxon distributed across a wide range of environmental gradients (e.g. temperature, pH, specific conductance, total phosphorus), although it is more common in slightly alkaline water (Siver, 1991a). The cyst has been found in slightly acidic to alkaline lakes (Rybak *et al.*, 1991; Zeeb & Smol, 1993a; Duff, 1994). It displayed a positive, but non-significant, correlation with lake size and total phosphorus in Duff's (1994) survey.

Stomatocyst 229, Duff & Smol, 1994 (5) Fig. 125

BIOLOGICAL AFFINITY: unknown

LOCALITY: Florence Lake (48°30'N; 123°30'W), British Columbia, Canada. Ekman dredge, surface (1 cm) sediment sample.

SEM DESCRIPTION: This stomatocyst is spherical (diameter 12.8-17.7 μm; $\bar{x}$ = 15.0 μm). The ornamentation is identical to that of stomatocyst 177, except that the pore (diameter 1.3-1.5 μm) is larger relative to cyst size and may be surrounded by a very low, rounded conical collar. The pore is set asymmetrically in a smooth or striated, roughly circular margin which may possess a swollen outer rim (diameter 4.3-5.7 μm).

LM DESCRIPTION: This cyst is recognizable by its reticulum, irregular pore margin, and large pore. It is smaller than stomatocyst 177.

REFERENCES: This stomatocyst has been found in British Columbia, Canada (Duff & Smol, 1994), Ontario, Canada (Rybak *et al.*, 1987 - cyst 30; Rybak *et al.*, 1991 - cyst 133), and New York, U.S.A. (Duff & Smol, unpubl.).

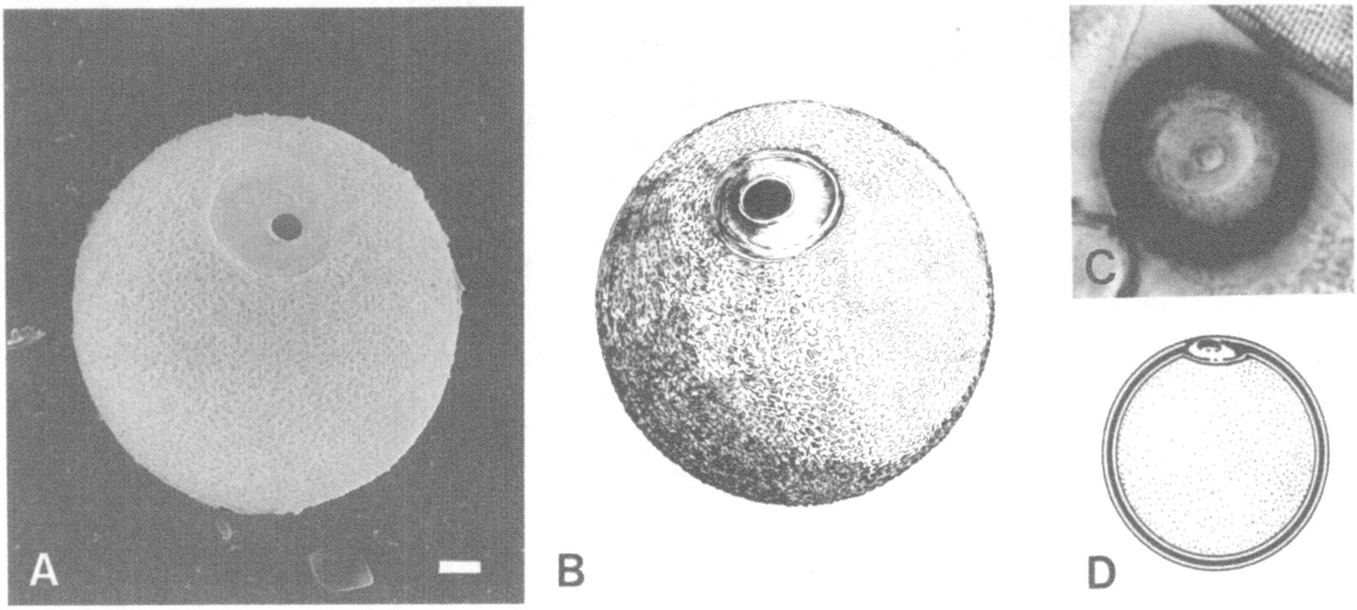

Fig. 125. Stomatocyst 229. A: SEM. B: Line drawing of SEM. C: LM. D: Line drawing of LM. Scale bar = 2 μm. LM x2000.

ECOLOGY: This cyst has been classified as eutrophic (Rybak *et al.*, 1987).

Stomatocyst 241, Duff *et al.* (this volume) (4) Fig. 126

BIOLOGICAL AFFINITY: unknown

NEGATIVE NUMBER: J.P. Smol 2998

LOCALITY: "Horseshoe Pond" (unofficial name), Cape Herschel (78°37'N; 74°42'W), Ellesmere Island, Northwest Territories, Canada. Moss sample, collected July 19, 1986.

SEM DESCRIPTION: This stomatocyst is spherical (diameter 7.2-8.5 μm). The collar is cylindrical with a widely rounded apex and a sloping inner margin (basal diameter 3.2-5.0 μm; collar diameter : cyst diameter ratio 0.44-0.59). The outer margin may be striated. There is a notch-like indentation in the collar (Fig. 126B), similar to that of stomatocyst 158 (Fig. 60A). On all our SEM specimens, the pore is obscured by a roughly spherical, convex siliceous plug (diameter 1.2-2.2 μm). The cyst surface ranges from very slightly verrucose to an irregular low reticulum with circular lacunae of variable diameter (0.1-0.5 μm) or polygonal lacunae (honeycomb pattern).

Stomatocyst 241 resembles stomatocysts 56 and 112. However, both these morphs are considerably larger and lack verrucose or reticular ornamentation.

LM DESCRIPTION: This stomatocyst is distinguishable from stomatocyst 56 using LM by its smaller size, relatively larger collar, and by the presence of ornamentation on stomatocyst 241.

REFERENCES: This stomatocyst has been observed in the Northwest Territories, Canada (Duff *et al.*, this volume; Webb & Smol, unpubl.), and Minnesota, U.S.A. (Zeeb & Smol, 1993a - misidentified as stomatocyst 56). Adam & Mahood (1979b - California, U.S.A.) type 85 has identical ornamentation, but is more similar in size to stomatocyst 56. The cyst identified by Rybak (1986 - Poland) as *cysta crassicollis* Nygaard (≠ Nygaard's (1956) specimens) is similar to stomatocyst 241, but is larger with a more acute collar apex.

ECOLOGY: Autecological data are not yet available.

Stomatocyst 6, Duff & Smol, 1988 (5) Fig. 127

BIOLOGICAL AFFINITY: unknown

LOCALITY: "Rock Basin Lake" (unofficial name) (78°29.5'N; 76°44.8'W), Ellesmere Island, Northwest Territories, Canada. Piston core, core depth 0-2 cm.

SEM DESCRIPTION: This is a spherical to widely oval stomatocyst (diameter 5.0-6.9 μm) with a complex collar.

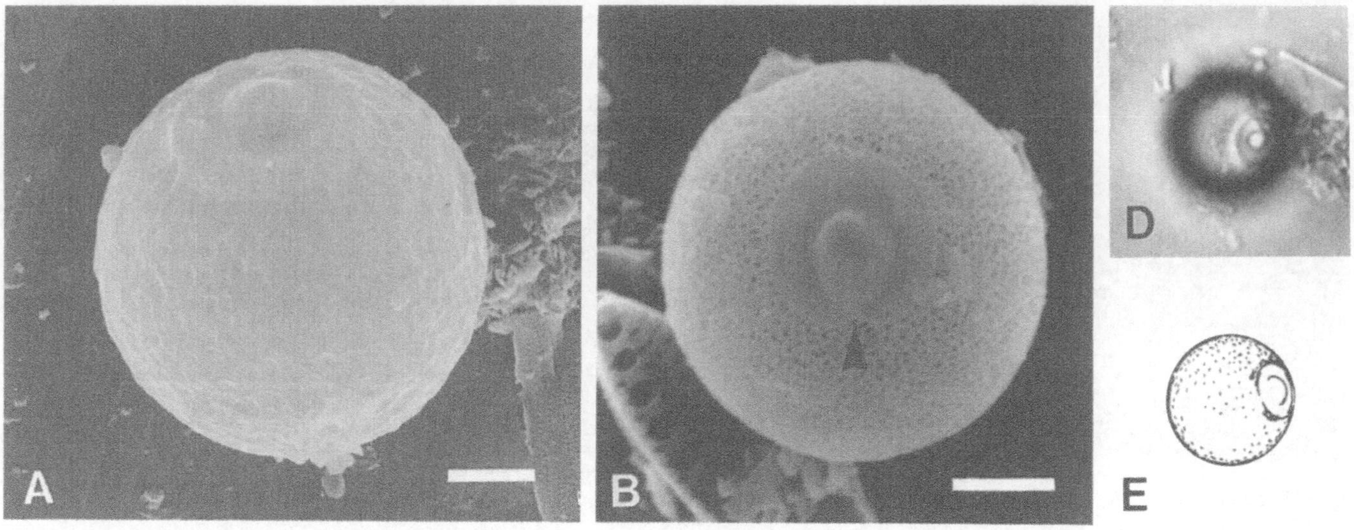

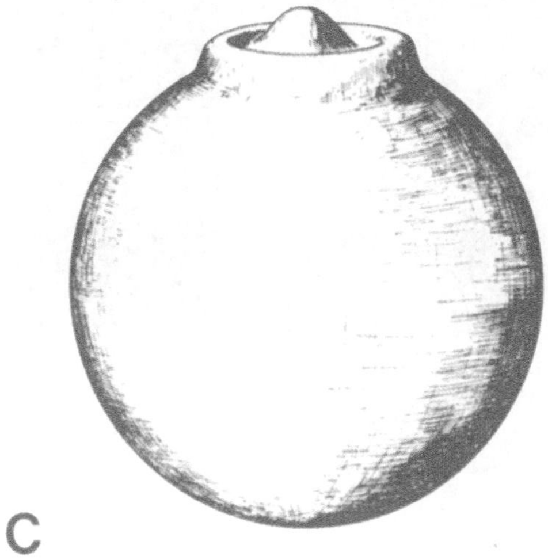

Fig. 126. Stomatocyst 241. A: SEM. B: SEM, apical view. Notch-like indentation in collar indicated by arrow. C: Line drawing of SEM, unornamented specimen. D: LM. E: Line drawing of LM. Scale bars = 2 μm. LM x2000.

The regular pore (diameter 0.6-0.8 μm) is usually surrounded by a low, acutely conical primary collar; the primary collar may be absent on undeveloped or degraded specimens. The secondary collar is also conical and identical in height to the primary collar (secondary collar diameter 0.9-1.6 μm; height 0.3-0.4 μm). The secondary collar is formed from one of the reticular lacunae. The surface is covered by a low, variable reticulum with fairly large, polygonal (usually pentagonal or hexagonal) lacunae (lacuna diameter 0.7-3.8). The reticular ridges are apically acute, and connect with the secondary collar.

LM DESCRIPTION: Stomatocyst 6 can be identified with LM by its reticulum. It is very similar to stomatocyst 86. The latter has an unusual complex collar, but if a good collar view is not obtained, stomatocyst 86 could be misidentified as stomatocyst 6. Stomatocyst 6 is also similar to stomatocyst 143, but is smaller; the reticular lacunae of stomatocyst 6 are usually larger than those of stomatocyst 143, and stomatocyst 6 never has circular lacunae or spines.

REFERENCES: Stomatocyst 6 has been observed in the Northwest Territories, Canada (Duff & Smol, 1988, 1989; Duff *et al.*, 1992), Ontario, Canada (Rybak *et al.*, 1991 - cyst 130), British Columbia, Canada (Zeeb, 1994), New York, U.S.A. (Duff & Smol, unpubl.), and Denmark (Nygaard, 1956 - *cysta areolata* Nygaard).

ECOLOGY: This cyst is likely produced by an acidophilic species (Nygaard, 1956; Duff & Smol, 1988, 1989; Rybak *et al.*, 1991), although it has been found in low abundances in circumneutral to alkaline arctic ponds (Duff *et al.*, 1992).

Stomatocyst 242, Duff & Smol in Duff *et al.* (this volume) (4) Fig. 128

BIOLOGICAL AFFINITY: unknown

NEGATIVE NUMBER: J.P. Smol 2994

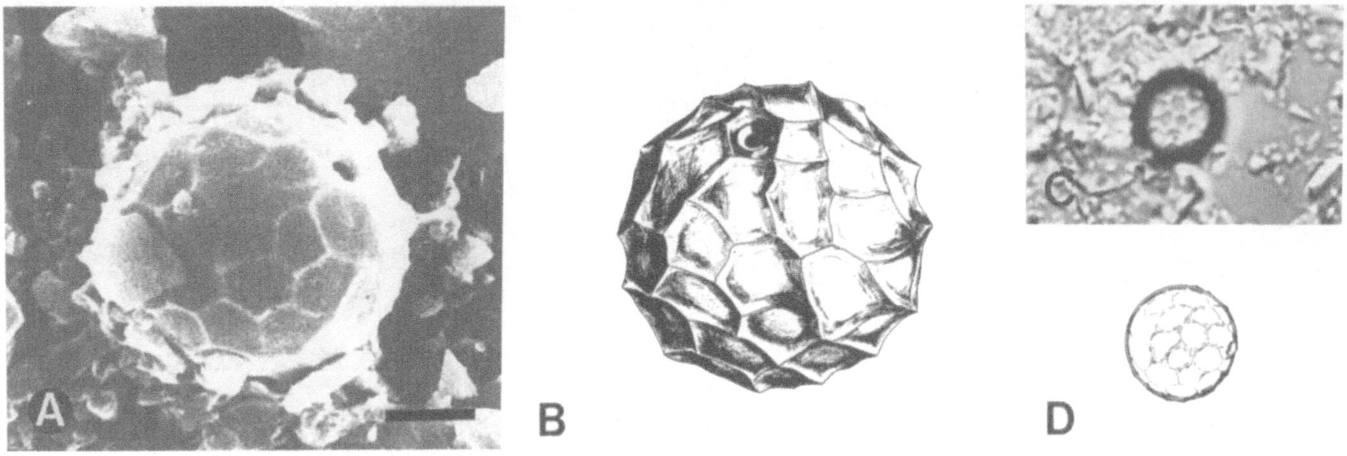

Fig. 127. Stomatocyst 6. A: SEM. B: Line drawing of SEM. C: LM. D: Line drawing of LM. Scale bar = 2 μm. LM x2000.

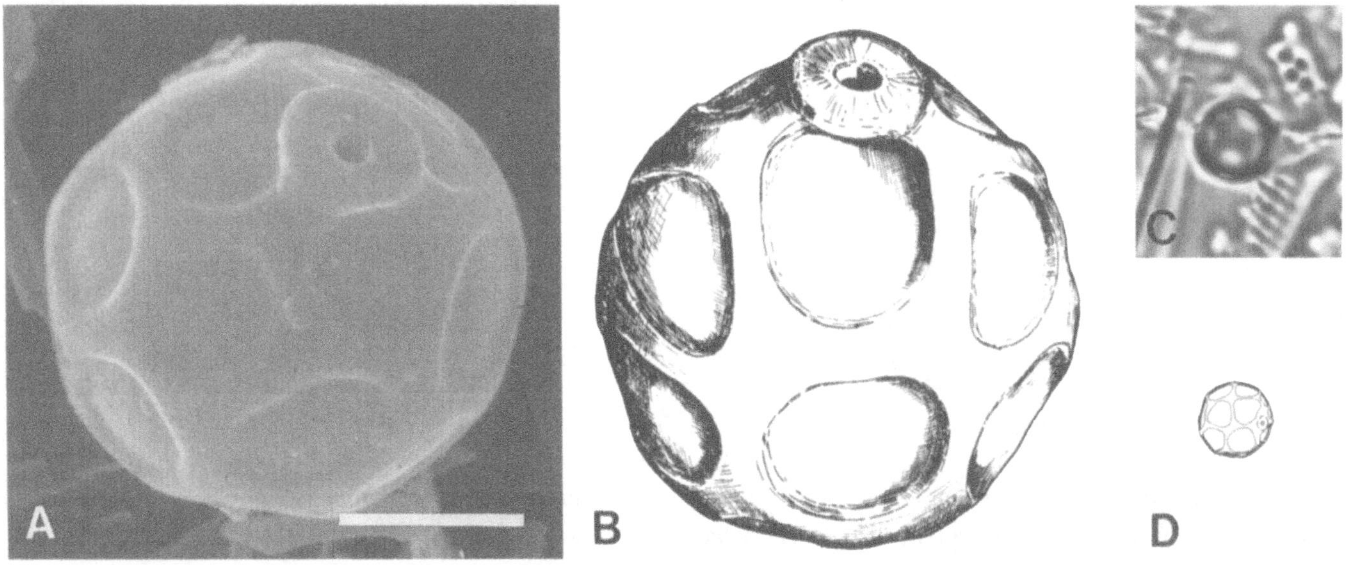

Fig. 128. Stomatocyst 242. A: SEM. B: Line drawing of SEM. C: LM. Collar not visible. D: Line drawing of LM. Scale bar = 2 μm. LM x2000.

LOCALITY: "Unnamed Pond" (unofficial name) (43°11'N; 74°37'W), Adirondack Park, New York, U.S.A. Gravity core, surface (0-0.25 cm) sediment sample.

SEM DESCRIPTION: This is a small, spherical stomatocyst (diameter 4.5-5.8 μm), ornamented with a very low reticulum with relatively large, roughly circular lacunae (lacuna diameter 0.9-2.3 μm). The reticulum ridges are flattened or rounded and are quite wide (ridge width 0.4-0.6 μm). The collar is low, wide, and cylindrical with a flattened apex (diameter 1.3-1.6 μm; height ca. 0.3 μm), and is connected to the reticulum at the base of its outer margin. It surrounds a regular pore (diameter 0.4 μm).

LM DESCRIPTION: This cyst is recognized by its size and the large lacunae of the reticulum, which give the cyst body a polygonal appearance.

REFERENCES: This stomatocyst has been observed in British Columbia, Canada (Zeeb, 1994), and New York, U.S.A. (Duff *et al.*, this volume).

ECOLOGY: The ecological optima and tolerances of stomatocyst 242 have not yet been determined.

Stomatocyst 179, Zeeb & Smol, 1993a (2) Fig. 129

BIOLOGICAL AFFINITY: This is probably the stomatocyst of *Chrysidiastrum catenatum* Lauterborn (Sandgren, 1983b).

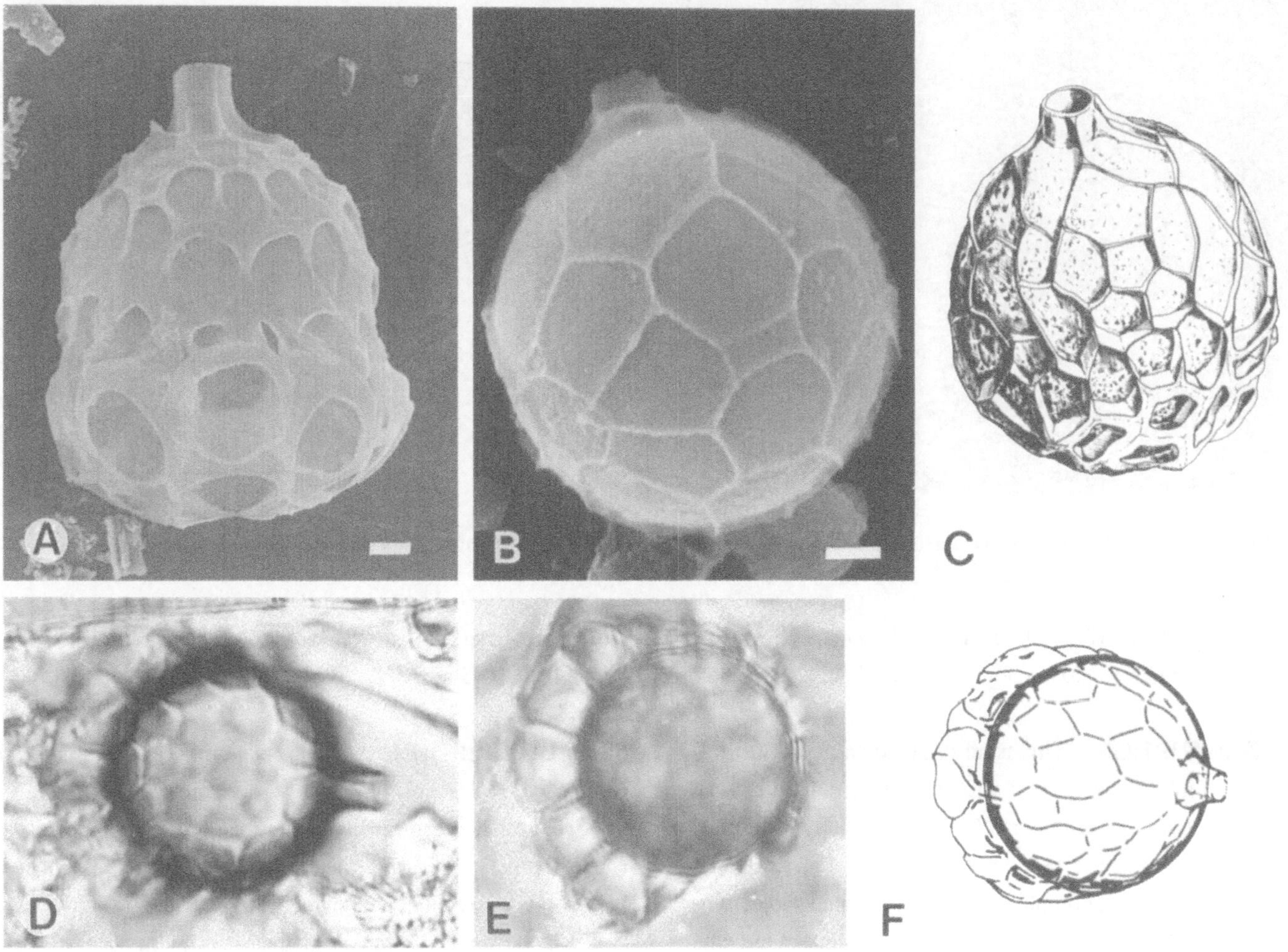

Fig. 129. Stomatocyst 179. A: SEM, mature specimen. B: SEM, immature specimen. C: Line drawing of SEM. D: LM, surface view. E: LM, cross-sectional view. Collar not visible. F: Line drawing of LM. Scale bars = 2 μm. LM x2000.

LOCALITY: Elk Lake (47°11'N; 95°13'W), Clearwater County, Minnesota, U.S.A. Sediment trap sample.

SEM DESCRIPTION: This is a large, spherical stomatocyst (diameter 12.1-16.8 μm). The collar is long, conical or cylindrical, and may be bent slightly (diameter 1.7-3.8 μm; height 1.7-4.2 μm). The cyst surface is covered by a wide reticulum with roughly polygonal lacunae (lacunae diameter 1.2-5.0 μm). On more developed specimens, the reticulum ridges are thicker and higher (≥ 1.3 μm), and may fold over in the posterior hemisphere to form nearly closed chambers (Fig. 129A). The reticulum may extend part way up the collar on mature specimens.

LM DESCRIPTION: This stomatocyst is easily recognizable by its size and ornamentation.

REFERENCES: This cyst has been found in Ontario, Canada (Rybak *et al.*, 1987 - cysts 24, 31 and 32; Smol, 1988a - Plate 1F; Rybak *et al.*, 1991 - cyst 131); British Columbia, Canada (Duff & Smol, 1994; Zeeb, 1994); Minnesota, U.S.A. (Zeeb & Smol, 1993a); Washington, U.S.A. (Smith & White, 1985 - fig. 16); California, U.S.A. (Adam & Mahood, 1980a - type 146; Carney *et al.*, 1992 - cyst 41); Denmark (*cysta polyedra* Nygaard (1956)); and Poland (Rybak, 1986 - *cysta polyedra*, *cysta arcuatus* Rybak).

ECOLOGY: This is a circumneutral to alkaliphilic taxon (Rybak *et al.*, 1991; Carney *et al.*, 1992; Duff, 1994), likely with a low tolerance to eutrophic water (Duff, 1994).

Stomatocyst 144, Duff & Smol in Duff *et al.*, 1992 (4) Fig. 130

BIOLOGICAL AFFINITY: unknown

LOCALITY: "Col Pond" (unofficial name), Cape Herschel

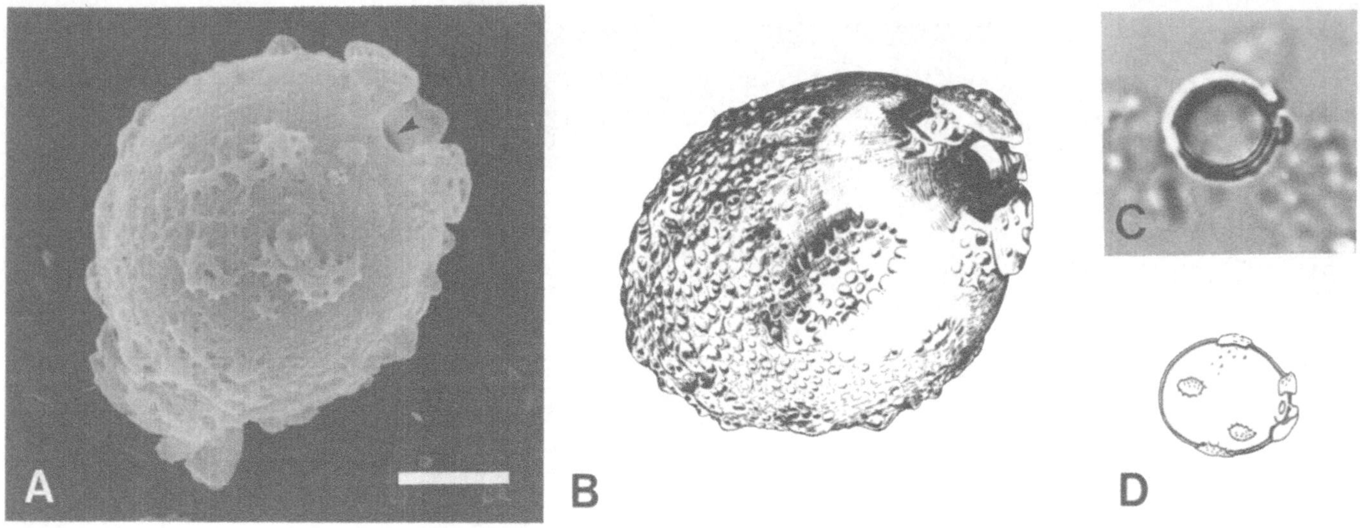

Fig. 130. Stomatocyst 144. A: SEM. Pore indicated by arrow. B: Line drawing of SEM. C: LM. D: Line drawing of LM. Scale bar = 2 μm. LM x2000.

(78°37'N; 74°42'W), Ellesmere Island, Northwest Territories, Canada. Moss sample, collected July 29, 1986.

SEM DESCRIPTION: This is an oval stomatocyst (5.3-6.0 x 6.0-7.2 μm). The surface ornamentation consists of an irregular reticulum with small, approximately spherical chambers. The pore is regular (diameter 0.7-1.1 μm), and surrounded by several complex, irregularly shaped projections of the surface reticulum, up to 1.5 μm in height. On some specimens, the surface reticulum is present as approximately oval patches, oriented longitudinally; on others, the reticulum forms widely spaced surface projections, similar to those surrounding the pore.

LM DESCRIPTION: This stomatocyst is distinguished from similar cysts (e.g. stomatocyst 136 and 137) by the presence of surface texture, although the individual ridges are usually not discernible, and by the irregular shape of its collar.

REFERENCES: Stomatocyst 144 has apparently not been observed elsewhere.

ECOLOGY: No autecological data are yet available for this rare morphotype.

Stomatocyst 101, Duff & Smol, 1991 (13) Fig. 131

BIOLOGICAL AFFINITY: unknown

LOCALITY: Upper Wallface Pond (44°08.8'N; 74°03.3'W), Adirondack Park, New York, U.S.A. Gravity core, core depth 10.0-10.5 cm.

SEM DESCRIPTION: This is a spherical stomatocyst (diameter 3.4-4.3 μm) with a low, obconical collar (diameter 0.9-1.6 μm; height 0.2-0.6 μm) surrounding a conical pore (inner diameter 0.3 μm). The entire cyst surface is covered by a highly irregular and overlapping reticulum with wavy ridges and roughly circular lacunae (reticulum ridge width <0.1-0.3 μm; lacunae diameter 0.4-0.8 μm). Fig. 131B shows an immature specimen of stomatocyst 101.

LM DESCRIPTION: Careful attention must be paid to differentiate stomatocyst 101 from similar cysts with LM. It is distinguished from stomatocyst 34 by its very small size and less obconical collar. It differs from stomatocysts 89 and 232 in that its ornamentation covers the entire cyst surface, and is not differentiated in the anterior and posterior hemispheres. Nevertheless, we recommend SEM analysis to confirm LM identification of these small cysts.

REFERENCES: This stomatocyst has been found in Ontario, Canada (Rybak *et al.*, 1991 - cysts 96 and 97); the Northwest Territories, Canada (Duff & Smol, 1988 - Fig 13, misidentified as stomatocyst 7); the Yukon Territory, Canada (Brown & Smol, unpubl.); New York, U.S.A. (Duff & Smol, 1991, unpubl.); and California, U.S.A. (Adam & Mahood, 1979b - type 61).

ECOLOGY: Stomatocyst 101 has been found in oligotrophic locations (Duff & Smol, 1988, 1991; Rybak *et al.*, 1991). It

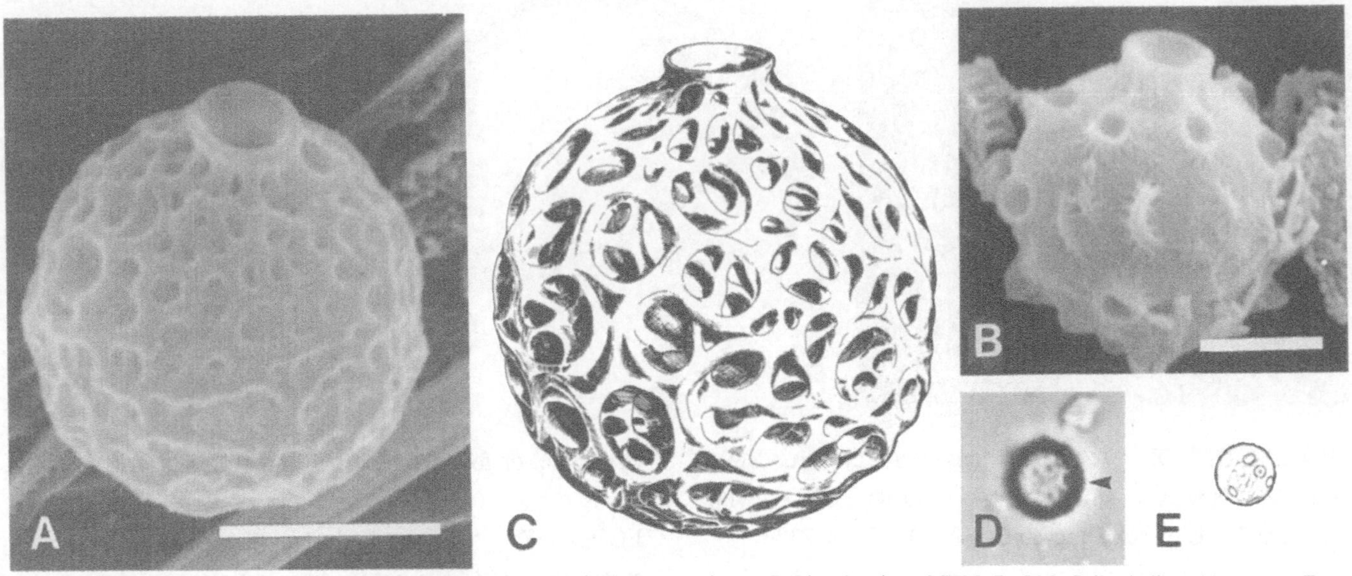

Fig. 131. Stomatocyst 101. A: SEM, mature specimen. B: SEM, immature specimen. C: Line drawing of SEM. D: LM. Collar indicated by arrow. E: Line drawing of LM. Scale bars = 2 μm. LM x2500.

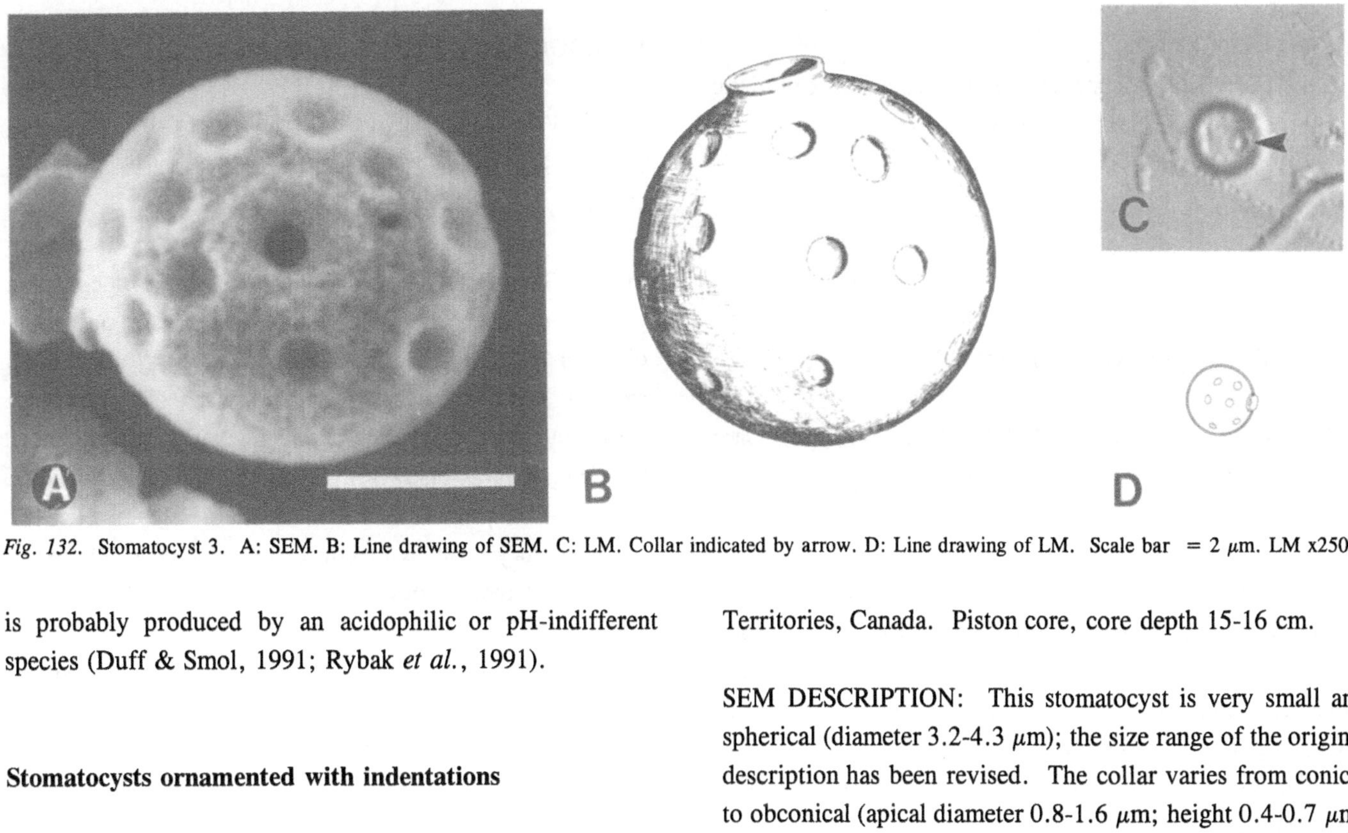

Fig. 132. Stomatocyst 3. A: SEM. B: Line drawing of SEM. C: LM. Collar indicated by arrow. D: Line drawing of LM. Scale bar = 2 μm. LM x2500.

is probably produced by an acidophilic or pH-indifferent species (Duff & Smol, 1991; Rybak *et al.*, 1991).

Stomatocysts ornamented with indentations

Stomatocyst 3, Duff & Smol, 1988 (8) Fig. 132

BIOLOGICAL AFFINITY: unknown

LOCALITY: "Rock Basin Lake" (unofficial name) (78°29.5'N; 76°44.8'W), Ellesmere Island, Northwest Territories, Canada. Piston core, core depth 15-16 cm.

SEM DESCRIPTION: This stomatocyst is very small and spherical (diameter 3.2-4.3 μm); the size range of the original description has been revised. The collar varies from conical to obconical (apical diameter 0.8-1.6 μm; height 0.4-0.7 μm) and surrounds a regular pore (diameter 0.3-0.5 μm). The ornamentation consists of shallow, circular depressions (diameter 0.3-0.7 μm), usually concentrated in the anterior hemisphere, but occasionally located also in the posterior hemisphere.

It is possible that this stomatocyst is the immature form of another cyst (e.g. stomatocyst 5, 88, 89, 101 or 232).

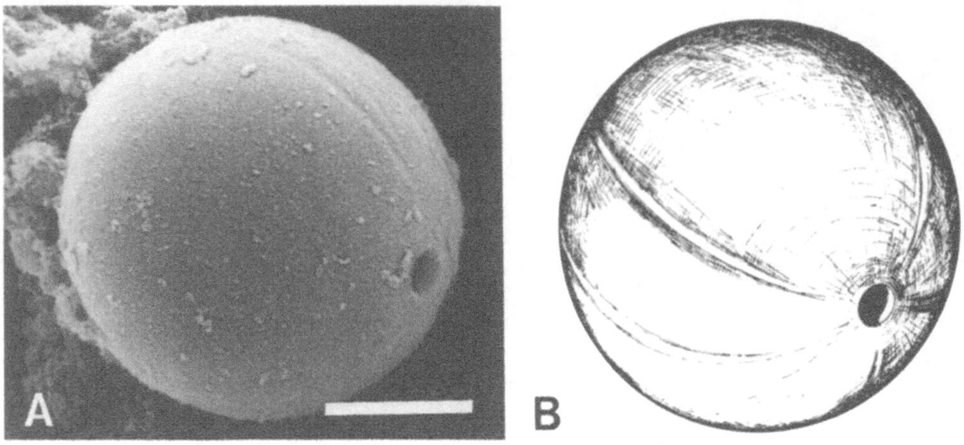

Fig. 133. Stomatocyst 148. A: SEM. B: Line drawing of SEM. Scale bar = 2 μm. See Fig. 6B for LM.

LM DESCRIPTION: The depressions on this cyst are visible with LM. However, stomatocyst 3 could be confused with immature forms of stomatocysts 5, 88, 89, 101 or 232.

REFERENCES: This stomatocyst has been observed in the Northwest Territories, Canada (Duff & Smol, 1988, 1989; Duff *et al.*, 1992), and New York, U.S.A. (Duff & Smol, unpubl.).

ECOLOGY: This stomatocyst may be characteristic of arctic or alpine sites (Duff & Smol, 1988, 1989, unpubl.; Duff *et al.*, 1992).

Stomatocyst 148, Zeeb & Smol, 1993a (2) Fig. 133

BIOLOGICAL AFFINITY: unknown

LOCALITY: Elk Lake (47°11'N; 95°13'W), Clearwater County, Minnesota, U.S.A. Piston core, varve year A.D. 1864.

SEM DESCRIPTION: This stomatocyst is spherical with a regular pore (cyst diameter 4.9-5.4 μm; pore diameter 0.5-0.6 μm; pore diameter : cyst diameter 0.10). The cyst surface is smooth, except for several shallow longitudinal fossae extending from the pore region.

This cyst is nearly identical to stomatocyst 149 but has a different pore morphology.

LM DESCRIPTION: Because the fossae of this stomatocyst cannot be observed with LM, it must be counted as a collective category with similar cysts (e.g. stomatocyst 1) if they are present (Fig. 6B).

REFERENCES: We have not observed this morphotype elsewhere, or found it reported in the literature.

ECOLOGY: The ecological optima and tolerances of stomatocyst 148 have not yet been estimated.

Stomatocyst 149, Zeeb & Smol, 1993a (17) Fig. 134

BIOLOGICAL AFFINITY: unknown

LOCALITY: Elk Lake (47°11'N; 95°13'W), Clearwater County, Minnesota, U.S.A. Piston core, varve year A.D. 1708.

SEM DESCRIPTION: This stomatocyst is spherical to widely oval (diameter 5.4-7.0 μm) with a concave pore (outer diameter 1.0-1.6 μm; inner diameter 0.5-0.7 μm) and a swollen pseudoannulus when mature. The surface is ornamented with several fairly regularly spaced fossae extending longitudinally from below the pore, and sometimes a few isolated psilae.

Stomatocyst 149 is identical to stomatocyst 148, except for the pore morphology.

LM DESCRIPTION: With LM, this cyst may be confused with stomatocyst 120. However, it can generally be distinguished by its shape, which is usually somewhat asymmetric, by its thicker cyst wall, and by the fossae if they are visible (Fig. 134D).

REFERENCES: This stomatocyst has been observed in Ontario, Canada (Rybak *et al.*, 1987 - cyst 6; Rybak *et al.*, 1991 - cyst 20); British Columbia, Canada (Duff & Smol, 1994; Zeeb, 1994); the Yukon Territory, Canada (Brown & Smol, unpubl.); Minnesota, U.S.A. (Zeeb & Smol, 1993a); and Michigan, U.S.A. (Sandgren & Carney, 1983 - cyst 8).

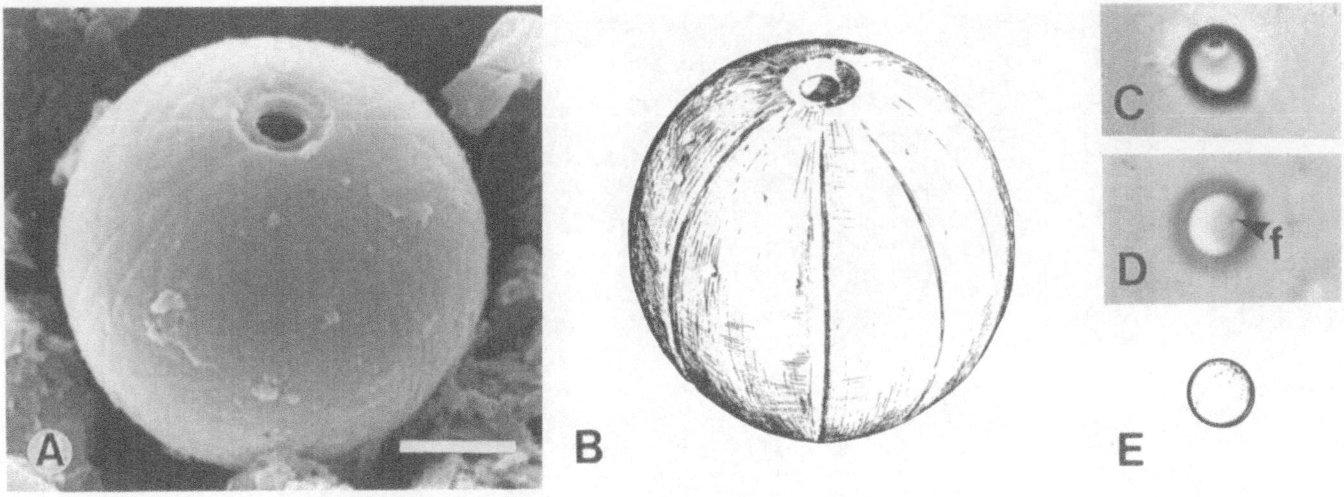

Fig. 134. Stomatocyst 149. A: SEM. B: Line drawing of SEM. C: LM, pore view. D: LM, view of single fossa (f). E: Line drawing of LM. Scale bar = 2 μm. LM x2000.

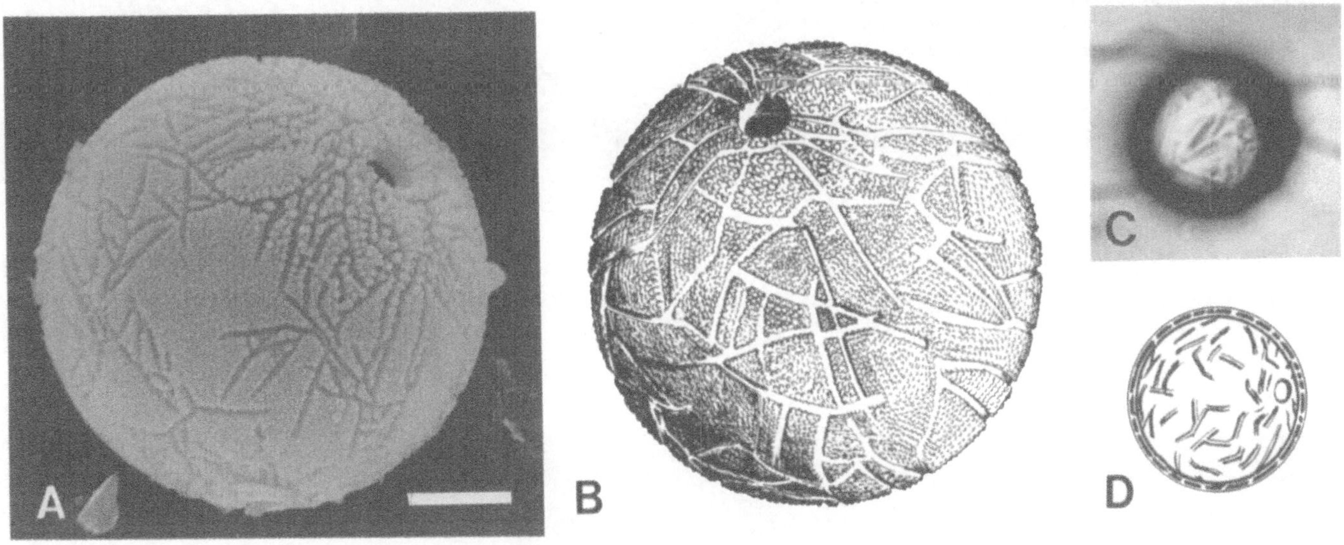

Fig. 135. Stomatocyst 145. A: SEM. B: Line drawing of SEM. C: LM. Collar not visible. D: Line drawing of LM. Scale bar = 2 μm. LM x2000.

ECOLOGY: Stomatocyst 149 has been found in slightly acidic (Rybak *et al.*, 1991) to alkaline lakes (Sandgren & Carney, 1983; Zeeb & Smol, 1993a; Duff & Smol, 1994). It was classified as a meso-eutrophic indicator by Rybak *et al.* (1987), and was highly positively correlated with total phosphorus in Duff's (1994) survey. It may be produced by a taxon with a preference for shallow water (Duff, 1994).

Stomatocyst 145, Duff & Smol in Duff *et al.*, 1992 (3) Fig. 135

BIOLOGICAL AFFINITY: unknown

LOCALITY: "Camp Pond" (unofficial name), Cape Herschel (78°37'N; 74°42'W), Ellesmere Island, Northwest Territories, Canada. Net tow, collected July 28, 1986.

SEM DESCRIPTION: This is a spherical stomatocyst (diameter 7.9-8.7 μm) with a conical pore (outer diameter 1.0-1.1 μm; inner diameter 0.6-0.7 μm) and a fossulate surface pattern of straight, intersecting fossae, mostly oriented longitudinally (fossa width ca. 0.1 μm). The cyst wall between the fossae, and especially around the collar, is ornamented with small, round, very densely spaced scabrae.

LM DESCRIPTION: This cyst is distinguished by its ornamentation and multiple cell walls. The fossae appear

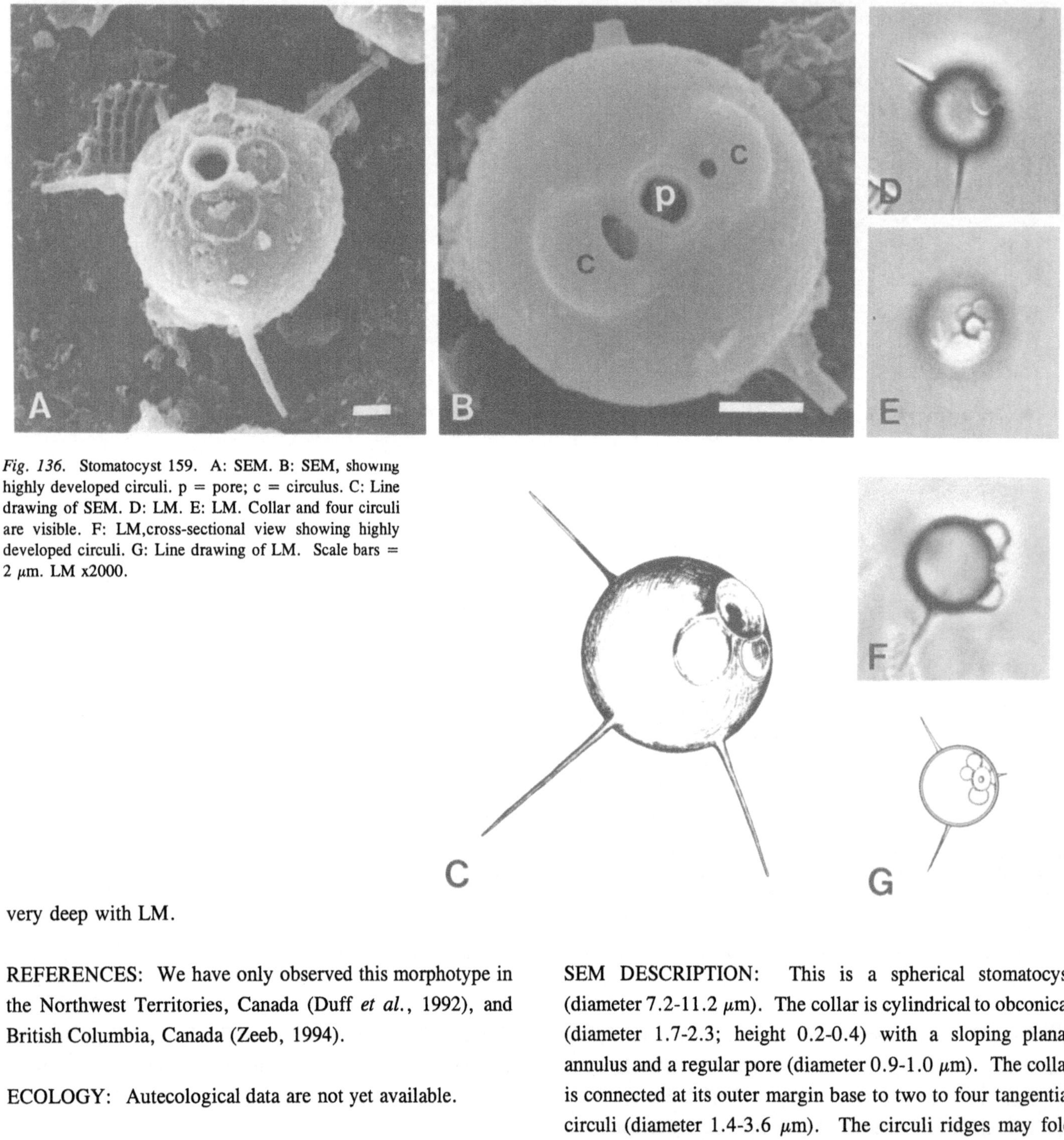

Fig. 136. Stomatocyst 159. A: SEM. B: SEM, showing highly developed circuli. p = pore; c = circulus. C: Line drawing of SEM. D: LM. E: LM. Collar and four circuli are visible. F: LM,cross-sectional view showing highly developed circuli. G: Line drawing of LM. Scale bars = 2 μm. LM x2000.

very deep with LM.

REFERENCES: We have only observed this morphotype in the Northwest Territories, Canada (Duff *et al.*, 1992), and British Columbia, Canada (Zeeb, 1994).

ECOLOGY: Autecological data are not yet available.

Stomatocysts ornamented with spines and ridges

Stomatocyst 159, Zeeb & Smol, 1993a (7) Fig. 136

BIOLOGICAL AFFINITY: unknown

LOCALITY: Elk Lake (47°11'N; 95°13'W), Clearwater County, Minnesota, U.S.A. Piston core, varve year A.D.

SEM DESCRIPTION: This is a spherical stomatocyst (diameter 7.2-11.2 μm). The collar is cylindrical to obconical (diameter 1.7-2.3; height 0.2-0.4) with a sloping planar annulus and a regular pore (diameter 0.9-1.0 μm). The collar is connected at its outer margin base to two to four tangential circuli (diameter 1.4-3.6 μm). The circuli ridges may fold over to form closed, or nearly closed, chambers. The cyst surface is smooth, except for three to four long, non-bifurcating echinate spines (length to 8.7 μm) located in the posterior hemisphere.

LM DESCRIPTION: This stomatocyst can be recognized by the circuli surrounding the collar, even when the spines are broken.

REFERENCES: Stomatocyst 159 has been observed in

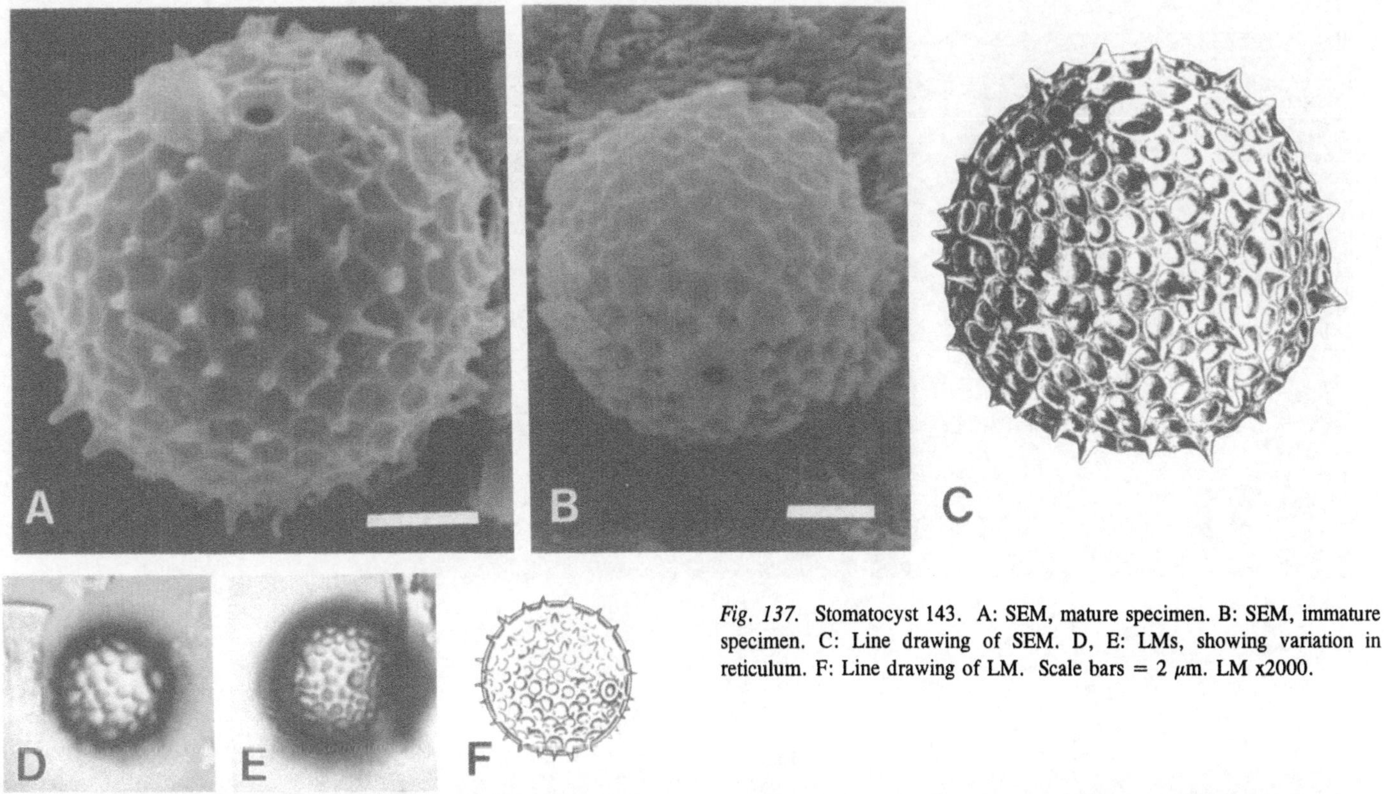

Fig. 137. Stomatocyst 143. A: SEM, mature specimen. B: SEM, immature specimen. C: Line drawing of SEM. D, E: LMs, showing variation in reticulum. F: Line drawing of LM. Scale bars = 2 μm. LM x2000.

British Columbia, Canada (Duff & Smol, 1994; Zeeb, 1994); Minnesota, U.S.A. (Zeeb & Smol, 1993a); Michigan, U.S.A. (Carney & Sandgren, 1983 - *cysta splendens* Carney & Sandgren; Sandgren & Carney, 1983 - cyst 27); New York, U.S.A. (Duff & Smol, unpubl.); and Poland (Rybak, 1986 - *cysta splendens*).

ECOLOGY: This cyst appears to be indifferent to lake trophic status (Carney & Sandgren, 1983; Rybak, 1986; Zeeb & Smol, 1993b).

Stomatocysts ornamented with spines and a reticulum

Stomatocyst 143, Duff & Smol in Duff *et al.*, 1992 (14) Fig. 137

BIOLOGICAL AFFINITY: unknown

LOCALITY: "Beach Ridge Pond" (unofficial name), Cape Herschel (78°37'N; 74°42'W), Ellesmere Island, Northwest Territories, Canada. Net tow, collected July 23, 1986

SEM DESCRIPTION: This is a spherical stomatocyst (diameter 7.7-11.3 μm) with a regular to variable reticulum. The reticulum lacunae are spherical to polygonal (lacuna diameter 0.3-1.5 μm). Small, echinate spines, which may be bent, are located at the reticular interstices of mature specimens (Fig. 137A). There is a small cylindrical collar (diameter 1.0-1.4 μm) which is connected to the reticulum on most specimens.

LM DESCRIPTION: Stomatocyst 143 could be confused using LM with stomatocyst 6, but the latter morph is smaller, with reticular lacunae that are usually larger than those of stomatocyst 143, and are never circular. Also, stomatocyst 6 never has spines. Stomatocyst 243 is very similar, but has larger circular lacunae and usually has more spines than stomatocyst 143.

REFERENCES: Stomatocyst 143 has been observed in the Northwest Territories, Canada (Duff *et al.*, 1992; Webb & Smol, unpubl.); Ontario, Canada (Rybak *et al.*, 1991 - cyst 45); British Columbia, Canada (Duff & Smol, 1994; Zeeb, 1994); California, U.S.A. (Adam & Mahood, 1980a - type 145; Carney *et al.*, 1992 - cyst 19; Zeeb *et al.*, subm.); Washington, U.S.A. (VanLandingham, 1964 - type 7); and Michigan, U.S.A. (Sandgren & Carney, 1983 - cyst 40).

ECOLOGY: This cyst is likely produced by an alkaliphilic (Duff *et al.*, 1992; Carney *et al.*, 1992; Sandgren & Carney, 1983) or pH-indifferent (Rybak *et al.*, 1991) taxon.

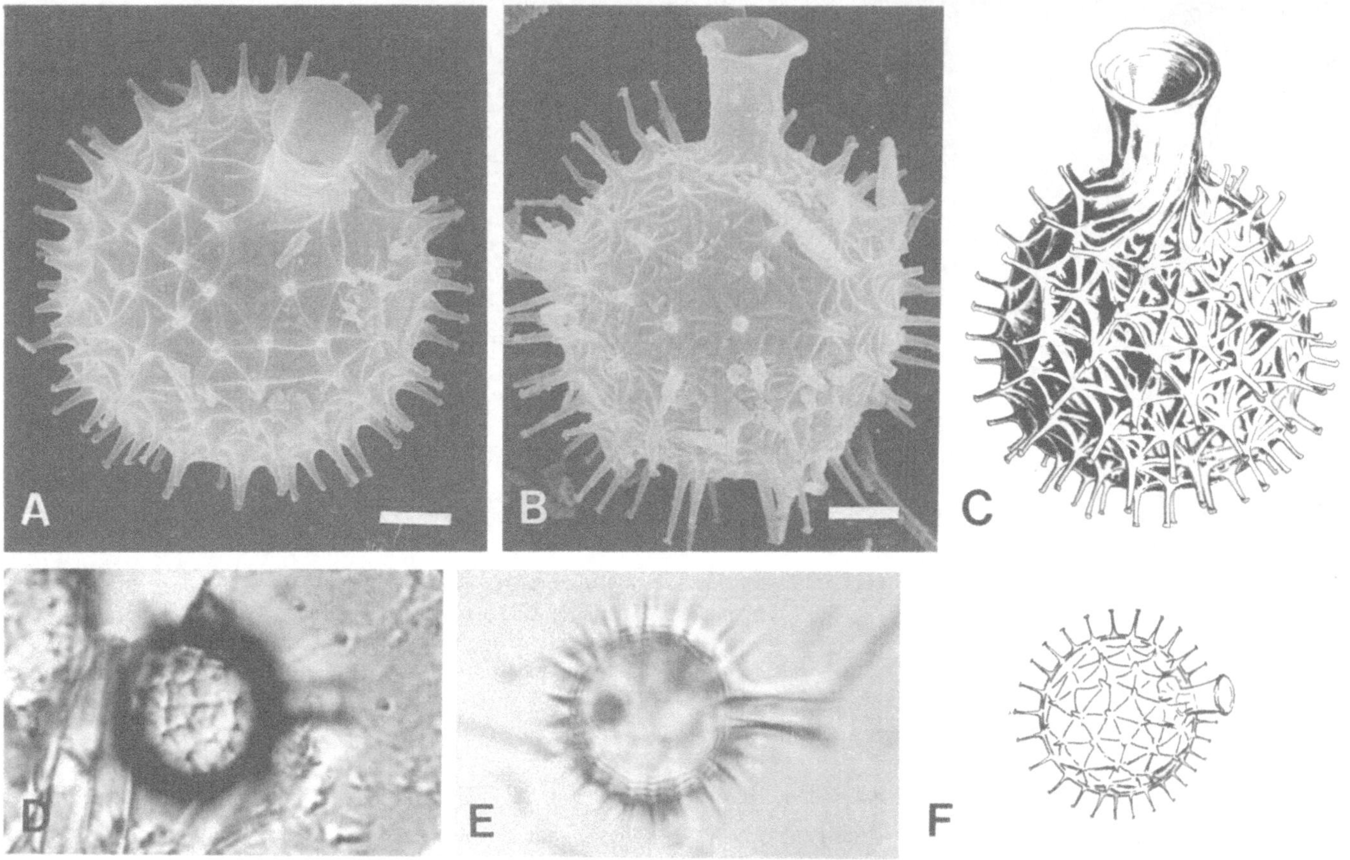

Fig. 138. Stomatocyst 35 *formae* A and B. A: *Forma* A, SEM. B: *Forma* B, SEM. C: *Forma* B, line drawing of SEM. D: LM, surface view. E: LM, cross-sectional view. F: Line drawing of LM. Scale bars = 2 μm. LM x2000.

Stomatocyst 35, Duff & Smol, 1989 (15) Fig. 138

BIOLOGICAL AFFINITY: This stomatocyst is possibly produced by *Mallomonas doignonii* Bourrelly *emend.* Asmund & Cronberg (H. Kling, pers. commun.)

LOCALITY: Tasikutaaq Lake (66°23'N; 65°57'W), Baffin Island, Northwest Territories, Canada. Percussion core, core depth 75 cm.

SEM DESCRIPTION: This is a spherical stomatocyst (diameter 6.9-12.4 μm) with a cylindrical to obconical collar (apical diameter 1.4-4.4 μm; height 1.1-4.1 μm) surrounding a regular pore (diameter 0.9-1.0 μm). The cyst body is ornamented with a low, roughly triangular reticulum, with echinate spines, often with hooked apices, located at fairly regularly spaced interstices (spine length 0.3-2.6 μm). There appear to be a number of developmental stages leading to at least two fully developed *formae*.

***Forma* A** - This *forma* has a basic triangular reticulum which ends at, or near to, the outer collar margin base (Fig. 138A).

***Forma* B** - The reticulum of this *forma* is more complex, with additional wavy ridges connecting the spines. The reticulum extends up the outer collar margin to approximately one-quarter of the collar height, sometimes in a spiralling pattern (Fig. 138B).

Stomatocyst 35 now includes several cysts previously described by us as separate morphs (see Chapter 5). These are: stomatocyst 22 (Duff & Smol, 1988), which was improperly described based on light microscopy only; stomatocyst 81 (Duff & Smol, 1991); and stomatocyst 82 (Duff & Smol, 1991), which is an immature morph.

LM DESCRIPTION: The two *formae* cannot be reliably distinguished using LM (Figs. 138D, E, and F). The triangular reticulum can be used to identify this morphotype, unless the ornamentation is undeveloped.

REFERENCES: This cyst has been found in the Northwest Territories, Canada (Duff & Smol, 1988 - stomatocyst 22; Duff & Smol, 1989); Ontario, Canada (Rybak *et al.*, 1991 - cyst 55); British Columbia, Canada (Duff & Smol, 1994);

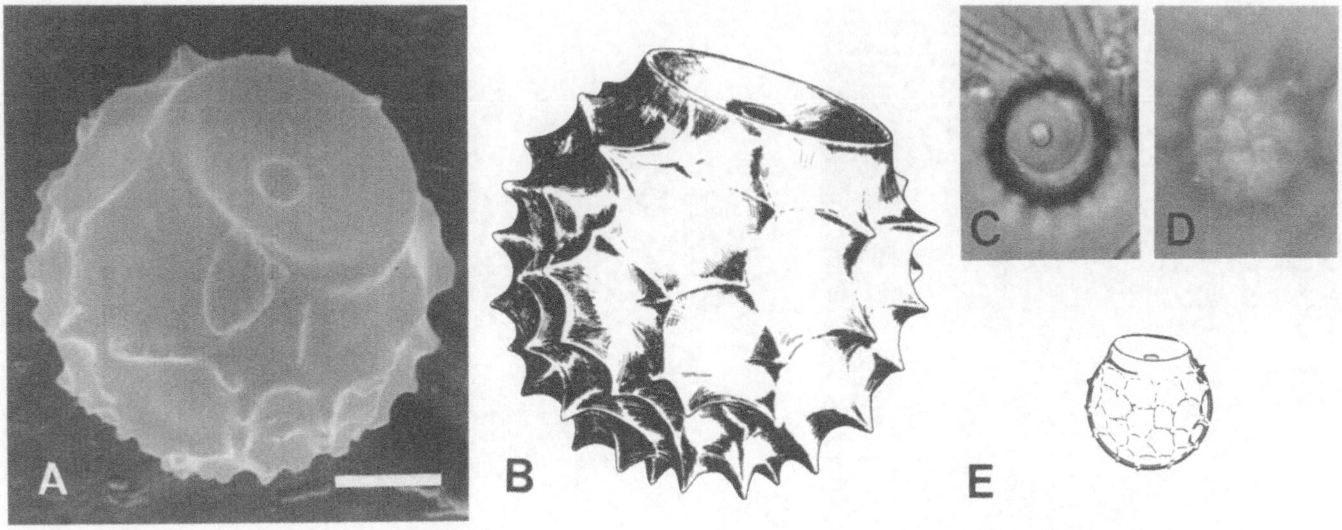

Fig. 139. Stomatocyst 85. A: SEM. B: Line drawing of SEM. C: LM, collar view. D: LM, view of reticulum. E: Line drawing of LM. Scale bar = 2 μm. LM x2000.

Connecticut, U.S.A. (Gritten, 1977 - *cysta hexagrammica* Gritten); New York, U.S.A. (Duff & Smol, 1991 - stomatocysts 81 and 82; Duff & Smol, unpubl.); Oregon, U.S.A. (Adam & Mehringer, 1980c - types 171 and 172); California, U.S.A. (Adam & Mahood, 1980b - type 171; Carney *et al.*, 1992 - cyst 7); and Denmark (Nygaard, 1956 - *cysta spinulosa* Nygaard).

ECOLOGY: This cyst appears to be produced by an acidophilic (Nygaard, 1956; Duff & Smol, 1989, 1991; Rybak *et al.*, 1991) or circumneutral (Carney *et al.*, 1992) taxon. It is likely tolerant of cold water temperatures and/or extended ice cover (Adam & Mehringer, 1980c; Duff & Smol, 1989). In this respect, *Mallomonas doignonii* is thought to be a coldwater species (Siver, 1991a).

Stomatocyst 85, Duff & Smol, 1991 (4) Fig. 139

BIOLOGICAL AFFINITY: unknown

LOCALITY: Upper Wallface Pond (44°08.8'N; 74°03.3'W), Adirondack Park, New York, U.S.A. Gravity core, core depth 0.0-0.5 cm.

SEM DESCRIPTION: This stomatocyst is spherical (diameter 4.5-8.0 μm) with a complex collar. The primary collar is low and conical, surrounding a regular pore (pore diameter 0.8-1.1 μm). On all but one SEM specimen, it is located centrally within the interannulus. The secondary collar is wide, low, and conical, with an abrupt outer margin, a gently sloping inner margin, and usually with a rounded apex (secondary collar diameter 3.2-6.1 μm). The ornamentation below the secondary collar consists of a variable polygonal reticulum (lacuna diameter 0.7-2.7 μm), with blunt conula located at the reticulum interstices.

LM DESCRIPTION: Stomatocyst 85 can be distinguished from cysts with similar ornamentation (e.g. stomatocysts 6 and 86) by the complex collar.

REFERENCES: This stomatocyst has been observed in Ontario, Canada (Rybak *et al.*, 1991 - cyst 58), and New York, U.S.A. (Duff & Smol, 1991, unpubl.). The cyst identified by Gritten (1977 - Connecticut, U.S.A.) as *cysta collaris* Nygaard is probably an undeveloped or degraded stomatocyst 85; it is not the same morph shown as *cysta collaris* by Nygaard (1956).

ECOLOGY: This morphotype may be produced by an acidophilic (Gritten, 1977; Duff & Smol, 1991) or pH-indifferent (Rybak *et al.*, 1991) chrysophyte. It appears to be associated with oligotrophic, and possibly humic, water (Gritten, 1977; Duff & Smol, 1991; Rybak *et al.*, 1991).

Stomatocyst 86, Duff & Smol, 1991 (3) Fig. 140

BIOLOGICAL AFFINITY: unknown

LOCALITY: Upper Wallface Pond (44°08.8'N; 74°03.3'W), Adirondack Park, New York, U.S.A. Gravity core, core

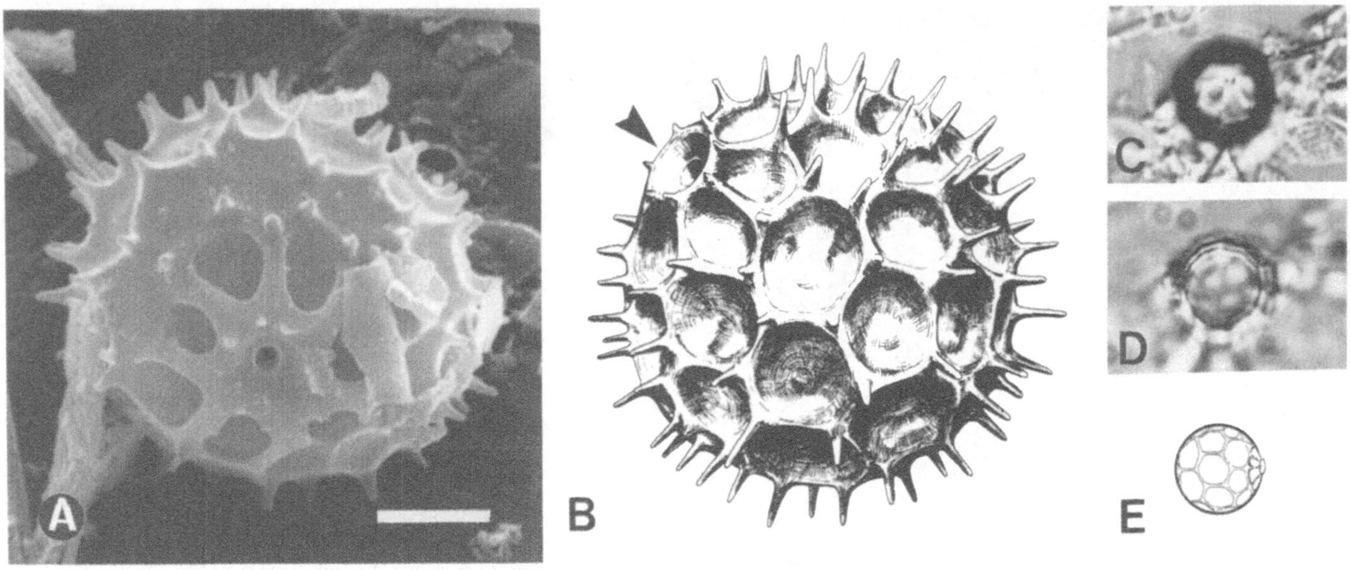

Fig. 140. Stomatocyst 86. A: SEM. B: Line drawing of SEM. Collar indicated by arrow. C: LM, collar view. D: LM, cross-sectional view showing reticulum. E: Line drawing of LM. Scale bar = 2 μm. LM x2000.

depth 10.0-10.5 cm.

SEM DESCRIPTION: This stomatocyst is spherical (diameter 6.6-6.7 μm) with an unusual collar. The pore is regular (diameter ca. 0.4 μm) and may have a slight, apically acute, conical primary collar. There is a cylindrical secondary collar, polygonal in outline (apical diameter 1.6-1.8 μm; height ca. 1.0 μm); from each corner a straight ridge radiates (six or seven total; length ca. 1.5 μm). On mature specimens the ridges are quite high, with flattened apices that fold over on each side of the ridge to form rounded, recessed lacunae (diameter 0.9-1.2 μm). Small echinate spines may also be found on the collar apex (height to 0.3 μm). The remainder of the cyst surface is ornamented with a low, variable, polygonal reticulum (lacuna diameter 1.4-2.1 μm) that is continuous with the collar ridges. Echinate spines (height 0.3-0.9 μm) are located at each reticular interstice.

LM DESCRIPTION: Usually, stomatocyst 86 can be distinguished from morphotypes with similar ornamentation (e.g. stomatocysts 6 and 85) by the presence of spines and the collar morphology. However, if the spines are undeveloped and a good view of the collar is not obtained, stomatocyst 86 could be confused with stomatocyst 6.

REFERENCES: Stomatocyst 86 has been observed in Ontario, Canada (Rybak *et al.*, 1991 - cyst 59), and New York, U.S.A. (Duff & Smol, 1991, unpubl.).

ECOLOGY: This stomatocyst is most likely produced by an acidophilic chrysophyte (Duff & Smol, 1991; Rybak *et al.*, 1991).

Stomatocyst 243, Duff & Smol in Duff *et al.* (this volume) (4) Fig. 141

BIOLOGICAL AFFINITY: unknown

NEGATIVE NUMBER: J.P. Smol 1447

LOCALITY: Lake Frances (44°45'N; 74°22'W), Adirondack Park, New York, U.S.A. Gravity core, surface (0-1 cm) sediment sample.

SEM DESCRIPTION: This stomatocyst is spherical (diameter 7.9-10.2 μm). It is ornamented with an irregular reticulum of smaller polygonal lacunae and larger circular lacunae (diameter 0.7-1.9 μm; Fig. 141B), with small, echinate spines at each interstice (spine height 0.3-0.9 μm). The circular lacunae appear to be shallow depressions in the cyst wall, which are slightly separated from the reticulum ridges surrounding them. The collar was visible on only one of our SEM specimens (Fig. 141A); it is cylindrical with a very rounded apex and a regular pore (collar diameter 1.0 μm; pore diameter 0.5 μm).

LM DESCRIPTION: Stomatocyst 243 can be distinguished from stomatocyst 143 by the presence of larger circular lacunae, and often by the greater number of spines.

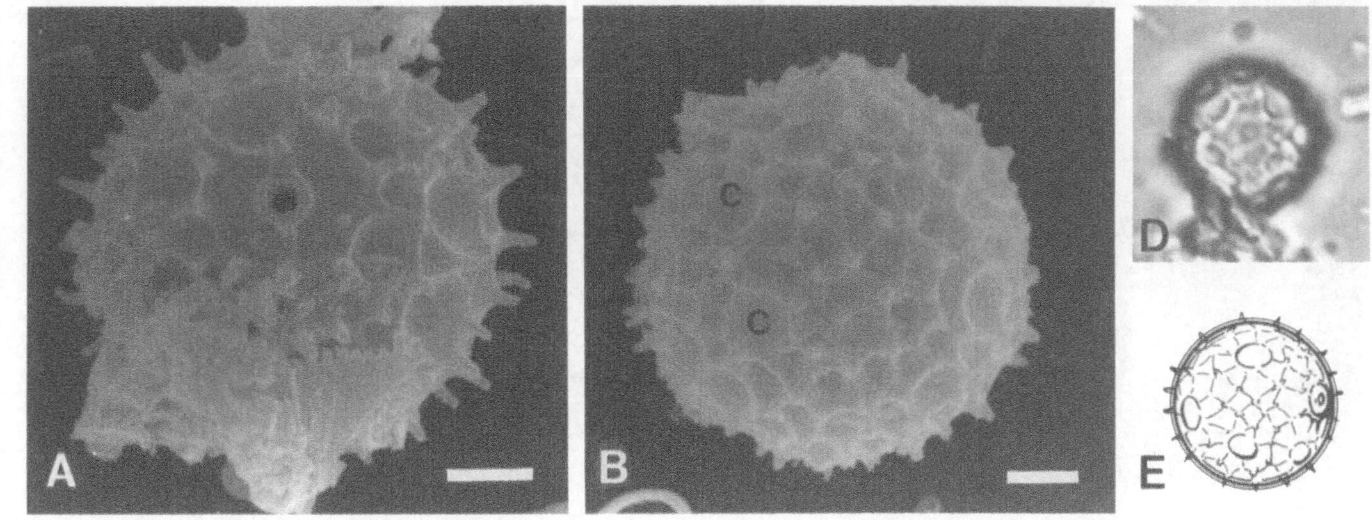

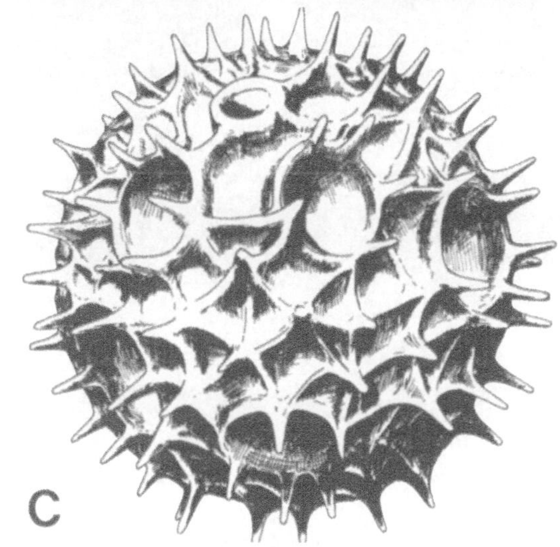

Fig. 141. Stomatocyst 243. A: SEM, apical view. B: SEM. Collar not visible. c = larger circular lacunae. C: Line drawing of SEM. D: LM. Collar not visible. E: Line drawing of LM. Scale bars = 2 μm. LM x2000.

Stomatocyst 243 is distinguished from stomatocyst 100 because: the latter morph lacks the polygonal reticulum; its depressions appear deeper with LM than do those of stomatocyst 243; and the spines of stomatocyst 100 never appear on the edge of the depressions.

REFERENCES: This cyst has been observed in Ontario, Canada (Smol, 1988a - Plate 1E), and New York, U.S.A. (Duff *et al.*, this volume). Adam & Mahood (1979b - California, U.S.A.) type 132 is similar.

ECOLOGY: We cannot yet estimate the ecological optima and tolerances of this morph.

Stomatocyst 98, Duff & Smol, 1991 (2) Fig. 142

BIOLOGICAL AFFINITY: unknown

LOCALITY: Upper Wallface Pond (44°08.8'N; 74°03.3'W), Adirondack Park, New York, U.S.A. Gravity core, core depth 0.0-0.5 cm.

SEM DESCRIPTION: This is a small, spherical to widely obovate stomatocyst (diameter 4.5-5.1 μm) with an obconical collar surrounding a conical pore (collar diameter 0.7-0.9 μm; inner pore diameter 0.3 μm). The surface is ornamented with a regular reticulum of hexagonal lacunae (lacuna diameter 0.6-0.7 μm). Inside each lacuna is a short baculate spine with an apical corona of finger-like projections (spine diameter 0.1-0.2 μm; height 0.1-0.2 μm). The surface immediately surrounding the collar is not ornamented.

LM DESCRIPTION: This stomatocyst has not yet been identified definitely with LM. Duff & Smol (1991 - Fig. 138) observed a similar sized cyst with a hexagonal reticulum; however, no spines were visible.

REFERENCES: To our knowledge, this stomatocyst has only been observed in New York, U.S.A. (Duff & Smol, 1991, unpubl.).

ECOLOGY: The two reports of stomatocyst 98 are both from acidic, oligotrophic lakes (Duff & Smol, 1991, unpubl.).

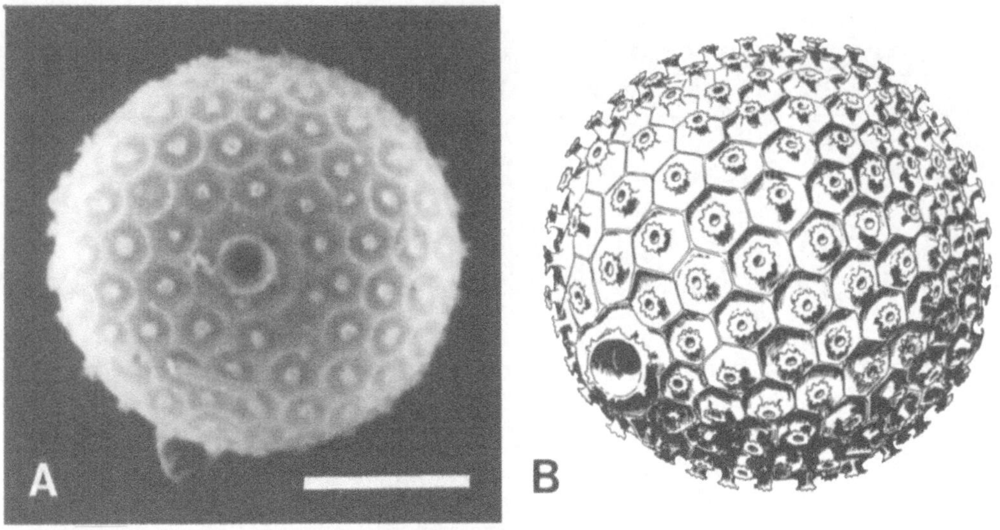

Fig. 142. Stomatocyst 98. A: SEM. B: Line drawing of SEM. Scale bar = 2 μm.

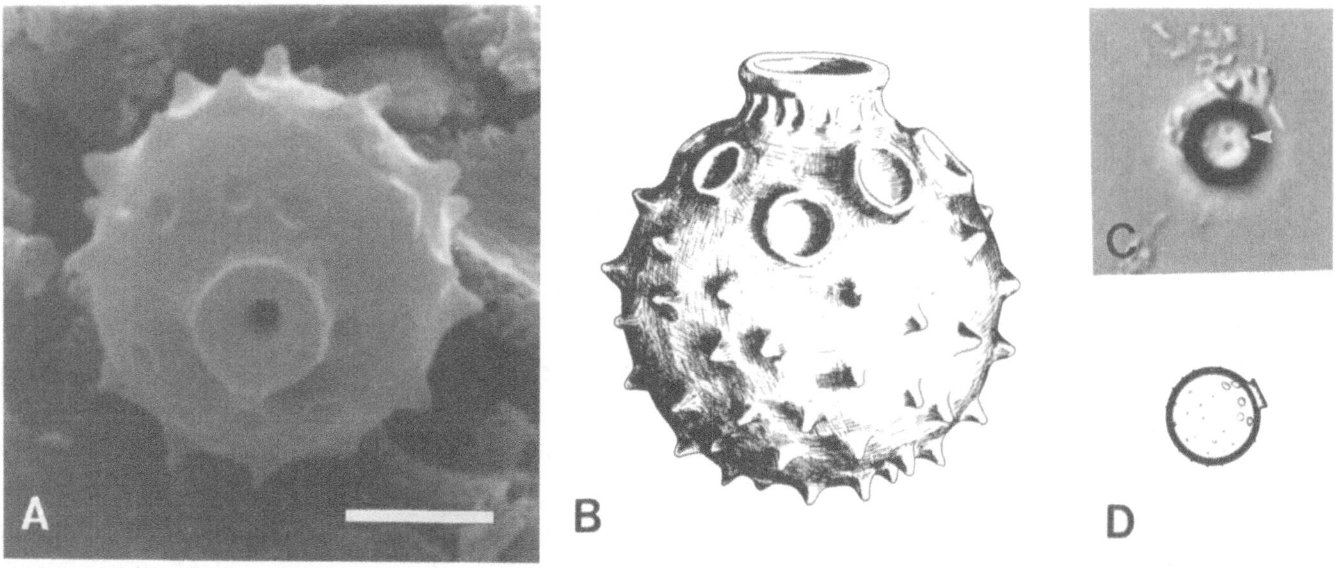

Fig. 143. Stomatocyst 5 *formae* A and B. A: *Forma* A, SEM. B: *Forma* B, line drawing of SEM. C: LM. Collar indicated by arrow. D: Line drawing of LM. Scale bar = 2 μm. LM x2000.

Stomatocysts ornamented with spines and indentations

Stomatocyst 5, Duff & Smol, 1988 *emend.* Duff & Smol, 1994 (27) Fig. 143

BIOLOGICAL AFFINITY: unknown

LOCALITY: "Rock Basin Lake" (unofficial name) (78°29.5'N; 76°44.8'W), Ellesmere Island, Northwest Territories, Canada. Piston core, core depth 15-16 cm.

SEM DESCRIPTION: This stomatocyst is small and spherical (diameter 3.0-5.2 μm), with a regular pore (diameter 0.3-0.6 μm) surrounded by a collar that varies from conical to obconical (apical diameter 1.0-1.8 μm; height 0.2-0.7 μm). Shallow circular depressions (diameter 0.3-0.8 μm) are found mostly in the anterior hemisphere, but sometimes slightly subequatorially. Short conula (base diameter 0.1-0.9 μm; height 0.1-0.8 μm) are located in the posterior hemisphere, or slightly supraequatorially.

There are two *formae*, differentiated by the presence or absence of a continuous rim surrounding each depression:

***Forma* A** - lacks a depression rim (Fig. 143A).
***Forma* B** - possesses a depression rim (Fig. 143B).

LM DESCRIPTION: The two *formae* of this stomatocyst cannot be differentiated reliably with LM. Although this stomatocyst was counted as a collective category with

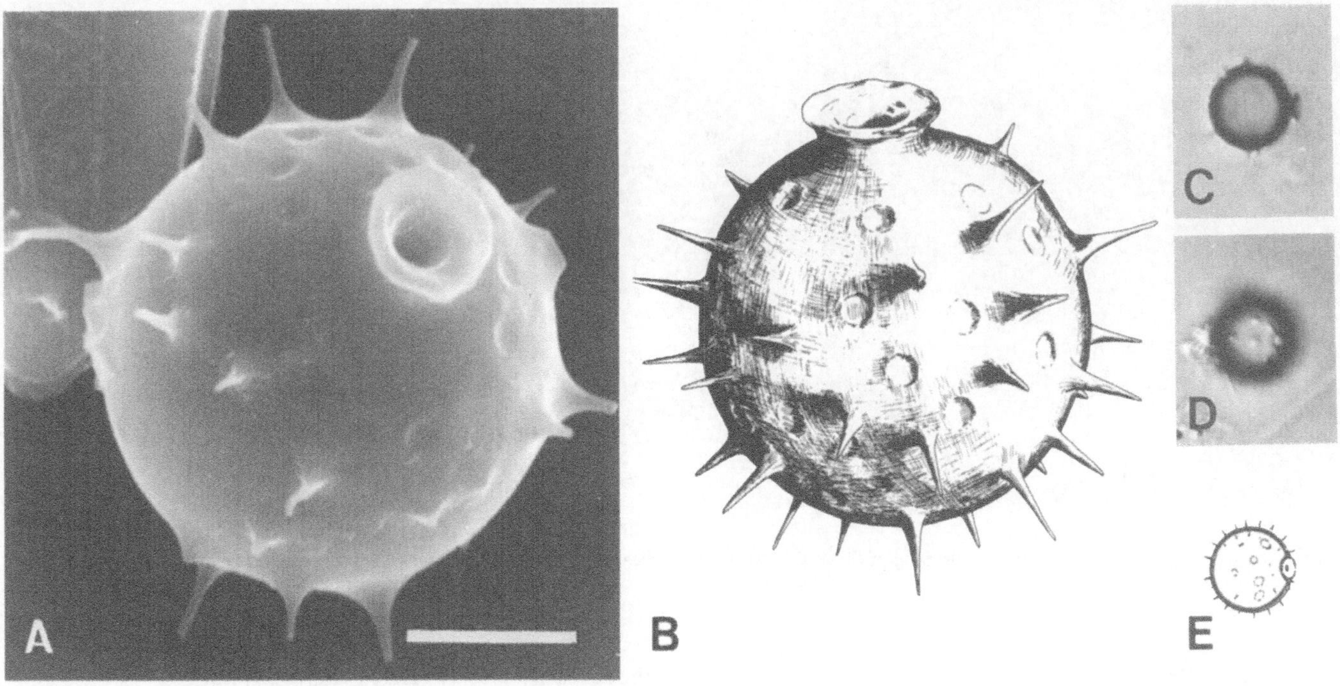

Fig. 144. Stomatocyst 88. A: SEM. B: Line drawing of SEM. C: LM, cross-sectional view. D: LM, surface view. Depressions not visible. E: Line drawing of LM. Scale bar = 2 μm. LM x2000.

stomatocyst 89 by Duff & Smol (1991), these two morphs are distinguishable if the microscopist can resolve between small spines and ridges. On this basis, stomatocyst 5 can also be distinguished from stomatocyst 232 *formae* A and B. Stomatocyst 88 is very similar, but has longer spines and randomly oriented depressions.

REFERENCES: Stomatocyst 5 *forma* A has been observed in the Northwest Territories, Canada (Duff & Smol, 1988, 1989); British Columbia, Canada (Duff & Smol, 1994); New York, U.S.A. (Duff & Smol, 1991); Minnesota, U.S.A. (Zeeb & Smol, 1993a); California, U.S.A. (Adam & Mahood, 1979b - type 45); and Washington, U.S.A. (Smith & White, 1985 - Figs. 12, 13).

Stomatocyst 5 *forma* B has been found in Ontario, Canada (Rybak *et al.*, 1987 - cyst 23; Rybak *et al.*, 1991 - cyst 95); British Columbia, Canada (Duff & Smol, 1994); New York, U.S.A. (Duff & Smol, 1991); Minnesota, U.S.A. (Zeeb & Smol, 1993a); and California, U.S.A. (Adam & Mahood, 1979b - type 45; Carney *et al.*, 1992 - cyst 5).

ECOLOGY: Stomatocyst 5 is often found in deep, cold, and oligotrophic lakes (Adam & Mahood, 1979; Smith & White, 1985; Duff & Smol, 1988, 1989, 1991; Duff, 1994). It appears to be tolerant of a wide range of pH (Duff & Smol, 1991; Rybak *et al.*, 1991), and may also be aluminum-tolerant (Smith & White, 1985; Duff & Smol, 1991; Duff, 1994).

Stomatocyst 88, Duff & Smol, 1991 (5) Fig. 144

BIOLOGICAL AFFINITY: unknown

LOCALITY: Upper Wallface Pond (44°08.8'N; 74°03.3'W), Adirondack Park, New York, U.S.A. Gravity core, core depth 0.0-0.5 cm.

SEM DESCRIPTION: This stomatocyst is spherical (diameter 3.1-5.5 μm) with a cylindrical to obconical collar (apical diameter 0.7-1.7 μm; height 0.3-0.6 μm) surrounding a conical pore (diameter ca. 0.7 μm). Shallow circular depressions (diameter 0.4-0.5 μm) without a rim are found scattered irregularly over the cyst body. Among the depressions are irregularly scattered echinate spines (base diameter 0.4-0.7 μm; length to 1.1 μm).

Stomatocyst 88 is very similar to stomatocyst 5 *forma* A, but has longer spines and more randomly placed ornamentation.

LM DESCRIPTION: With LM, this stomatocyst is differentiated from stomatocyst 5 by its longer spines and randomly oriented depressions.

REFERENCES: This stomatocyst has been observed in Ontario, Canada (Zeeb *et al.*, 1994), the Yukon Territory, Canada (Brown & Smol, unpubl.), and New York, U.S.A.

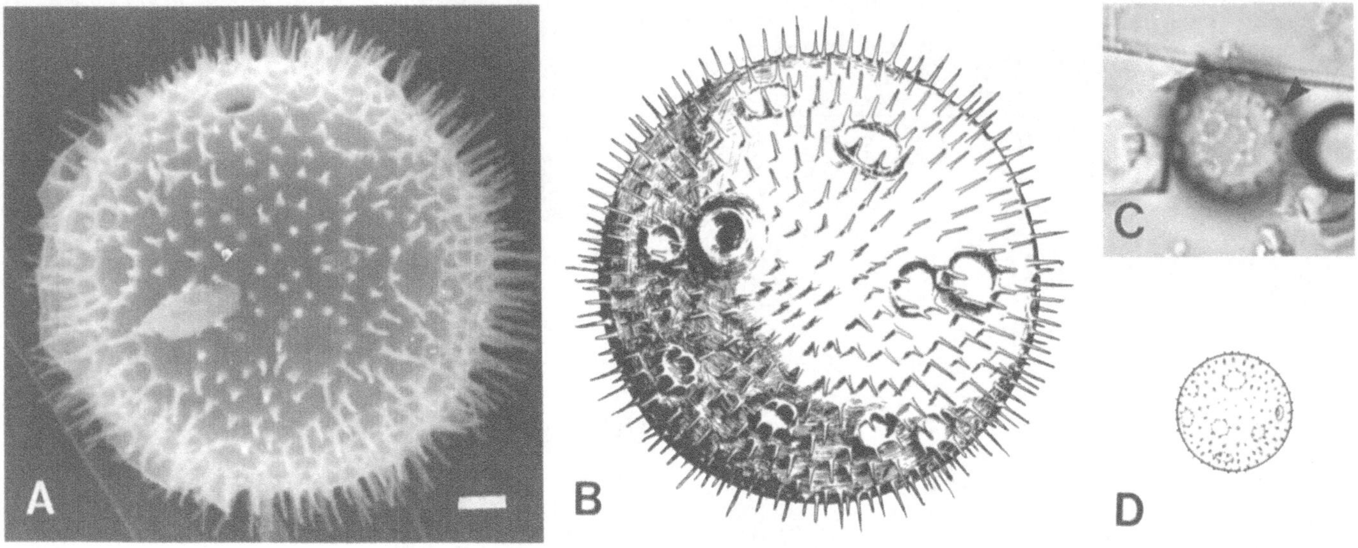

Fig. 145. Stomatocyst 76. A: SEM. B: Line drawing of SEM. C: LM. Collar indicated by arrow. D: Line drawing of LM. Scale bar = 2 μm. LM x2000.

(Duff & Smol, 1991).

ECOLOGY: This cyst increased in abundance in an artificially eutrophied lake after the fertilizer used was changed from an N:P ratio of 15:1 to 5:1 (Zeeb *et al.*, 1994).

Stomatocyst 76, Duff & Smol, 1991 (7) Fig. 145

BIOLOGICAL AFFINITY: unknown

LOCALITY: Upper Wallface Pond (44°08.8'N; 74°03.3'W), Adirondack Park, New York, U.S.A. Gravity core, core depth 10.0-10.5 cm.

SEM DESCRIPTION: This is a spherical stomatocyst (diameter 8.1-9.9 μm). It has a very small, cylindrical to conical collar with an abrupt outer margin, sloping inner margin, and rounded apex (apical collar diameter 1.3-1.8 μm) surrounding a regular pore (diameter 0.4-0.6 μm). The cyst surface is ornamented with long, narrow, echinate spines (basal spine diameter 0.1-0.3 μm; length to 1.1 μm), either scattered randomly, or arranged in a ring around shallow depressions of the cyst surface (depression diameter 1.1-1.5 μm). Low ridges may connect the spines encircling the depressions (ridge height 0.3-0.4 μm). One broken specimen revealed a thinly silicified cyst wall.

Stomatocyst 76 can be distinguished from stomatocyst 171 by its smaller size, narrower spine bases, and more densely spaced spines. However, it is probable that these cysts are produced by closely related chrysophytes.

LM DESCRIPTION: The features used to distinguish stomatocysts 76 and 171 with SEM are also visible with LM. Stomatocyst 76 is differentiated from stomatocyst 243 by the absence of a reticulum.

REFERENCES: This cyst has been observed in Ontario, Canada (Rybak *et al.*, 1991 - cyst 52; Zeeb *et al.*, 1994), the Yukon Territory, Canada (Brown & Smol, unpubl.), British Columbia, Canada (Zeeb, 1994), and New York, U.S.A. (Duff & Smol, 1991, unpubl.). Adam & Mehringer (1980c - Oregon, U.S.A.) type 170 is probably also identical.

ECOLOGY: This cyst appears to be acidophilic (Duff & Smol, 1991) or pH-indifferent (Rybak *et al.*, 1991). In Zeeb *et al.*'s (1994) paleolimnological study of an artificially eutrophied lake, it was common in pre-manipulation sediments, declined abruptly with P and N additions, but returned to its original levels following a reduction in the N:P ratio from 15:1 to 5:1.

Stomatocyst 171, Zeeb & Smol, 1993a (21) Fig. 146

BIOLOGICAL AFFINITY: unknown

LOCALITY: Elk Lake (47°11'N; 95°13'W), Clearwater County, Minnesota, U.S.A. Piston core, varve year 8720 B.P.

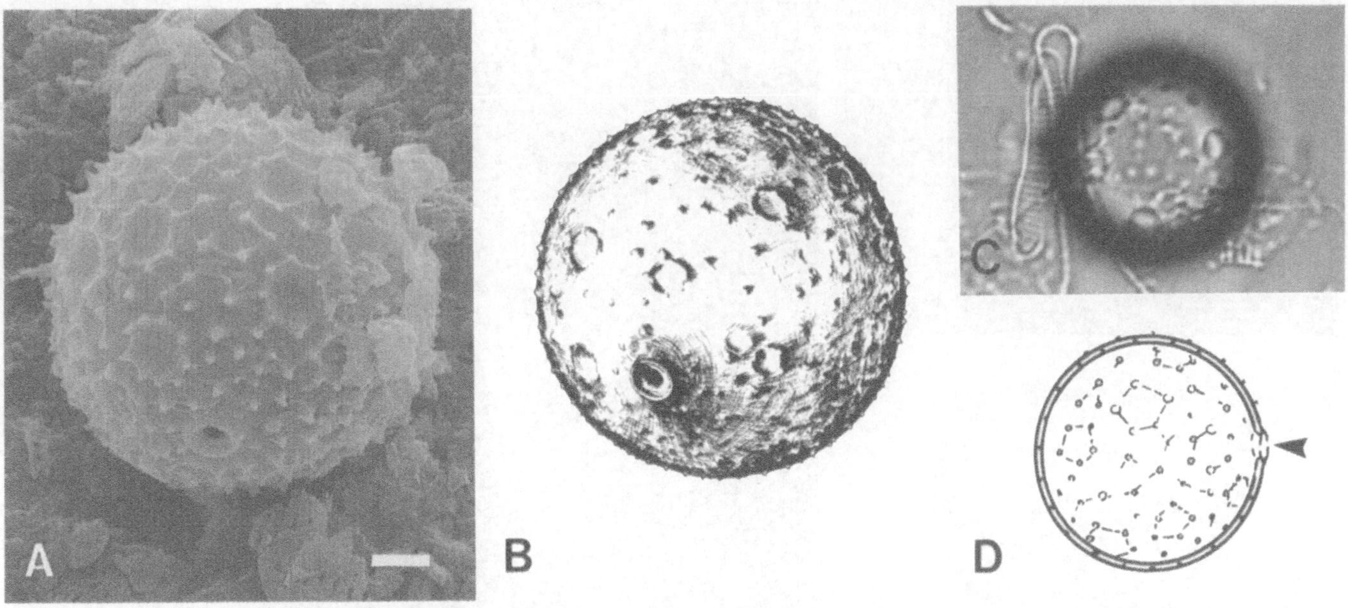

Fig. 146. Stomatocyst 171. A: SEM. B: Line drawing of SEM. C: LM. Collar not visible. D: Line drawing of LM. Collar indicated by arrow. Scale bar = 2 μm. LM x2000.

SEM DESCRIPTION: This is a large spherical stomatocyst (diameter 11.4-14.4 μm). The collar is low, cylindrical to conical with a sloping inner margin, and possibly a narrow, sloping planar annulus; the apex is usually rounded (apical collar diameter 1.4-1.8 μm, height 0.3-0.7 μm). The pore is regular (diameter 0.6-0.8 μm). The surface is ornamented with short conula or echinate spines, either scattered randomly, or arranged in a ring around shallow circular depressions in the cyst wall (basal spine diameter 0.4-0.8; height 0.2-1.1; depression diameter 1.3-2.2). The spines surrounding the depressions are often connected by low ridges (ridge height 0.2-0.7 μm).

Stomatocyst 171 can be distinguished from stomatocyst 76 in that the latter cyst is smaller and has narrower spine bases, making the spines appear almost needle-like.

LM DESCRIPTION: Stomatocysts 171 and 76 are distinguished with LM on the basis of size and spine length.

REFERENCES: This stomatocyst has been observed in Ontario, Canada (Rybak *et al.*, 1987 - cyst 25; Rybak *et al.*, 1991 - cyst 53); British Columbia, Canada (Duff & Smol, 1994; Zeeb, 1994); the Yukon Territory, Canada (Brown & Smol, unpubl.); Minnesota, U.S.A. (Zeeb & Smol, 1993a); Denmark (Nygaard, 1956 - *cysta polygonata* Nygaard); and Poland (Rybak, 1986 - *cysta polygonata*).

ECOLOGY: Stomatocyst 171 has been commonly found in slightly acidic to circumneutral water (Rybak *et al.*, 1991; Duff, 1994); it was classified as pH-indifferent by Rybak *et al.* (1991). It may be produced by a eutrophic taxon (Rybak *et al.*, 1987). It was more common in large, deep lakes in British Columbia, Canada (Duff, 1994).

"Stomatocyst" 17, Duff & Smol, 1988 (9) Fig. 147

BIOLOGICAL AFFINITY: Although originally described as a stomatocyst, this morph is actually the silicified vegetative cell of *Chrysococcus furcatus* (Dolgoff) Nicholls (Nicholls, 1981). The stomatocyst produced by this species is stomatocyst 130.

LOCALITY: "Rock Basin Lake" (unofficial name) (78°29.5'N; 76°44.8'W), Ellesmere Island, Northwest Territories, Canada. Piston core, core depth 0-2 cm.

SEM DESCRIPTION: *Chrysococcus furcatus* is pyramidal in shape (corner-to-corner length 9.2-11.7 μm). The collar is low, cylindrical to slightly obconical (diameter 1.1-2.0 μm) with a thick, flattened apex and a thinly silicified, flat planar annulus surrounding a regular pore (diameter 0.5-0.7 μm). The ornamentation consists of irregularly scattered shallow depressions (diameter 1.7-2.4 μm) covering the entire cyst surface. A thin circulus encircling the base of the characteristic specimen (Duff & Smol, 1988 - Fig. 36) is probably an artefact of that specimen.

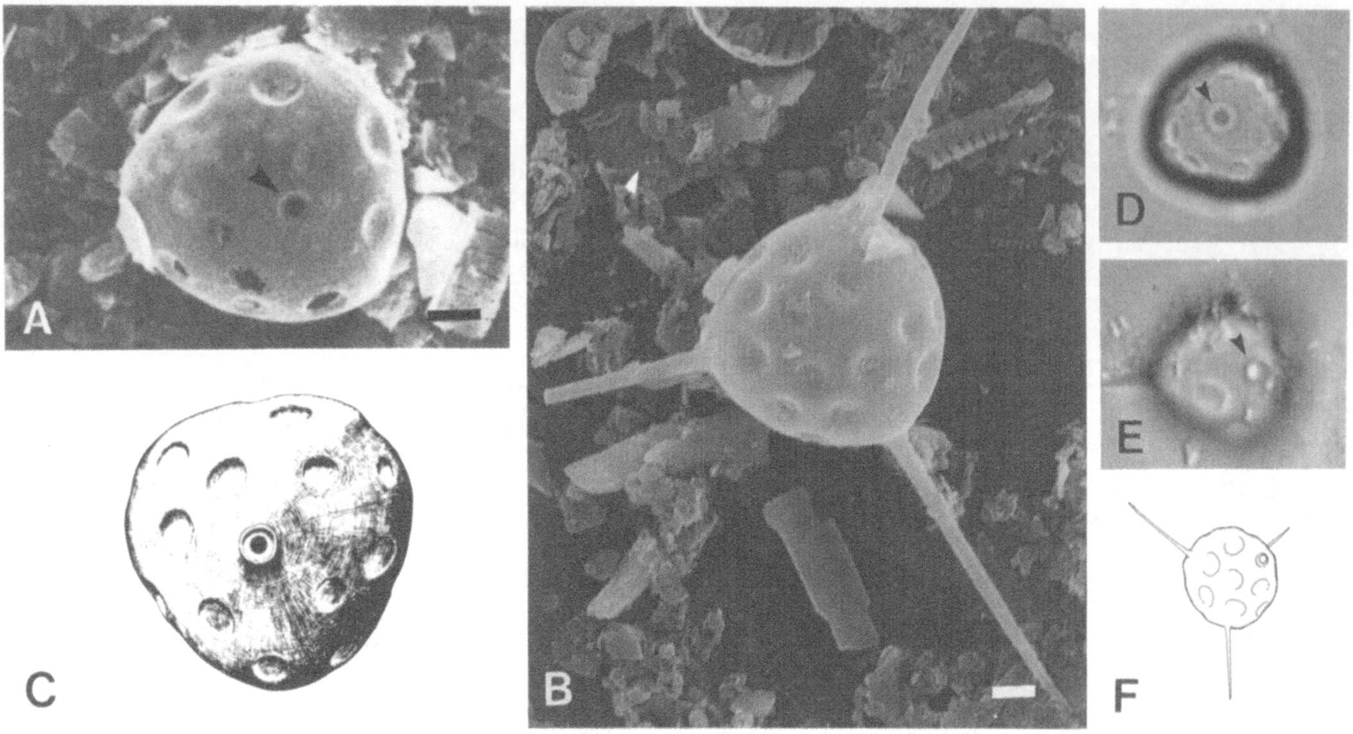

Fig. 147. "Stomatocyst" 17 *formae* A and B. A: *Forma* A, SEM. Collar indicated by arrow. B: *Forma* B, SEM. Collar not visible. C: *Forma* A, line drawing of SEM. D: *Forma* A, LM. Collar indicated by arrow. E: *Forma* B, LM. Collar indicated by arrow. F: *Forma* B, line drawing of LM. Scale bars = 2 μm. LM x2000.

Two *formae* are recognized, based on the presence or absence of corner spines:

***Forma* A** - spines are absent (Figs. 147A, C, and D).

***Forma* B** - narrow echinate spines (basal diameter 0.4-1.2 μm; length to 14.4 μm) are found on the basal corners (Figs. 147B, E, and F).

LM DESCRIPTION: This species is easily recognized by its shape and ornamentation.

REFERENCES: See the description of stomatocyst 130 for details.

ECOLOGY: See the description of stomatocyst 130 for details.

Stomatocysts ornamented with ridges and a reticulum

Stomatocyst 176, Zeeb & Smol, 1993a (5) Fig. 148

BIOLOGICAL AFFINITY: unknown

LOCALITY: Elk Lake (47°11'N; 95°13'W), Clearwater County, Minnesota, U.S.A. Piston core, varve year 5488 B.P.

SEM DESCRIPTION: This is a spherical to slightly oval stomatocyst (diameter 6.5-7.7 μm) with a short cylindrical collar (diameter 0.6-0.9 μm; height 0.7-1.0 μm; Fig. 148A). Four ridges or two circuli, more or less equidistantly placed, extend longitudinally from the collar; ridge height undulates from 0.4-1.5 μm, and occasionally a circulus may be discontinuous. The surface between the ridges may have very low, irregular reticular patches with circular lacunae.

LM DESCRIPTION: Stomatocyst 176 is identified with LM by its ridges and small collar.

REFERENCES: This cyst has been observed in Minnesota, U.S.A. (Zeeb & Smol, 1993a), Michigan, U.S.A. (Sandgren & Carney, 1983 - cyst 33), and New York, U.S.A. (Duff & Smol, unpubl.).

ECOLOGY: Autecological details are not yet available.

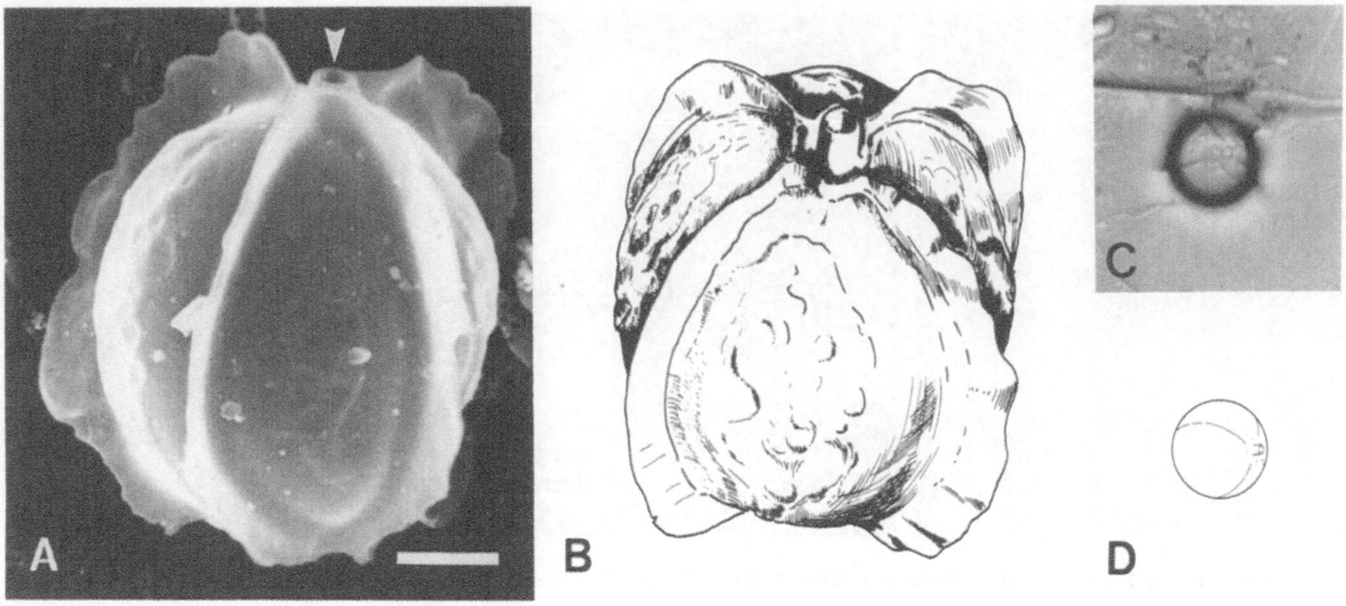

Fig. 148. Stomatocyst 176. A: SEM. Collar indicated by arrow. B: Line drawing of SEM. C: LM. D: Line drawing of LM. Scale bar = 2 μm. LM x2000.

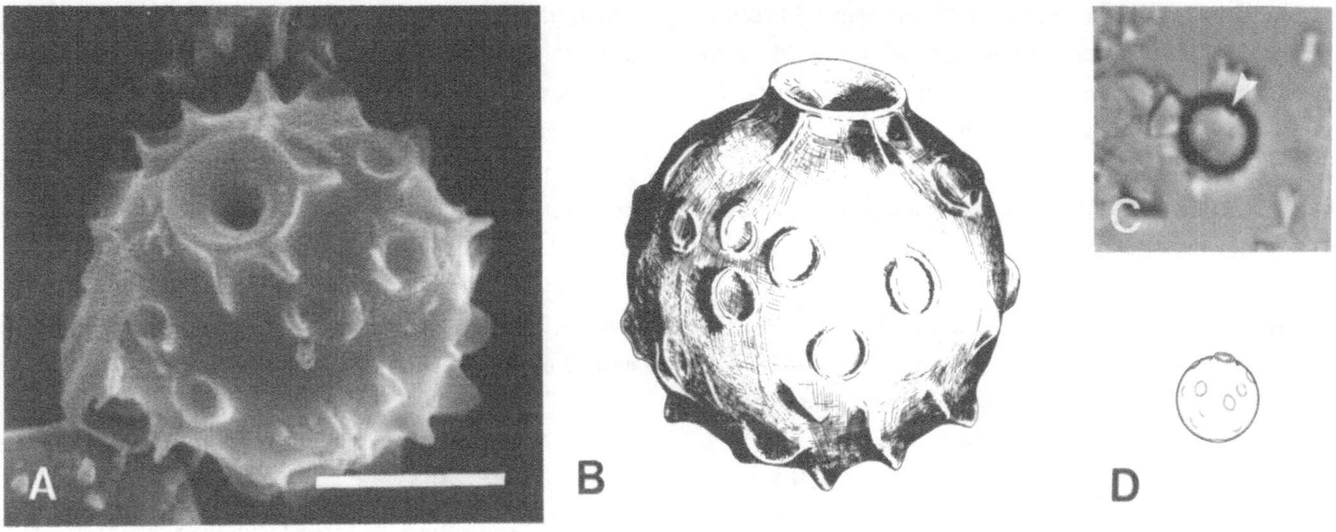

Fig. 149. Stomatocyst 89. A: SEM. B: Line drawing of SEM. C: LM. Collar indicated by arrow. Depressions and ridges also visible. D: Line drawing of LM. Scale bar = 2 μm. LM x2500.

Stomatocysts ornamented with ridges and indentations

Stomatocyst 89, Duff & Smol, 1991 (6) Fig. 149

BIOLOGICAL AFFINITY: unknown

LOCALITY: Upper Wallface Pond (44°08.8'N; 74°03.3'W), Adirondack Park, New York, U.S.A. Gravity core, core depth 10.0-10.5 cm.

SEM DESCRIPTION: This is a small, spherical stomatocyst (diameter 3.3-4.0 μm) with a cylindrical to obconical collar (diameter 1.2-1.5 μm; height 0.4 μm) surrounding a conical pore (inner pore diameter 0.3-0,4 μm; depth ca. 0.2 μm). The collar base is often strutted. Shallow circular depressions (diameter 0.4-0.7 μm) are located in the anterior hemisphere, or slightly subequatorially. Each depression has a rim formed by the joining (usually incomplete) of two or more lunate, curved ridges. Short, straight lunate ridges are found in the posterior hemisphere or slightly supraequatorially (ridge length 0.3-1.4 μm; height 0.1-0.6 μm; width 0.1-0.2 μm).

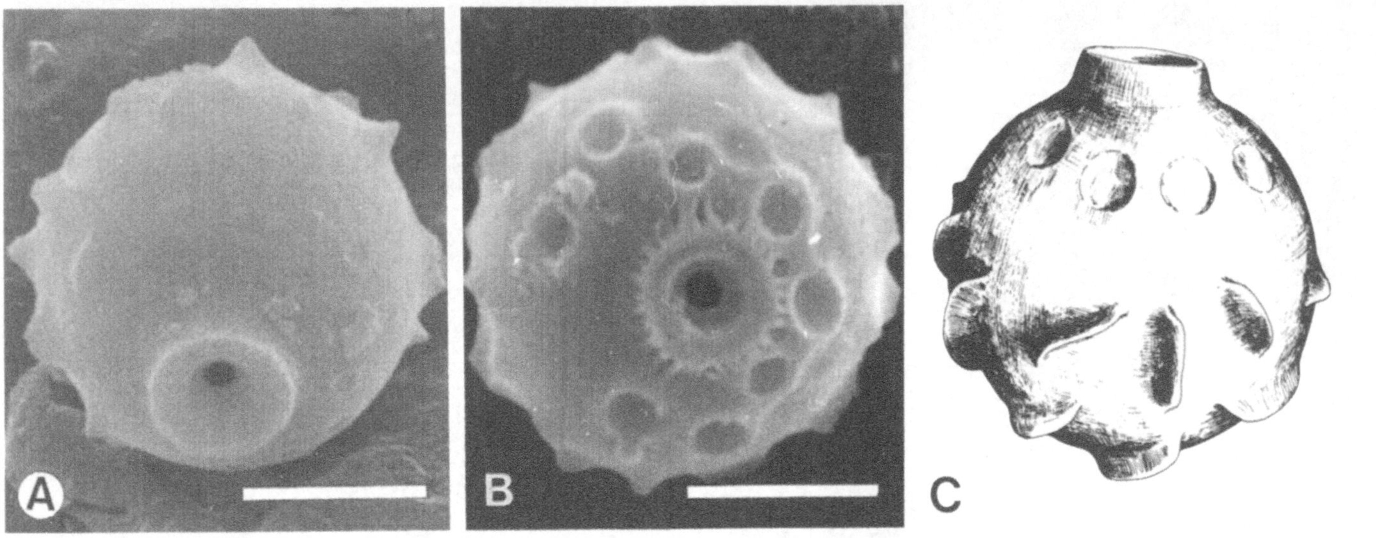

Fig. 150. Stomatocyst 232 *formae* A and B. A: *Forma* A, SEM. B: *Forma* B, SEM. C: *Forma* A, line drawing of SEM. Scale bars = 2 μm. See Figs. 149C, D for LM.

Stomatocyst 89 resembles both stomatocysts 5 and 232, but differs in the way the depression rim is formed. However, it is still possible that these morphologically similar cysts are produced by the same (or closely related) species. Stomatocyst 90 (Chapter 5) possesses identical ridges to stomatocyst 89, but lacks depressions; it also may be produced by a closely related species.

LM DESCRIPTION: With high resolution optics, the ridges of this stomatocyst can be differentiated from the spines of stomatocyst 5. However, stomatocyst 89 cannot be distinguished from stomatocyst 232 using LM.

REFERENCES: Stomatocyst 89 has been observed in Ontario, Canada (Rybak *et al.*, 1991 - cyst 92), and New York, U.S.A. (Duff & Smol, 1991, unpubl.). Rybak *et al.* (1991) cyst 89 appears to be intermediate between our stomatocysts 5 and 89; its depressions are identical to those of stomatocyst 89, but it has spines rather than ridges.

ECOLOGY: This cyst has only been found in acidic, oligotrophic lakes (Duff & Smol, 1991; Rybak *et al.*, 1991).

Stomatocyst 232, Duff & Smol, 1994 (18) Fig. 150

BIOLOGICAL AFFINITY: unknown

LOCALITY: Lang Lake (52°00'N; 121°00'W), British Columbia, Canada. Ekman dredge, surface (1 cm) sediment sample.

SEM DESCRIPTION: This stomatocyst is small and spherical (diameter 3.6-4.6 μm). The conical pore (inner diameter 0.3-0.4 μm) is surrounded by a highly variable collar, ranging from conical to obconical (diameter 1.1-1.7 μm; height 0.3-0.8 μm). Shallow circular depressions (diameter 0.3-0.7 μm) are located in the anterior hemisphere, but apparently not subequatorially. Short ridges, which vary greatly in length and height, are randomly oriented in the posterior hemisphere, and often supraequatorially (ridge length 0.4-2.0 μm; height 0.1-0.7 μm).

Like stomatocyst 5, this stomatocyst is divided into two *formae*, based on the presence or absence of a continuous rim around each depression:

***Forma* A** - lacks a depression rim; the depressions are often very faint (Fig. 150A).

***Forma* B** - possesses a depression rim; the collar of this *forma* usually has basal struts (Fig. 150B).

LM DESCRIPTION: The two *formae* cannot be differentiated reliably with LM. Stomatocyst 232 is indistinguishable from stomatocyst 89 (Figs. 149C, D), but can be distinguished from stomatocyst 5 if the resolution is sufficient to distinguish between spines and ridges. If the faint depressions of *forma* A were not visible, this morphotype could be misidentified as stomatocyst 113 (Figs. 102C, D).

REFERENCES: Stomatocyst 232 *forma* A has been observed in British Columbia, Canada (Duff & Smol, 1994), Ontario, Canada (Rybak *et al.*, 1991 - cysts 90, 92, 94), and

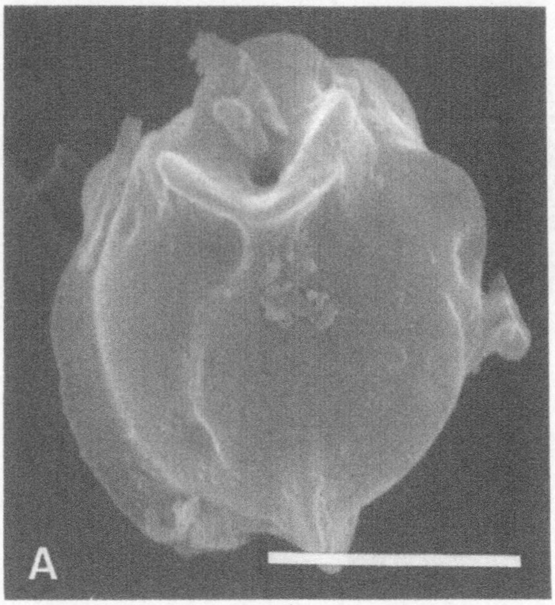

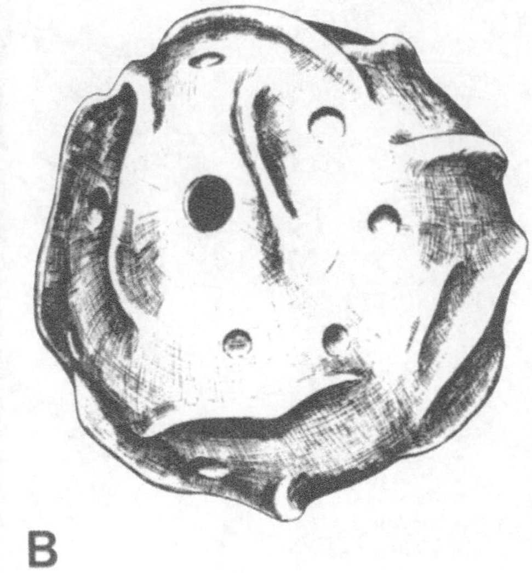

Fig. 151. Stomatocyst 94. A: SEM. B: Line drawing of SEM. Scale bar = 2 μm. See Figs. 111D, E, and F for LM.

Michigan, U.S.A. (Sandgren & Carney, 1983 - cyst 38). We have not found other references to *forma* B.

ECOLOGY: Rybak *et al.* (1991) classified this morphotype as either acidophilic or pH-indifferent. In Duff's (1994) study, its distribution was uncorrelated with pH, but was significantly negatively correlated with total phosphorus.

Stomatocyst 94, Duff & Smol, 1991 (15) Fig. 151

BIOLOGICAL AFFINITY: unknown

LOCALITY: Upper Wallface Pond (44°08.8'N; 74°03.3'W), Adirondack Park, New York, U.S.A. Gravity core, core depth 0.0-0.5 cm.

SEM DESCRIPTION: This is a small, spherical stomatocyst (diameter 3.0-4.8 μm) with a regular pore (diameter 0.2-0.4 μm). It is ornamented with randomly oriented ridges. There is no true collar, but usually two or more ridges located in the pore region give the appearance of a collar. A few to numerous small, shallow, circular depressions (diameter 0.2-0.5 μm) are scattered irregularly on the cyst surface between the ridges or on the ridges themselves.

This cyst is identical to stomatocyst 239 except for the presence of depressions.

LM DESCRIPTION: Stomatocysts 94 and 239 are indistinguishable with LM (Figs. 111D, E, and F).

REFERENCES: This stomatocyst has been observed in Ontario, Canada (Rybak *et al.*, 1991 - cysts 110 and 117); the Yukon Territory, Canada (Brown & Smol, unpubl.); California, U.S.A. (Adam & Mahood, 1979b - type 112); Washington, U.S.A. (Smith & White, 1985 - Fig. 17); and New York, U.S.A. (Duff & Smol, 1991, unpubl.).

ECOLOGY: This stomatocyst appears to be common in arctic or alpine environments (Adam & Mahood, 1979b; Smith & White, 1985; Duff & Smol, 1991, unpubl.; Brown & Smol, unpubl.). It has been classified variously as an acidophil (Duff & Smol, 1991; Rybak *et al.*, 1991; Duff, 1994) to an alkaliphil (Rybak *et al.*, 1991). In one study (Duff & Smol, 1991), stomatocyst 94 appeared intolerant of very low pH (<5.0) and/or elevated aluminum concentrations. Stomatocyst 94 may also be associated with eutrophic conditions or watershed disturbances (Smith & White, 1985).

Stomatocyst 34, Duff & Smol, 1989 *emend.* Duff & Smol, 1991 (10) Fig. 152

BIOLOGICAL AFFINITY: unknown

LOCALITY: Upper Wallface Pond (44°08.8'N; 74°03.3'W), Adirondack Park, New York, U.S.A. Gravity core, core depth 10.0-10.5 cm.

SEM DESCRIPTION: This stomatocyst is spherical (diameter 3.9-6.2 μm). The collar is cylindrical with an obconical apex, which may be reflexed back towards the cyst body (collar base diameter 0.7-1.0 μm; apical diameter 0.8-

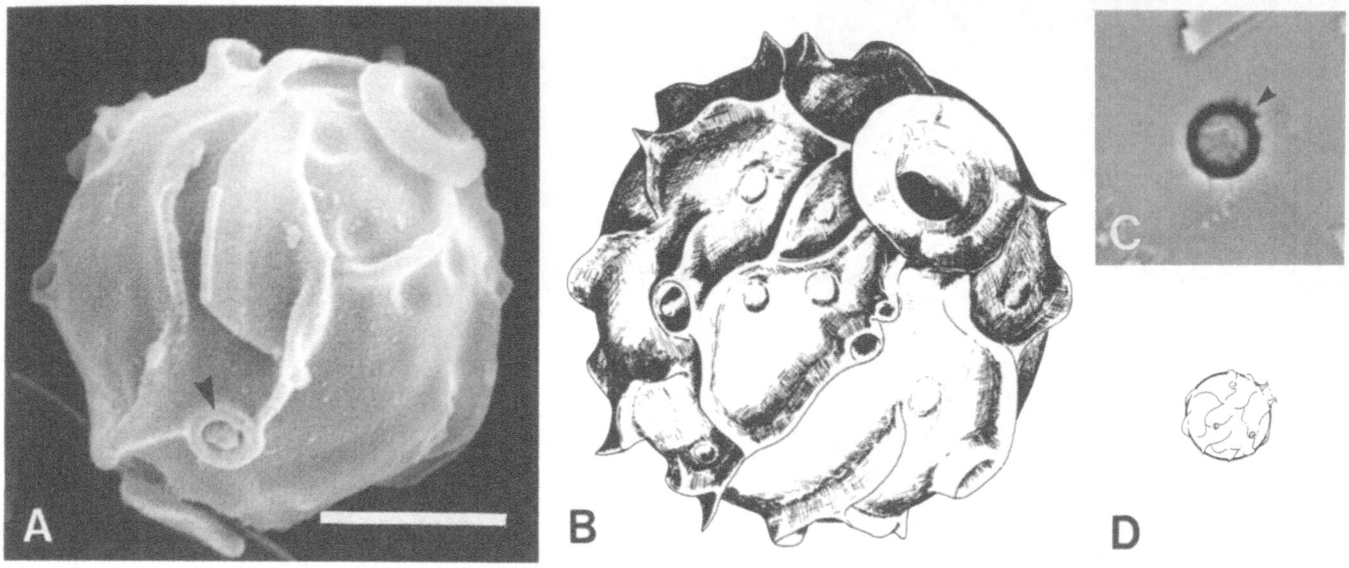

Fig. 152. Stomatocyst 34. A: SEM. Siliceous ring indicated by arrow. B: Line drawing of SEM. C: LM. Collar indicated by arrow. D: Line drawing of LM. Scale bar = 2 μm. LM x2000.

2.0 μm; height 0.2-0.6 μm), a sloping inner margin, and a regular pore (diameter 0.4-0.6 μm). The ornamentation consists of low, wavy ridges that occasionally extend into long, curved, echinate spines; the ridges are variable in length and height, and the ridge apices may curl over (ridge width 0.1-0.2 μm; height 0.1-0.8 μm). Between the ridges, particularly in the anterior hemisphere, are shallow circular depressions (diameter 0.3-0.7 μm). An occasional feature is a siliceous ring (diameter 0.7-1.0 μm) surrounding a plug-like siliceous deposition (Fig. 152A).

LM DESCRIPTION: Stomatocyst 34 resembles stomatocyst 93, but can be distinguished with LM by its shorter, wavier ridges, the presence of occasional spines, and the collar, which is often shorter and more reflexed. Stomatocyst 93 also tends to be larger than stomatocyst 34.

REFERENCES: Stomatocyst 34 has been found in the Northwest Territories, Canada (Duff & Smol, 1989), Ontario, Canada (Rybak *et al.*, 1991 - cysts 100, 107 and 108), New York, U.S.A. (Duff & Smol, 1991, unpubl.), and California, U.S.A. (Adam, 1980a - type 214; Adam & Mahood, 1979b - types 102, 103 and 115; Adam & Mahood, 1980b - types 214 and 218; Adam & Mahood, 1981b - types 214 and 290). The collar shown for Rybak *et al.* (1991) cyst 107 is the siliceous ring described above (Fig. 152A).

ECOLOGY: This cyst appears to be common in acidic, oligotrophic lakes (Duff & Smol, 1989, 1991; Rybak *et al.*, 1991); it has been classified as acidophilic or acidobiontic (Rybak *et al.*, 1991). Adam type 214 has been collected from periphytic samples of montane lakes (Adam, 1980; Adam & Mahood, 1981).

Stomatocysts ornamented with more than two types of ornamentation

Stomatocyst 100, Duff & Smol, 1991 (8) Fig. 153

BIOLOGICAL AFFINITY: unknown

LOCALITY: Upper Wallface Pond (44°08.8'N; 74°03.3'W), Adirondack Park, New York, U.S.A. Gravity core, core depth 20.0-20.5 cm.

SEM DESCRIPTION: This is a spherical stomatocyst (diameter 8.6-12.6 μm) with a small conical collar with a rounded apex, and a thin, planar annulus around a regular pore (collar diameter ca. 1.2 μm; annulus diameter ca. 0.9 μm; pore diameter 0.6 μm). The surface ornamentation is variable. All specimens have roughly spherical depressions scattered irregularly across the cyst surface (depression diameter 1.1-2.5 μm); the depressions may have a raised rim. Echinate spines are usually found scattered on the cyst surface between the depressions (spine height 0.4-1.6 μm). The remainder of the cyst body may also be ornamented with shallow fossae (Fig. 153A), or with low, rounded ridges forming a triangular or wavy reticulum (Fig. 153B). It is

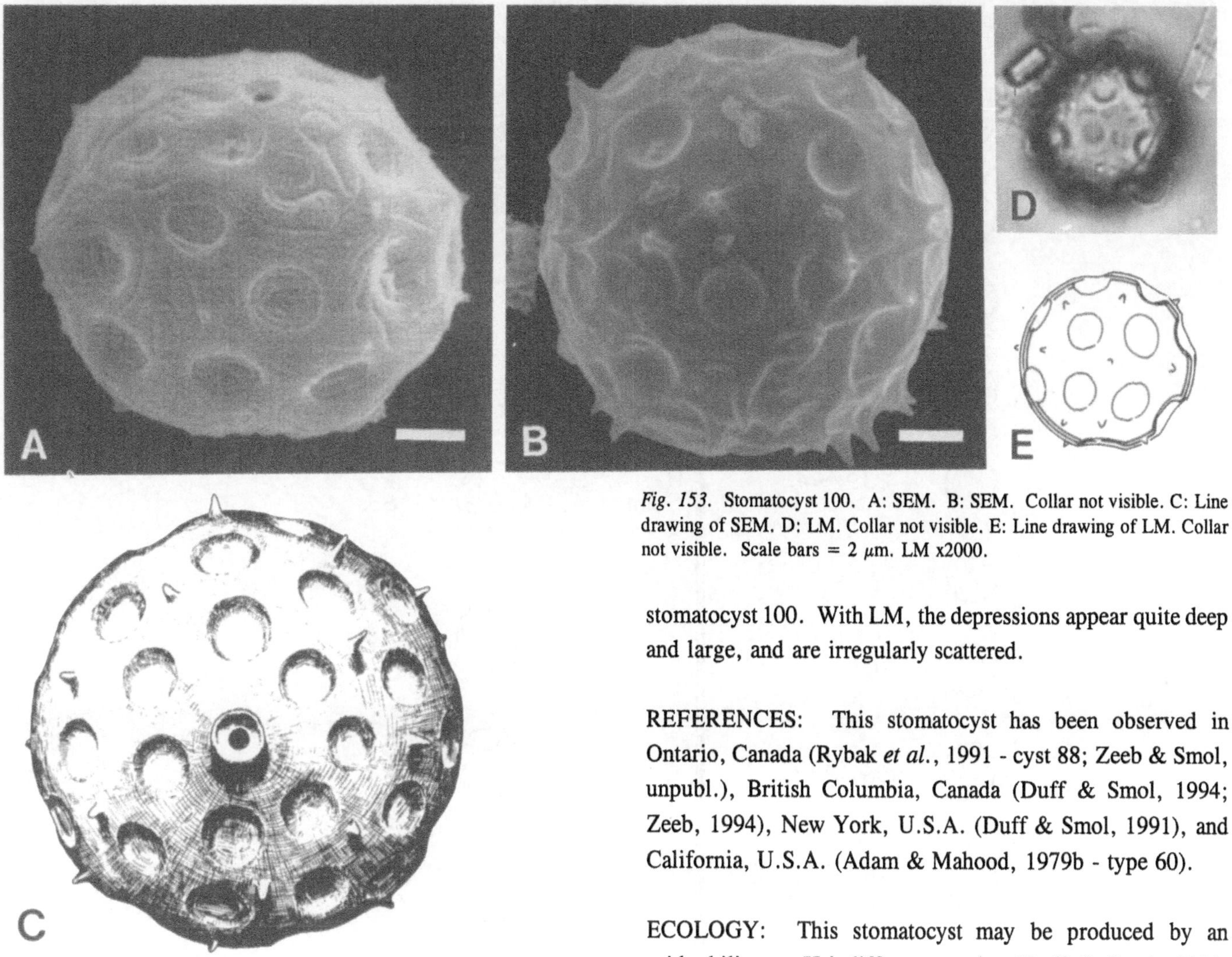

Fig. 153. Stomatocyst 100. A: SEM. B: SEM. Collar not visible. C: Line drawing of SEM. D: LM. Collar not visible. E: Line drawing of LM. Collar not visible. Scale bars = 2 μm. LM x2000.

possible that the different variations of stomatocyst 100 that we have described are produced by different taxa.

This morph can be distinguished from stomatocyst 99 by the fewer, and more irregularly scattered, depressions, and from stomatocyst 243 by the placement of the spines; on stomatocyst 100, spines are never found on the edge of the depressions.

LM DESCRIPTION: Duff & Smol (1991) grouped stomatocyst 100 with stomatocyst 99 as a collective category, although we now believe these two morphs could be differentiated with LM. The light micrograph given for stomatocysts 99/100 (Duff & Smol, 1991 - Fig. 141) is a misidentification; it is actually a picture of stomatocyst 243, although no SEM specimens of this cyst were found in their study. The LM tentatively attributed to stomatocyst 105 by Duff & Smol (1991 - Fig. 150), however, is probably stomatocyst 100. With LM, the depressions appear quite deep and large, and are irregularly scattered.

REFERENCES: This stomatocyst has been observed in Ontario, Canada (Rybak *et al.*, 1991 - cyst 88; Zeeb & Smol, unpubl.), British Columbia, Canada (Duff & Smol, 1994; Zeeb, 1994), New York, U.S.A. (Duff & Smol, 1991), and California, U.S.A. (Adam & Mahood, 1979b - type 60).

ECOLOGY: This stomatocyst may be produced by an acidophilic or pH-indifferent species (Duff & Smol, 1991; Rybak *et al.*, 1991).

Size continuum of cyst morphotypes

In Fig. 154, we have depicted those morphotypes or collective categories that are distinguishable using LM, drawn to scale. They are ordered according to increasing average cyst body dimensions. This figure serves to illustrate the large range in cyst size that may be encountered, from morphotypes that are less than 4 μm to morphotypes that are greater than 20 μm in diameter, as well as the great variety in cyst ornamentation. We expect that Fig. 154 will assist researchers to rapidly distinguish between different stomatocysts with LM.

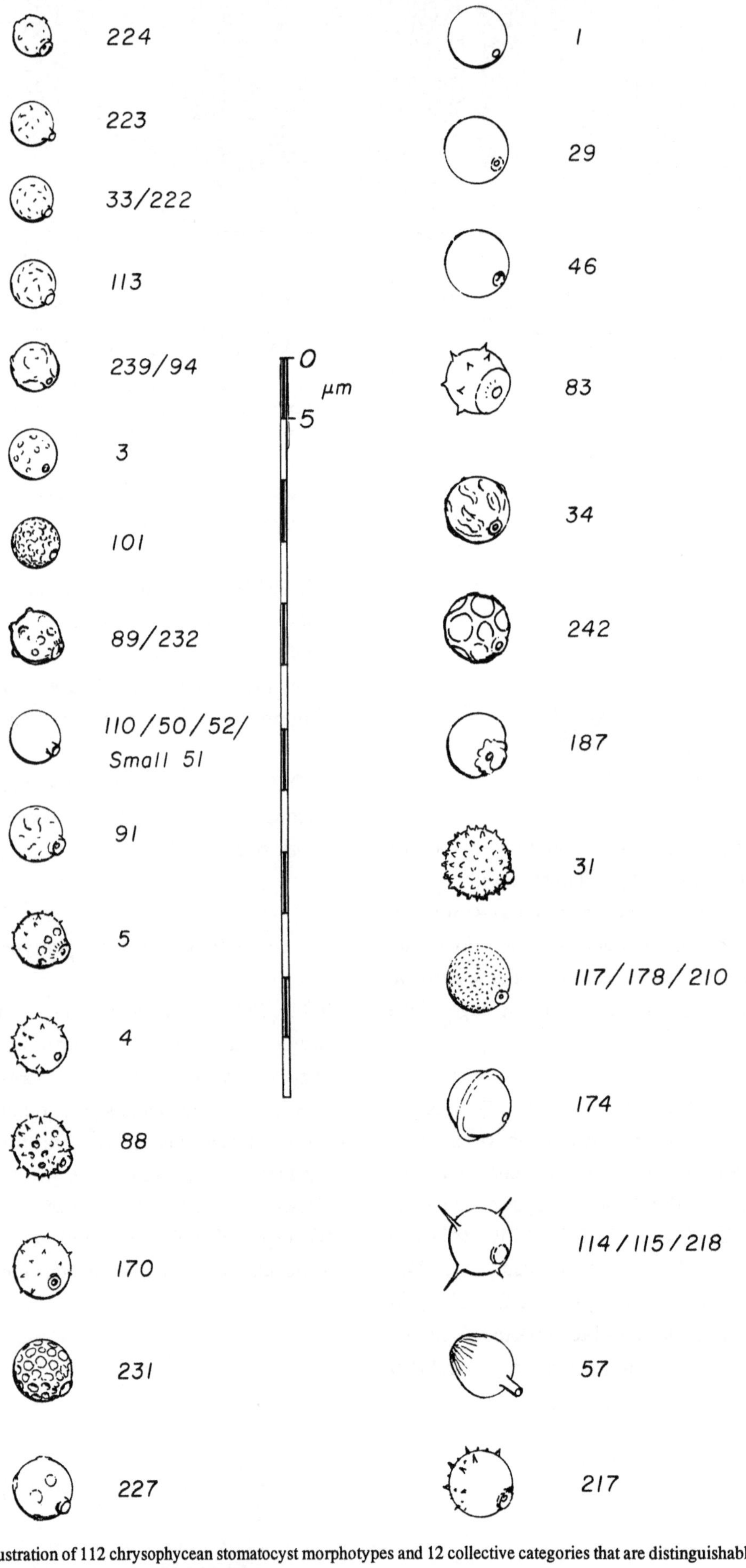

Fig. 154. Size continuum illustration of 112 chrysophycean stomatocyst morphotypes and 12 collective categories that are distinguishable using LM. Morphotypes are ordered according to increasing average cyst body dimensions.

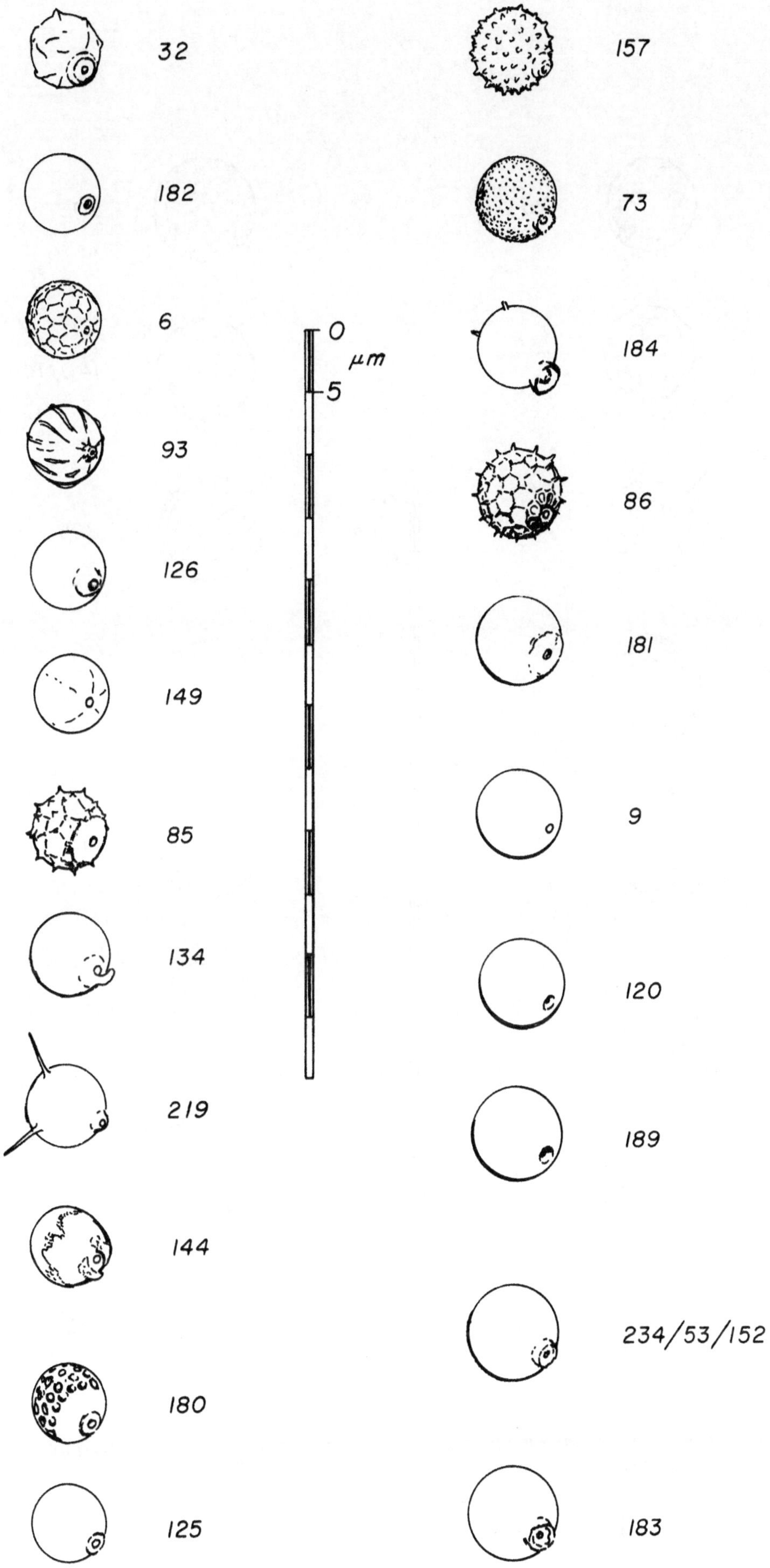

Fig. 154. continued

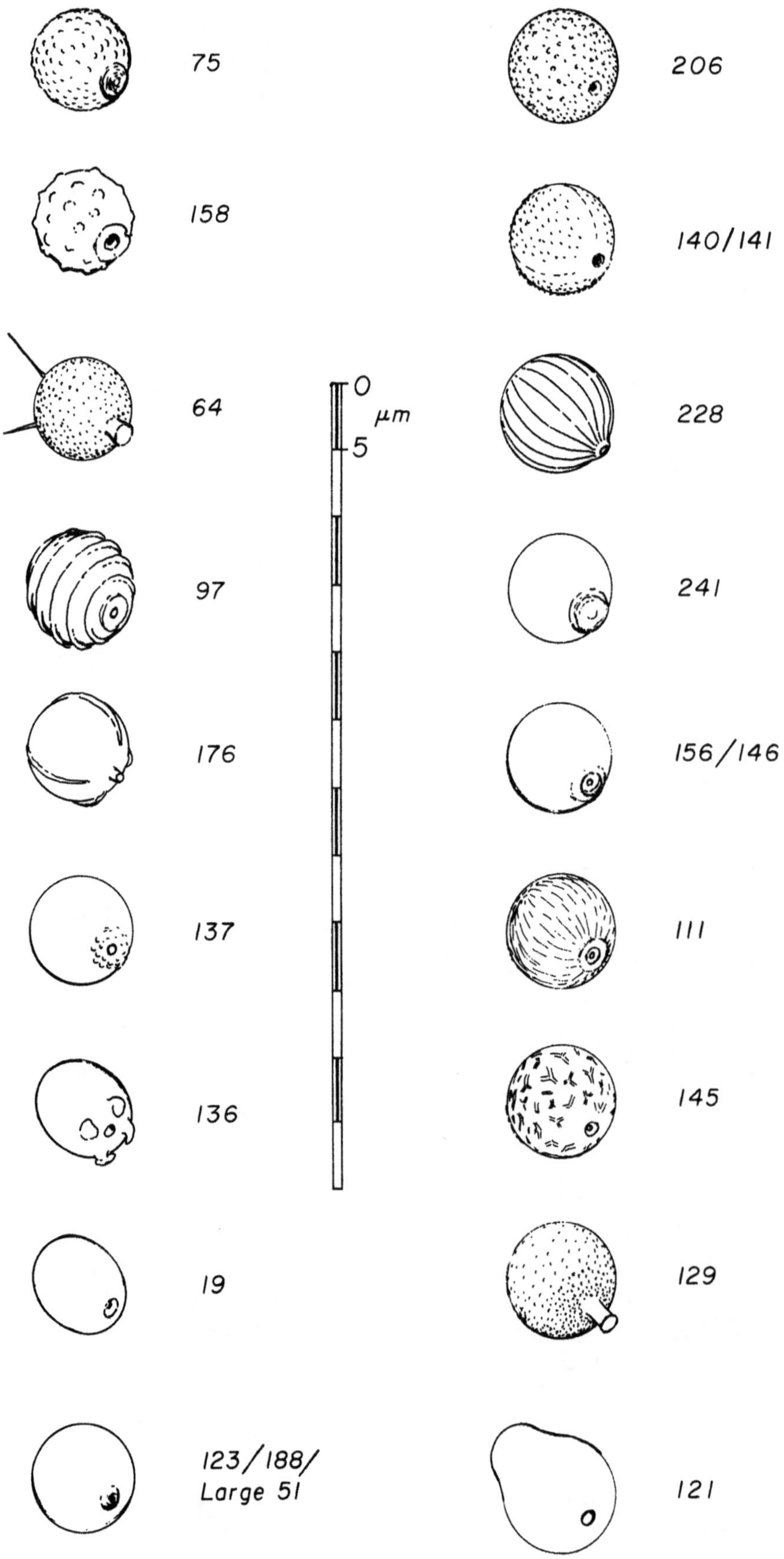

Fig. 154. continued

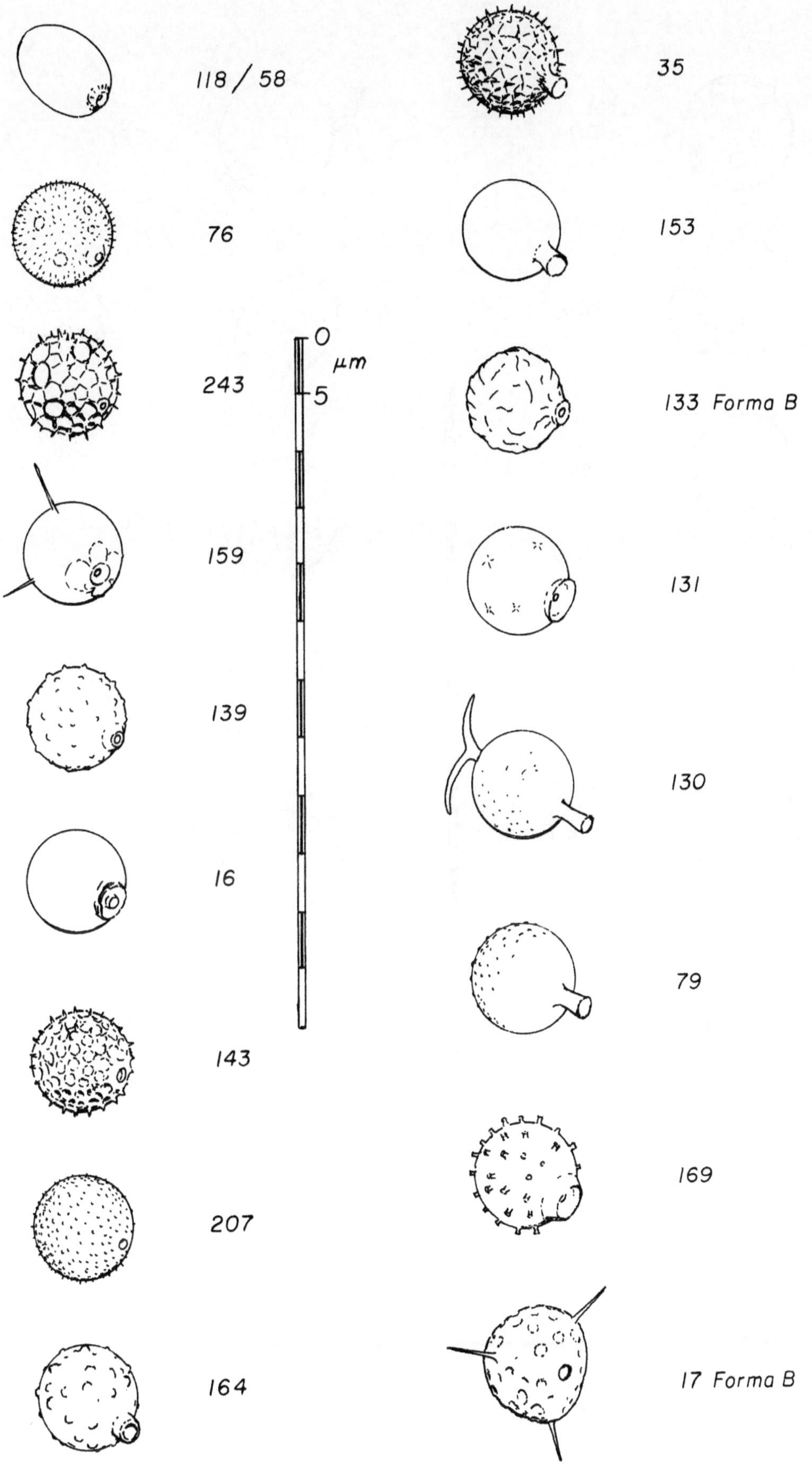

Fig. 154. continued

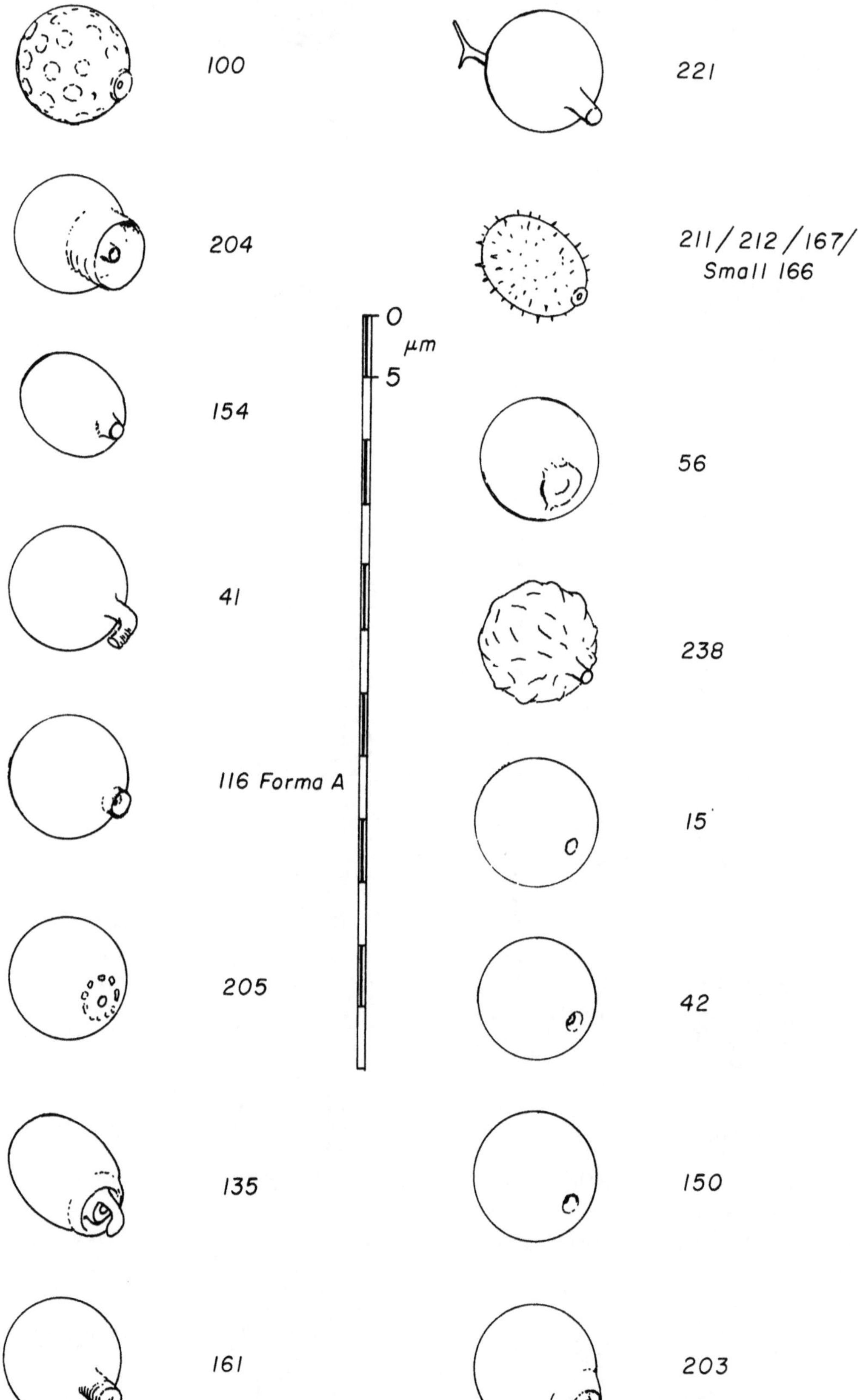

Fig. 154. continued

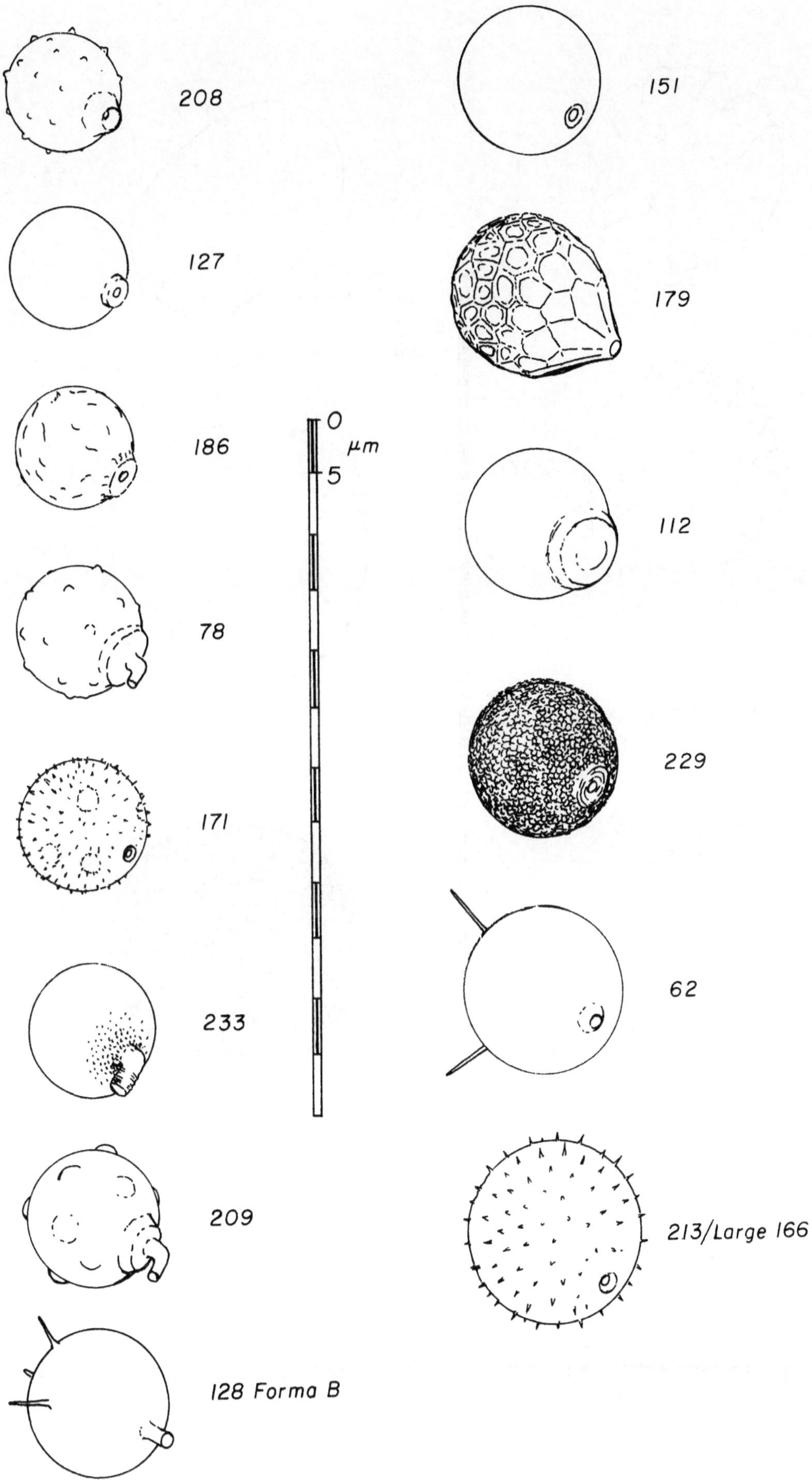

Fig. 154. continued

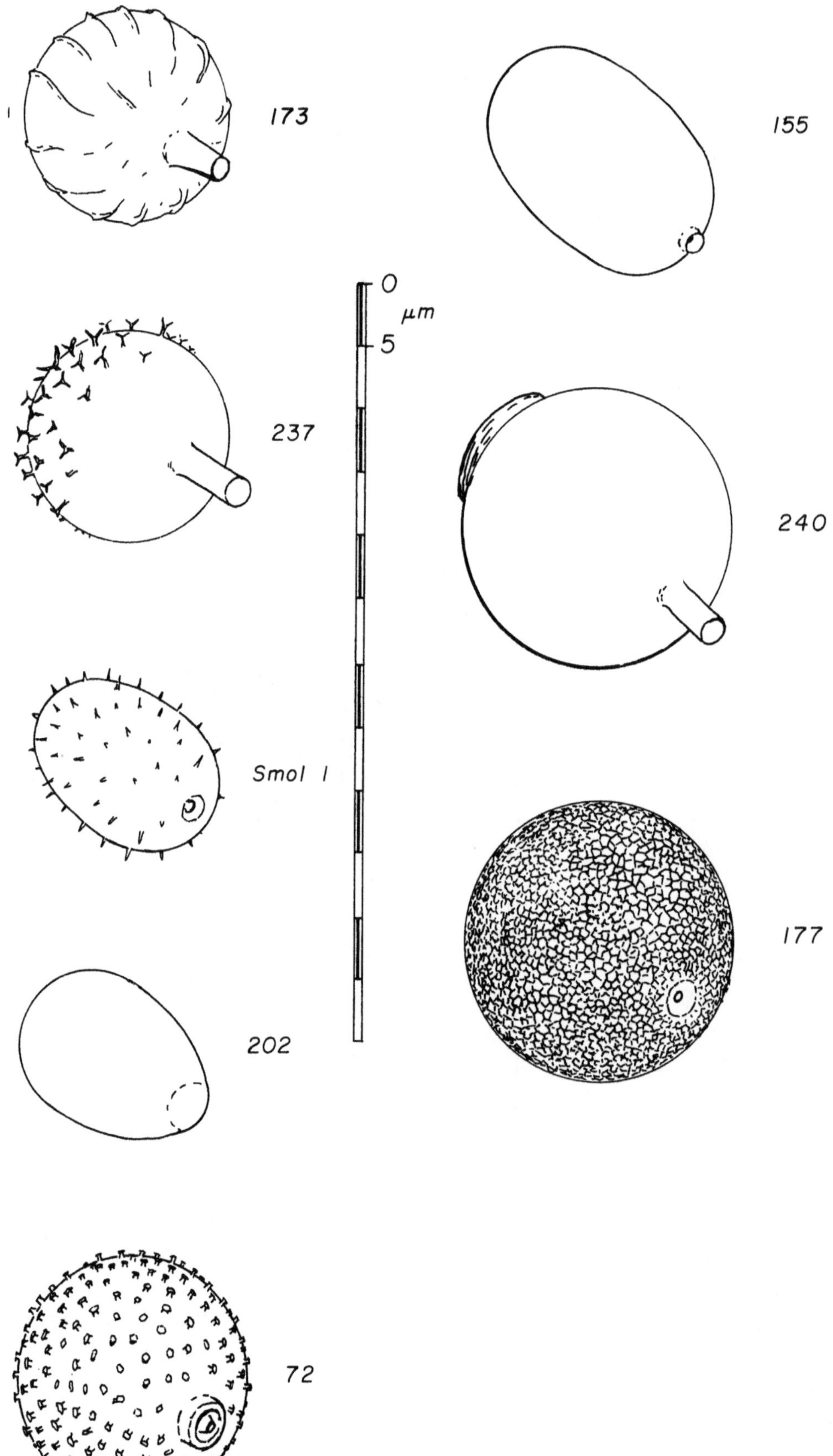

Fig. 154. continued

CHAPTER 5

Additional Morphotypes

The stomatocysts presented in this chapter are included for the sake of completion. They were described in our previous papers, but they are not considered fully in this Atlas for one of three reasons. The first group of morphotypes lacked sufficient SEM specimens of sufficient quality for a full description. Instead, we have presented a brief description, including the authors and year of the original publication, and a single scanning electron micrograph. The second group are cysts that we observed with light microscopy only. We have included a brief description of these morphotypes. The third group consists of morphotypes whose specimens have been reassigned to other stomatocyst numbers within our scheme (see Chapter 4).

Group 1: Morphotypes lacking sufficient SEM specimens

Stomatocyst 2, Duff & Smol, 1988 (Fig. 155A)

This is a small (diameter ca. 3.6 μm) spherical stomatocyst with a conical collar. With LM, this cyst would be counted with stomatocysts 110/50/52/small (< 5.0 μm) 51 (Figs. 12C, D).

Stomatocyst 8, Duff & Smol, 1988 (Fig. 155B)

This is a spherical stomatocyst (diameter 3.8-5.6 μm) with one to five long posterior spines; the pore/collar complex was not observed.

Stomatocyst 10, Duff & Smol, 1988 (Fig. 155C)

This is a spherical or slightly oblate stomatocyst (diameter 5.2-7.8 μm) with a low, obconical collar.

Stomatocyst 11, Duff & Smol, 1988 (Fig. 155D)

This stomatocyst is spherical (diameter ca. 8.9 μm) with a low, cylindrical collar and a single ridge extending longitudinally from the collar.

Stomatocyst 12, Duff & Smol, 1988 (Fig. 155G)

This is a spherical or slightly oblate stomatocyst (diameter 5.6-6.7 μm) with scattered verrucae, and possibly an obconical collar (not visible).

Stomatocyst 13, Duff & Smol, 1988 (Fig. 155E)

This is a spherical stomatocyst (diameter 7.9-92 μm) with a conical collar and a few conula or baculate spines.

Stomatocyst 14, Duff & Smol, 1988 (Fig. 155F)

This is a large spherical stomatocyst (diameter 11.2-11.5 μm) with a concave pore and irregularly scattered scabrae. This cyst may be an immature stomatocyst 142 (Fig. 160E).

Stomatocyst 18, Duff & Smol, 1988 (Fig. 155I)

This is an oval stomatocyst (6.8-7.0 x 9.6-10.4 μm) with a low cylindrical collar. This may be an undeveloped stomatocyst 118 (*Mallomonas akrokomos* Ruttner in Pascher - Fig. 39), although chrysophyte scales were never observed at

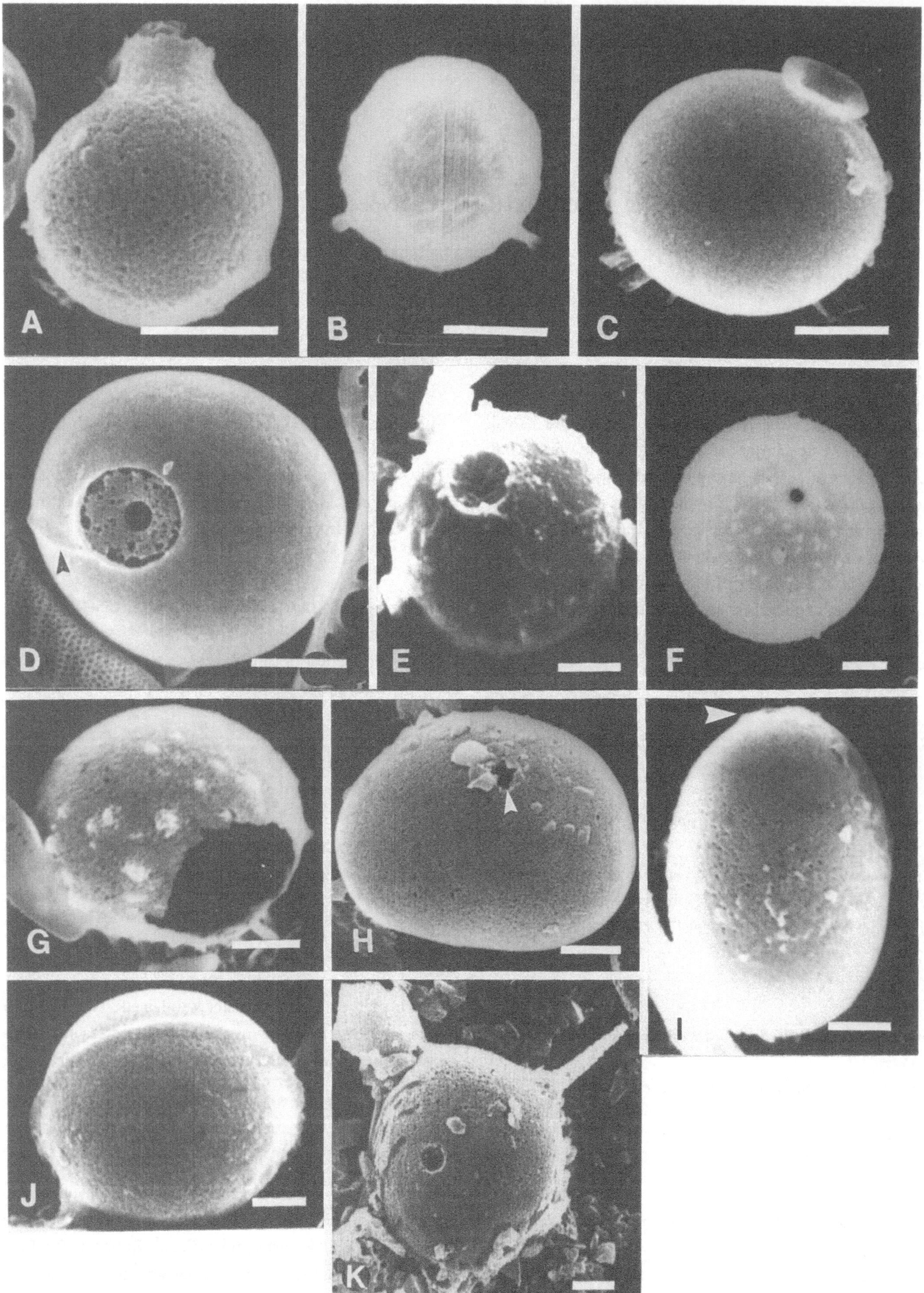
A
B
C
D
E
F
G
H
I
J
K

this site.

Stomatocyst 20, Duff & Smol, 1988 (Fig. 155H)

This stomatocyst is depressed ovate (5.3-7.6 x 6.9-8.0 μm) with a regular pore and no collar. This morph may be equivalent to stomatocyst 122 (Fig. 160B) but we lack enough SEM evidence to confirm this.

Stomatocyst 21, Duff & Smol, 1988 (Fig. 155J)

This is a spherical stomatocyst (diameter 4.8-8.9 μm) with a single circulus; the pore location is unknown. This morphotype may be stomatocyst 174 (Fig. 115).

Stomatocyst 27, Duff & Smol, 1989 (Fig. 155K)

This stomatocyst is pyramidal (corner-to-corner length 9.0-11.3 μm) with a small, low collar and long echinate spines at the base corners. It is very similar to the vegetative cell of *Chrysococcus furcatus* (Dolgoff) Nicholls ("stomatocyst" 17; Fig. 147).

Stomatocyst 28, Duff & Smol, 1989 (Fig. 156A)

This unornamented stomatocyst is oval (4.8-5.5 μm) with a complex collar. The primary collar is low and conical and is separated from the secondary conical collar by a sloping interannulus. Stomatocyst 85 (Fig. 139) has a similar collar morphology.

Stomatocyst 30, Duff & Smol, 1989 (Fig. 156B)

This stomatocyst is spherical (diameter ca. 5.5 μm) with a conical pore. Irregular verrucae surround the pore. This may be an incompletely developed specimen of a more ornamented morphotype.

Stomatocyst 36, Duff & Smol, 1989 (Fig. 156C)

This is a spherical stomatocyst (diameter 5.6-6.0 μm) with a wide, low, obconical collar. Possibly, it possesses a complex collar, previously described as siliceous deposition (Duff & Smol, 1989).

Stomatocyst 37, Duff & Smol, 1989 (Fig. 156D)

This is a smooth, spherical stomatocyst (diameter 10.4 μm) with a swollen annulus and a thin, conical collar.

Stomatocyst 38, Duff & Smol, 1989 (Fig. 156E)

This stomatocyst is spherical to slightly oblate (diameter ca. 7.9 μm) with a low, obconical collar and a few widely scattered conula.

Stomatocyst 39, Duff & Smol, 1989 (Fig. 156G)

This stomatocyst is spherical (diameter ca. 10.3 μm). It has a small conical collar with a rounded apex and possibly a thin, planar annulus. It has a sparsely scabrate surface.

Stomatocyst 40, Duff & Smol, 1989 (Fig. 156F)

This stomatocyst is slightly oblate (9.0 x 10.0 μm) with a small, conical collar and irregularly scattered scabrae. This may be an undeveloped form of stomatocyst 207 (Fig. 65).

Stomatocyst 47, Duff & Smol, 1991 (Fig. 156H)

This is a spherical stomatocyst (diameter ca. 6.5 μm) with a deep conical pore and a very slight conical collar.

Fig. 155. Stomatocyst morphotypes that lack sufficient SEM specimens for a full description. A: Stomatocyst 2. B: Stomatocyst 8. Collar not visible. C: Stomatocyst 10. D: Stomatocyst 11. Single ridge indicated by arrow. E: Stomatocyst 13. F: Stomatocyst 14. G: Stomatocyst 12. Collar not visible. H: Stomatocyst 20. Pore indicated by arrow. I: Stomatocyst 18. Collar indicated by arrow. J: Stomatocyst 21. Pore not visible. K: Stomatocyst 27. Scale bars = 2 μm.

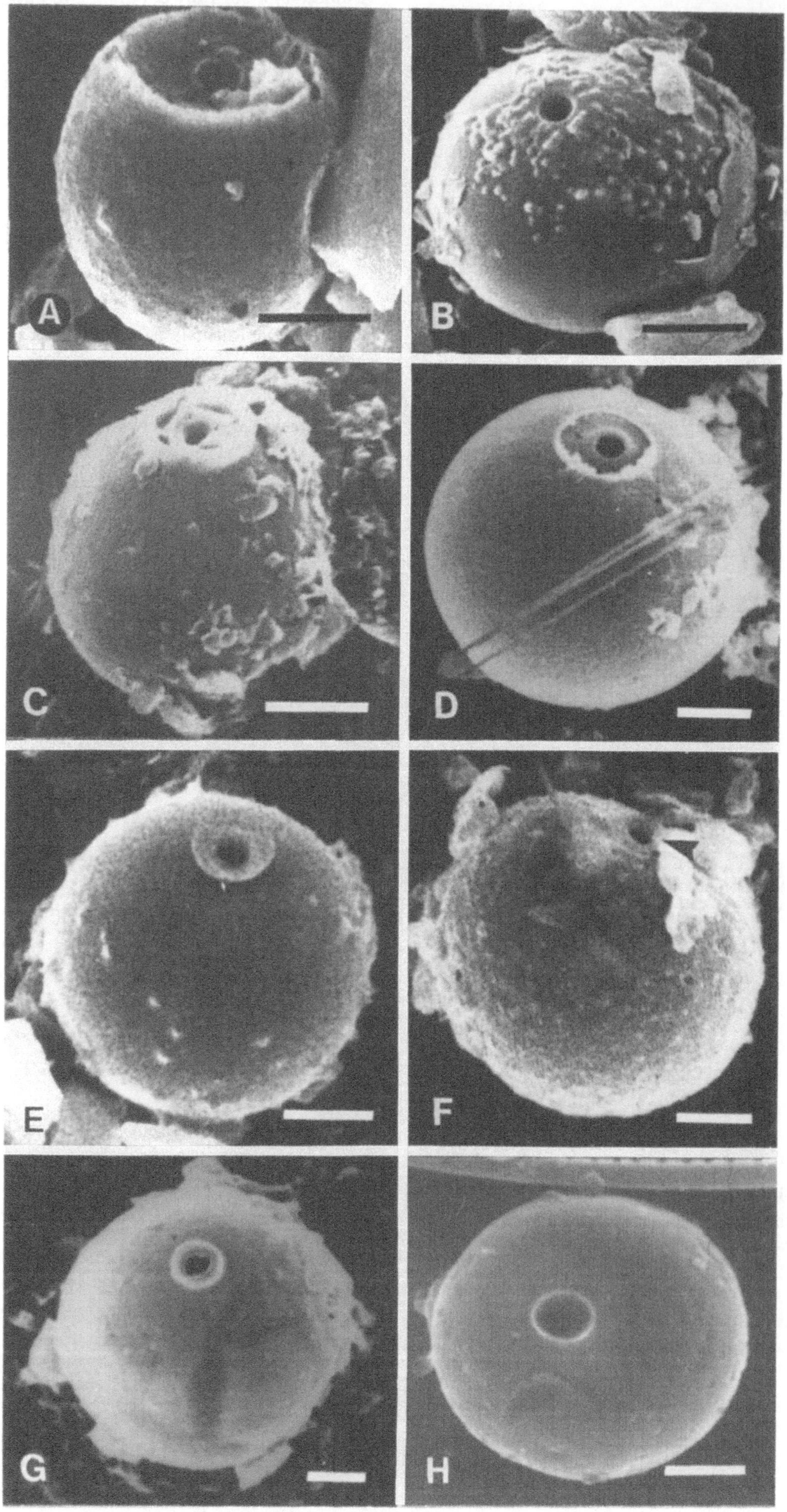
A
B
C
D
E
F
G
H

Stomatocyst 55, Duff & Smol, 1991 (Fig. 157A)

This stomatocyst is spherical (diameter ca. 10.9 μm) with a small conical collar and a slightly irregular, possibly scabrate, surface texture.

Stomatocyst 59, Duff & Smol, 1991 (Fig. 157B)

This stomatocyst is depressed ovate in shape (4.8 x 6.4 μm) with a low cylindrical collar and a flat planar annulus. It may be a variant of stomatocyst 50 (Fig. 24).

Stomatocyst 60, Duff & Smol, 1991 (Fig. 157C)

This is an oblate stomatocyst (4.2-6.4 μm x 4.5-7.1 μm) with a low, conical collar and approximately 10 to 20 spines in the posterior hemisphere.

Stomatocyst 61, Duff & Smol, 1991 (Fig. 157E)

This is a spherical stomatocyst (diameter ca. 5.7 μm) with a regular pore and either no collar or a short conical collar. The posterior hemisphere is ornamented with approximately four long, thin, non-bifurcating, echinate spines.

Stomatocyst 67, Duff & Smol, 1991 (Fig. 157D)

This is a spherical stomatocyst (diameter ca. 7.9 μm) with a concave pore. The ornamentation consists of verrucae with flattened apices scattered over the entire cyst surface.

Stomatocyst 68, Duff & Smol, 1991 (Fig. 157F)

This is an oblate stomatocyst (5.1 x 5.6 μm) with a regular pore and several low, rounded verrucae.

Stomatocyst 69, Duff & Smol, 1991 (Fig. 157G)

This stomatocyst is spherical (diameter ca. 6.2 μm) with a small conical collar. Low verrucae are regularly spaced over the entire cyst surface.

Stomatocyst 70, Duff & Smol, 1991 (Fig. 157H)

This stomatocyst, as originally described, displayed considerable variability. After further consideration, we believe that at least two distinct morphotypes have been included within stomatocyst 70; however, we do not have sufficient SEM specimens for full descriptions of the individual morphs. The cysts in this group are spherical (diameter 3.4-5.5 μm) with a cylindrical or obconical collar and approximately equidistant conula or echinate spines. They are likely indistinguishable from stomatocyst 31 (Figs. 70D, E, and F) with LM.

Stomatocyst 71, Duff & Smol, 1991 (Fig. 157I)

This stomatocyst is oval (12.0 x 13.7 μm) with a low, conical collar and evenly spaced conula or short echinate spines with flattened apices.

Stomatocyst 74, Duff & Smol, 1991 (Fig. 158A)

This is a spherical stomatocyst (diameter ca. 9.1 μm) with a low, widely obconical collar. The cyst surface is ornamented with low, irregularly scattered echinate spines with hooked apices.

Stomatocyst 77, Duff & Smol, 1991 (Fig. 158B)

This stomatocyst is spherical (diameter ca. 5.5 μm) with a low conical collar. The cyst surface is densely scabrate.

Fig. 156. Stomatocyst morphotypes that lack sufficient SEM specimens for a full description. A: Stomatocyst 28. B: Stomatocyst 30. C: Stomatocyst 36. D: Stomatocyst 37. E: Stomatocyst 38. F: Stomatocyst 40. Collar indicated by arrow. G: Stomatocyst 39. H: Stomatocyst 47. Scale bars = 2 μm.

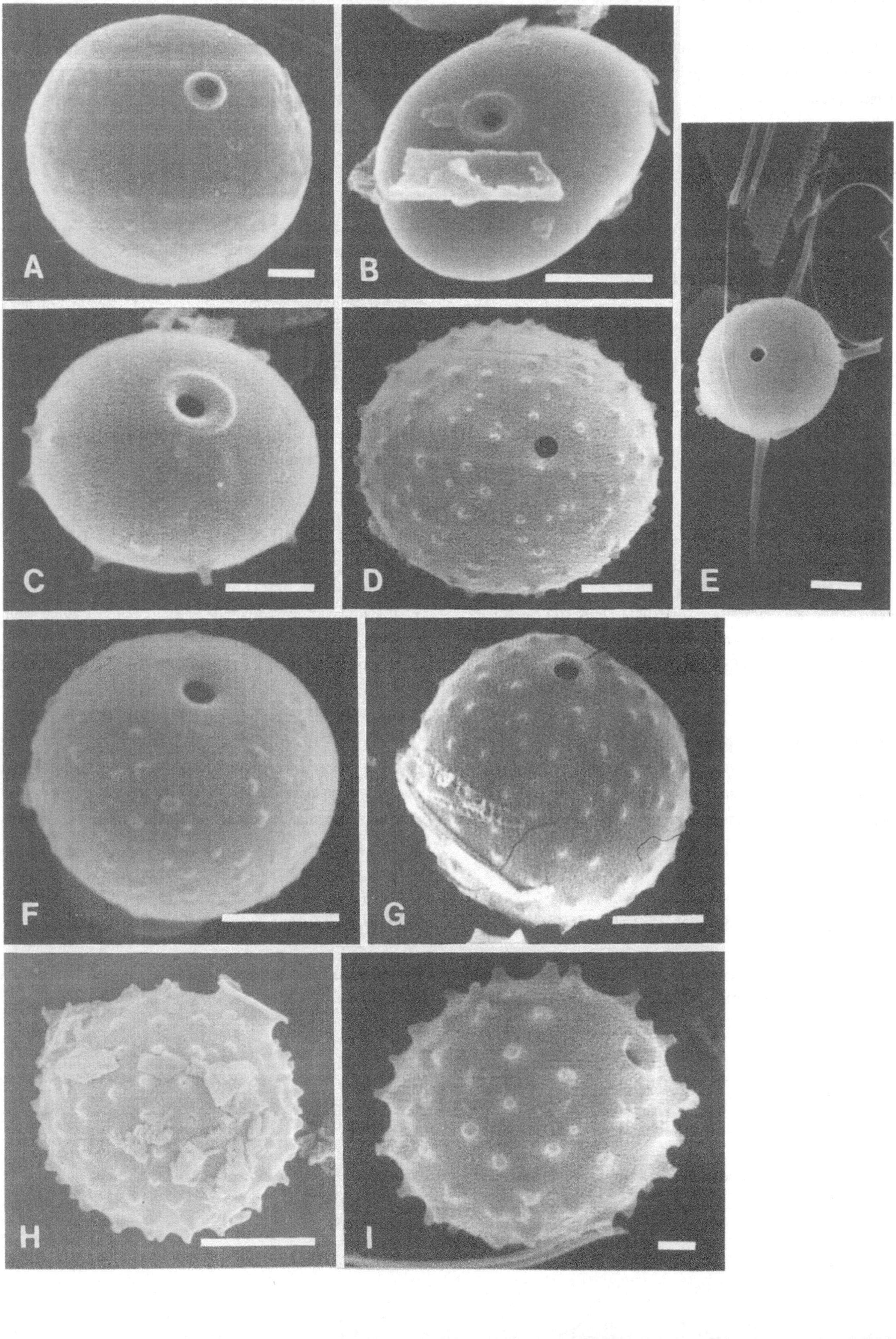
A
B
C
D
E
F
G
H
I

Stomatocyst 80, Duff & Smol, 1991 (Fig. 158C)

This stomatocyst is large and widely oval (10.0 x 10.5 μm) with a very low, broadly rounded, conical collar and several widely scattered baculate spines.

Stomatocyst 87, Duff & Smol, 1991 (Fig. 158D)

This stomatocyst closely resembles stomatocyst 88 (Fig. 144). It is small and spherical (diameter ca. 4.5 μm) with scattered shallow depressions and long echinate spines. It has a conical pore but no collar.

Stomatocyst 90, Duff & Smol, 1991 (Fig. 158E)

This is a small, spherical stomatocyst (diameter 3.1-5.6 μm) with a conical collar and short lunate ridges. It is identical to stomatocyst 89 (Fig. 149) but lacks depressions.

Stomatocyst 92, Duff & Smol, 1991 (Fig. 158F)

This is a very small, spherical stomatocyst (diameter ca. 3.0 μm) with a low, wide, obconical collar. The surface is ornamented with wavy ridges with struts; one or more ridges may connect with the collar base. This morph has not been identified definitely with LM, and is probably indistinguishable from stomatocyst 91 (Fig. 112).

Stomatocyst 95, Duff & Smol, 1991 (Fig. 158G)

This stomatocyst is small and spherical (diameter ca. 3.2 μm) with a regular pore and two to three longitudinal ridges. This morph is most likely an immature morph of another cyst.

Stomatocyst 96, Duff & Smol, 1991 (Fig. 159A)

This stomatocyst is spherical (diameter ca. 6.1 μm) with a short cylindrical collar and a single longitudinal circulus that is continuous with the collar. Stomatocyst 175 (Fig. 161E) is similar but ornamented.

Stomatocyst 102, Duff & Smol, 1991 (Fig. 159B)

This spherical stomatocyst (diameter ca. 8.0 μm) has a low, wide, rounded conical collar. The pore is not visible on our single SEM specimen (Fig. 159B). It is ornamented with both low ridges and a reticulum with roughly circular lacunae.

Stomatocyst 103, Duff & Smol, 1991 (Fig. 159C)

This is a spherical stomatocyst (diameter ca. 9.5 μm) with a short conical collar and several scattered shallow depressions.

Stomatocyst 104, Duff & Smol, 1991 (Fig. 159D)

This stomatocyst is spherical (diameter ca. 13.4 μm) with sparsely scattered shallow circular depressions. The large hole in the cyst wall probably represents the future location of the collar. It is likely that stomatocysts 103 (Fig. 159C) and 104 represent the same morphotype.

Stomatocyst 105, Duff & Smol, 1991 (Fig. 159E)

Our only SEM specimen of this stomatocyst (Fig. 159E) is large, spherical (diameter ca. 12.6 μm), and broken. The pore/collar was not observed. The surface is set with a few deep depressions.

Fig. 157, facing page. Stomatocyst morphotypes that lack sufficient SEM specimens for a full description. A: Stomatocyst 55. B: Stomatocyst 59. C: Stomatocyst 60. D: Stomatocyst 67. E: Stomatocyst 61. F: Stomatocyst 68. G: Stomatocyst 69. H: Stomatocyst 70. I: Stomatocyst 71. Scale bars = 2 μm.

Fig. 158, next page. Stomatocyst morphotypes that lack sufficient SEM specimens for a full description. A: Stomatocyst 74. B: Stomatocyst 77. C: Stomatocyst 80. Spine indicated by arrow. D: Stomatocyst 87. E: Stomatocyst 90. Collar indicated by arrow. F: Stomatocyst 92. Collar indicated by arrow. G: Stomatocyst 95. Scale bars = 2 μm.

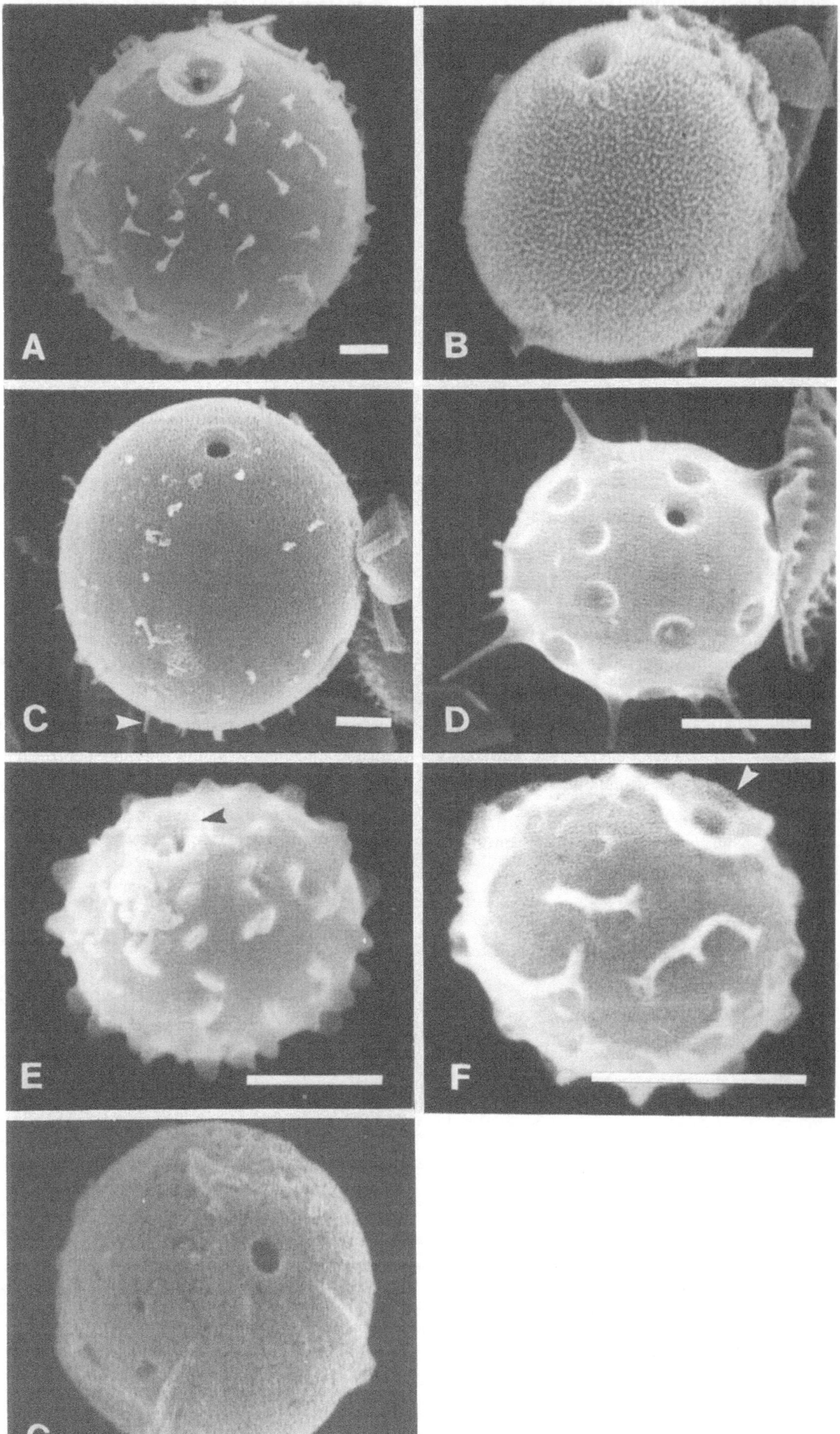
A
B
C
D
E
F
G

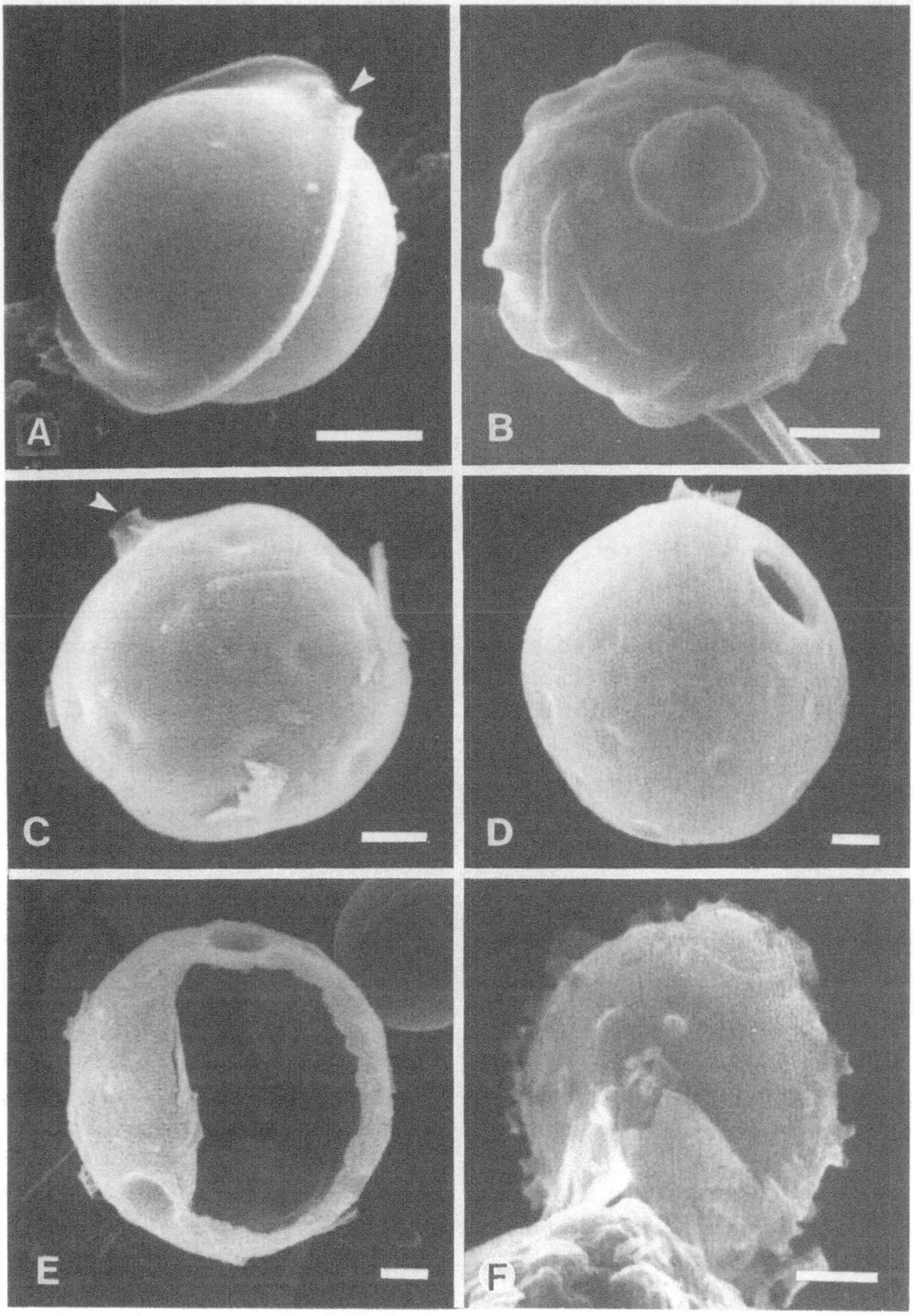
A
B
C
D
E
F

Stomatocyst 106, Duff & Smol, 1991 (Fig. 159F)

This stomatocyst is spherical (diameter ca. 8.5 μm) with a wide, cylindrical collar. The pore is obscured by a siliceous cap. This morphotype is ornamented with low baculate spines, many of which bifurcate. It may be equivalent to stomatocyst 169 (Fig. 93).

Stomatocyst 119, Zeeb *et al.*, 1990 (Fig. 160A)

This is a small, pyramidal cyst (corner-to-corner length 5.5-9.2 μm) with a small conical collar surrounding a concave pore. The surface is smooth or ornamented with several conula.

Stomatocyst 122, Duff & Smol in Duff *et al.*, 1992 (Fig. 160B)

This stomatocyst is obovate (4.1-4.6 x 5.5-6.4 μm) with a subpolar concave pore. This morph may be equivalent to stomatocyst 20 (Fig. 155H) but we lack sufficient SEM specimens to confirm this.

Stomatocyst 132, Duff & Smol in Duff *et al.*, 1992 (Fig. 160C)

This stomatocyst is spherical (diameter 5.5-5.7 μm) with a large obconical collar, a swollen annulus and a regular pore. It is similar to stomatocyst 163 (Fig. 161A).

Stomatocyst 138, Duff & Smol in Duff *et al.*, 1992 (Fig. 160D)

This is an oval to widely ovate stomatocyst (ca. 8.1 x 8.4 μm) with a wide, conical collar and a few conula located in the posterior hemisphere.

Stomatocyst 142, Duff & Smol in Duff *et al.*, 1992 (Fig. 160E)

This stomatocyst is spherical to slightly oblate (diameter 11.4-12.5 μm) with a low conical collar. It is ornamented with very densely spaced verrucae. Stomatocyst 207 (Fig. 65) is similar but smaller with differing pore/collar features.

Stomatocyst 147, Zeeb & Smol in Pienitz *et al.*, 1992 (Fig. 160F)

This stomatocyst is spherical (diameter ca. 5.5 μm) with a small conical pore, a planar annulus, and a wide conical collar.

Stomatocyst 160, Zeeb & Smol, 1993a (Fig. 160G)

This stomatocyst is spherical (diameter ca. 5.6 μm) with a conical pore surrounded by a very thick siliceous spiralling ridge.

Stomatocyst 162, Zeeb & Smol, 1993a (Fig. 160H)

This is a spherical stomatocyst (diameter 5.0-6.2 μm) with a cylindrical collar. The cyst surface is smooth or ornamented with a few conula.

Stomatocyst 163, Zeeb & Smol, 1993a (Fig. 161A)

This cyst is spherical (diameter ca. 5.4 μm) with a regular pore surrounded by a sloping planar annulus and a low, conical collar. The posterior hemisphere may be ornamented with low scabrae or verrucae, but additional SEM evidence is needed to confirm this. This morph resembles stomatocyst 132 (Fig. 160C).

Fig. 159, previous page. Stomatocyst morphotypes that lack sufficient SEM specimens for a full description. A: Stomatocyst 96. Collar indicated by arrow. B: Stomatocyst 102. C: Stomatocyst 103. Collar indicated by arrow. D: Stomatocyst 104. E: Stomatocyst 105. F: Stomatocyst 106. Scale bars = 2 μm.

Fig. 160, facing page. Stomatocyst morphotypes that lack sufficient SEM specimens for a full description. A: Stomatocyst 119. Conule indicated by arrow. B: Stomatocyst 122. C: Stomatocyst 132. D: Stomatocyst 138. Conule indicated by arrow. E: Stomatocyst 142. F: Stomatocyst 147. G: Stomatocyst 160. H: Stomatocyst 162. Scale bars = 2 μm.

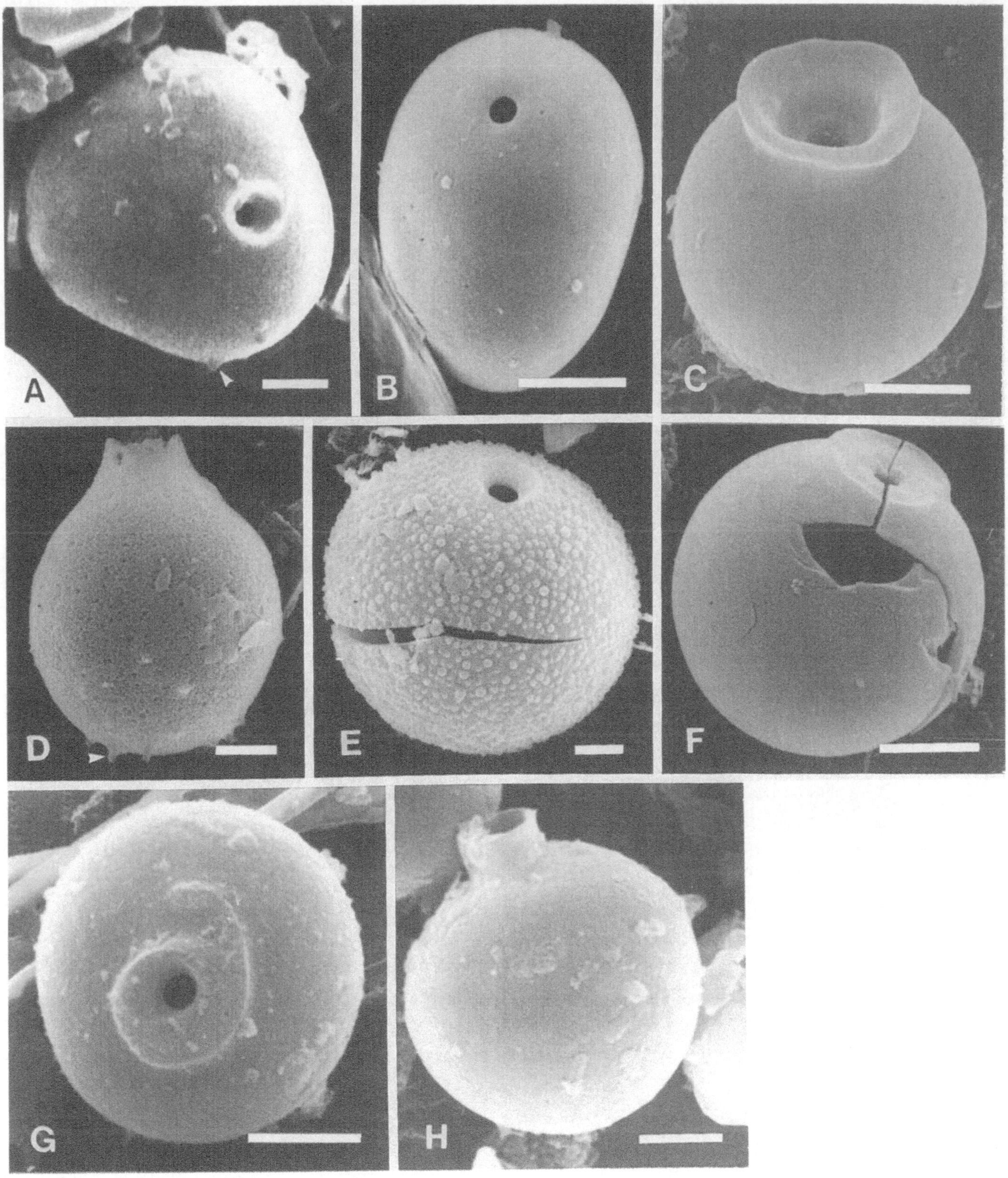
A
B
C
D
E
F
G
H

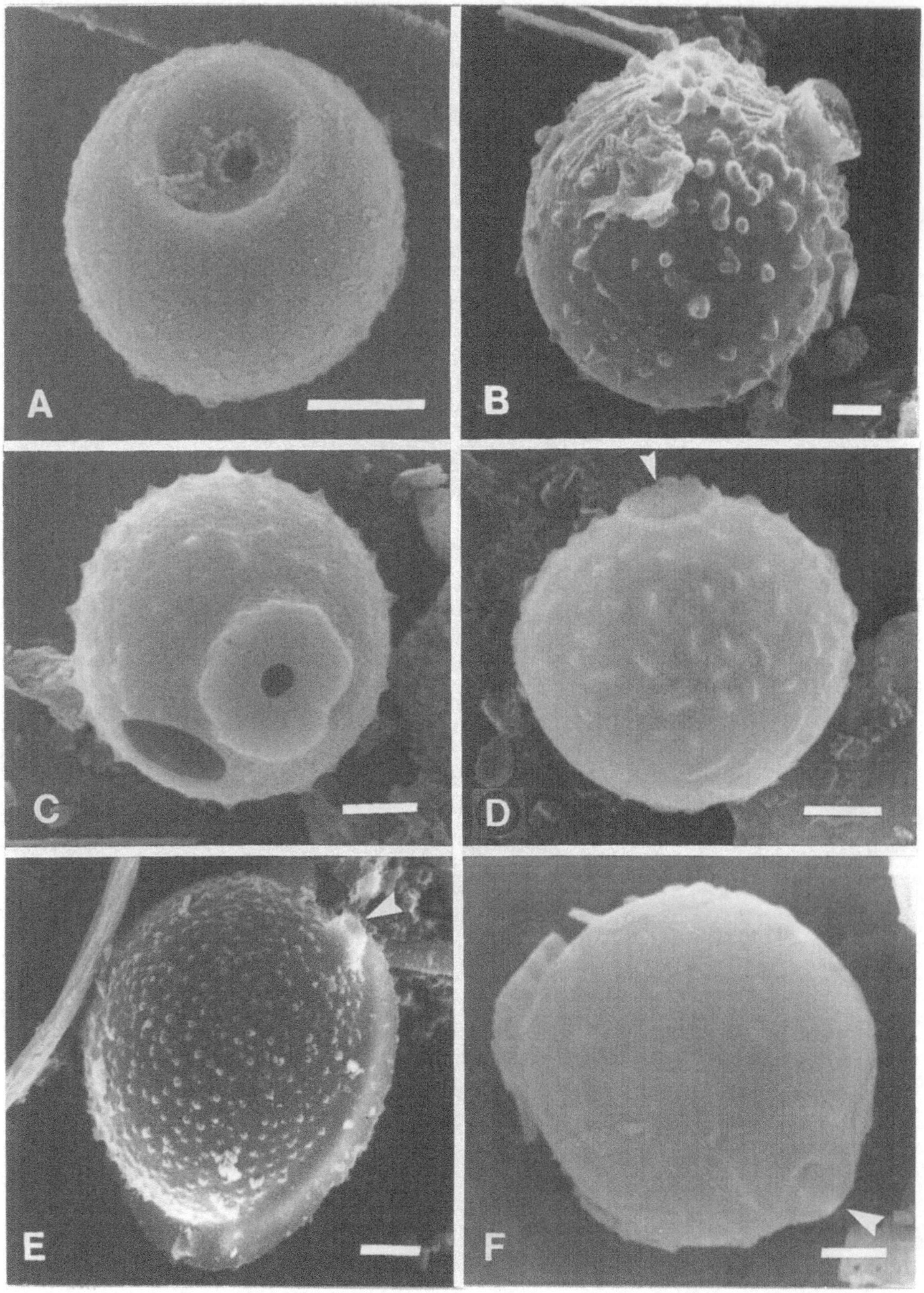
A
B
C
D
E
F

Stomatocyst 165, Zeeb & Smol, 1993a (Fig. 161B)

This cyst is large and spherical (diameter ca. 13 μm) with a cylindrical collar. Short thick conula are found over the entire cyst surface, but are concentrated most densely in the collar region.

Stomatocyst 168, Zeeb & Smol, 1993a (Fig. 161C)

This stomatocyst is spherical (diameter 6.6-7.9 μm) with a wide low cylindrical collar and irregularly scattered conula.

Stomatocyst 172, Zeeb & Smol, 1993a (Fig. 161D)

This is a spherical stomatocyst (diameter 7.9-8.0 μm) with a low, cylindrical to obconical collar. The cyst surface is ornamented with numerous low ridges of varying lengths. Stomatocyst 172 is similar to stomatocyst 111 (*Spiniferomonas trioralis* (Takahashi) Preisig & Hibberd - Fig. 106).

Stomatocyst 175, Zeeb & Smol, 1993a (Fig. 161E)

This stomatocyst is spherical (diameter ca. 9.5 μm) with a conical collar. It is ornamented with a single incomplete longitudinal circulus and numerous verrucae. It resembles stomatocyst 96 (Fig. 159A), an unornamented morphotype.

Stomatocyst 190, Zeeb & Smol in Zeeb *et al.*, subm. (Fig. 161F)

This stomatocyst is smooth and spherical (diameter 8.4-10.5 μm) with a cylindrical to slightly conical collar. It is similar to stomatocyst 126 (Fig. 18) but is larger.

Stomatocyst 191, Zeeb & Smol in Zeeb *et al.*, subm. (Fig. 162A)

This is a large spherical stomatocyst (diameter 9.5-9.7 μm) with a cylindrical collar and a smooth surface. A wide, low ridge running latitudinally below the collar may be a developing secondary collar.

Stomatocyst 192, Zeeb & Smol in Zeeb *et al.*, subm. (Fig. 162C)

This stomatocyst is spherical to oval (diameter ca. 8.7-9.2 μm). It has a high obconical collar with an uneven, corona-like apex.

Stomatocyst 193, Zeeb & Smol in Zeeb *et al.*, subm. (Fig. 162D)

This stomatocyst is spherical (diameter ca. 5.2 μm) with a complex collar. The primary collar is short and cylindrical, and the secondary collar is cylindrical with a slightly obconical apex. The cyst is ornamented with short echinate spines and lunate ridges.

Stomatocyst 195, Zeeb & Smol in Zeeb *et al.*, subm. (Fig. 162B)

This is a large, spherical to oblate stomatocyst (diameter ca. 13.5 μm) with a regular pore and a low, acute, conical collar. It is densely ornamented with thick, short, baculate spines, except in the immediate pore region.

Stomatocyst 196, Duff & Smol, 1994 (Fig. 162E)

This is a large, smooth, obovate stomatocyst (9.9-12.5 x 13.4-15.4 μm) with a shallow concave pore.

Stomatocyst 197, Duff & Smol, 1994 (Fig. 162F)

This stomatocyst is large and spherical or widely oval (diameter >15.0 μm) with a low, cylindrical collar, a gently sloping to flat planar annulus, and a regular pore. It is similar to stomatocyst 127 (Fig. 27), except that it is larger.

Fig. 161. Stomatocyst morphotypes that lack sufficient SEM specimens for a full description. A: Stomatocyst 163. B: Stomatocyst 165. C: Stomatocyst 168. D: Stomatocyst 172. Collar indicated by arrow. E: Stomatocyst 175. Collar indicated by arrow. F: Stomatocyst 190. Collar indicated by arrow. Scale bars = 2 μm.

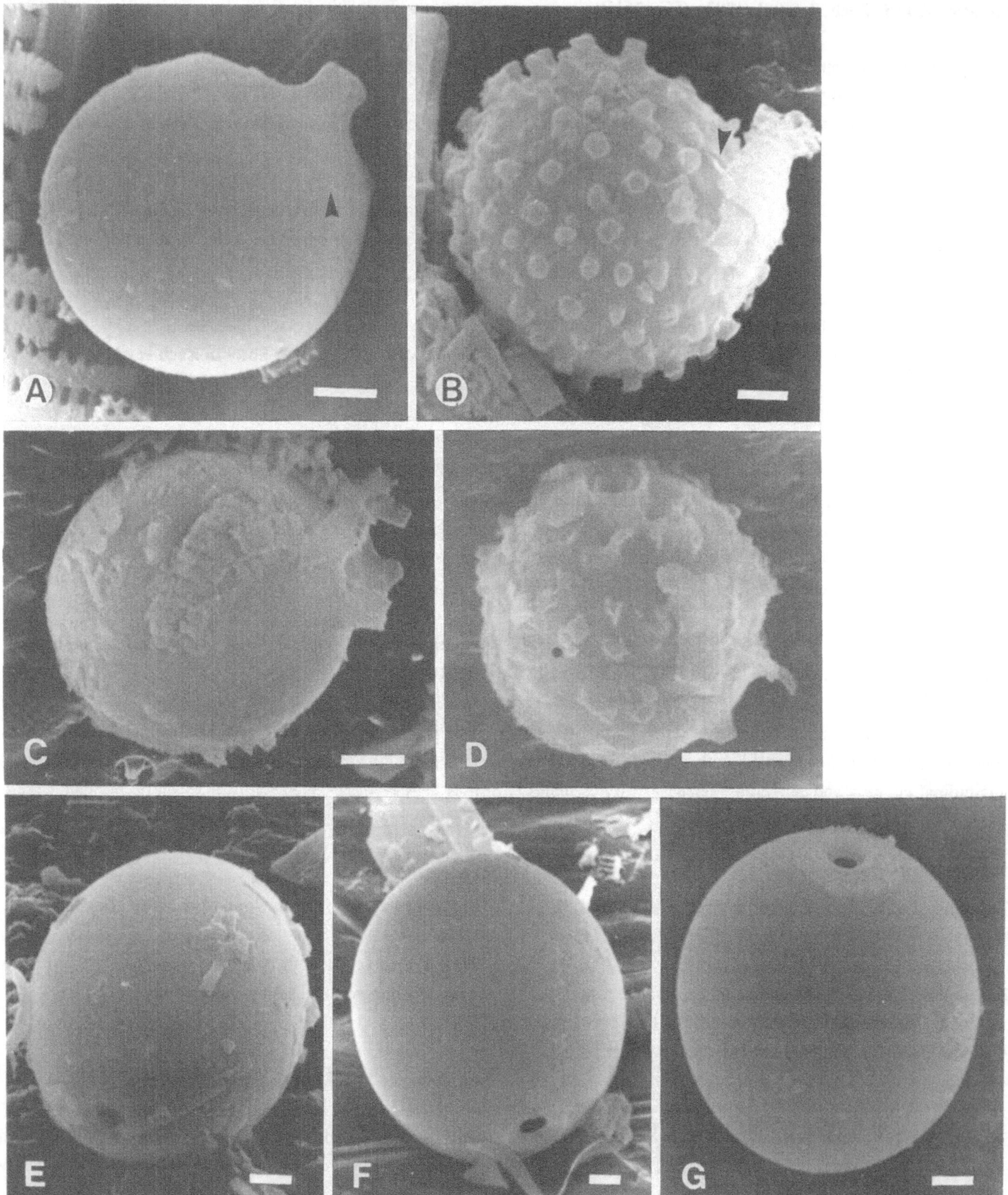
A
B
C
D
E
F
G

Stomatocyst 200, Duff & Smol, 1994 (Fig. 162G)

This is an oval stomatocyst (13.0-13.3 x 14.6-15.7 μm). The collar is low, wide, and roughly cylindrical with an irregular outline and texture.

Stomatocyst 201, Duff & Smol, 1994 (Fig. 163A)

This stomatocyst is large and oval (ca. 12.9 x 20.3 μm) with a cylindrical or slightly obconical collar. It is probably produced by *Mallomonas lychenensis* Conrad (Harris, 1953; Nygaard, 1956 - as *M. tesselata* Nygaard).

Stomatocyst 214, Duff & Smol, 1994 (Fig. 163B)

This stomatocyst is oval (ca. 9.4 x 11.7 μm) with a wide, obconical collar. The collar apex and the cyst surface are ornamented with echinate spines with corona-like apices. This morphotype is probably the cyst of *Mallomonas torquata* Asmund & Cronberg (Cronberg, 1980), although our specimens have longer collar spines than she described.

Stomatocyst 216, Duff & Smol, 1994 (Fig. 163C)

This is a large, spherical stomatocyst (11.6-13.0 μm) ornamented with echinate spines with spiralling basal striations, and several circuli, also ornamented with striations, that may be developing spines. The pore/collar morphology is unknown.

Stomatocyst 220, Duff & Smol, 1994 (Fig. 163D)

This is a spherical stomatocyst (diameter ca. 5.7 μm) with a conical primary collar, and a cylindrical to obconical secondary collar. A low ridge runs latitudinally below the secondary collar, and several echinate, non-bifurcating spines are found in the posterior hemisphere.

Stomatocyst 225, Duff & Smol, 1994 (Fig. 163E)

This is a spherical stomatocyst (diameter ca. 7.2 μm) with a cylindrical collar, and straight lunate ridges randomly oriented over the entire cyst surface.

Stomatocyst 226, Duff & Smol, 1994 (Fig. 163F)

This is a large, spherical stomatocyst (diameter ca. 13.6 μm) with a narrow cylindrical collar, and long, wavy, generally longitudinal ridges.

Stomatocyst 230, Duff & Smol, 1994 (Fig. 163G)

This is an oval stomatocyst (ca. 9.7 x 10.3 μm) with a low conical collar. It is ornamented with a regular reticulum with small, circular lacunae; conula are found at many of the reticular interstices. It is most likely produced by *Mallomonas areolata* Nygaard (Vigna, 1984).

Group 2: Morphotypes described using light microscopy only

Three of these morphotypes (stomatocysts 23, 24, and 25) were initially given ISWG numbers by Duff & Smol (1988). However, ISWG guidelines now require that descriptions be based on SEM specimens, and these numbers have been discontinued. In our papers following Duff & Smol (1988), cysts that were observed with LM only have been referred to as "Unidentified stomatocysts" and numbered consecutively.

Stomatocyst 23, Duff & Smol, 1988

This is a very large, oval stomatocyst (18.0 x 21. 6 μm) with regularly spaced, wide echinate spines.

Stomatocyst 24, Duff & Smol, 1988

This is a large, spherical stomatocyst (diameter 12.0-18.4 μm)

Fig. 162. Stomatocyst morphotypes that lack sufficient SEM specimens for a full description. A: Stomatocyst 191. Ridge indicated by arrow. B: Stomatocyst 195. Collar indicated by arrow. C: Stomatocyst 192. D: Stomatocyst 193. E: Stomatocyst 196. F: Stomatocyst 197. G: Stomatocyst 200. Scale bars = 2 μm.

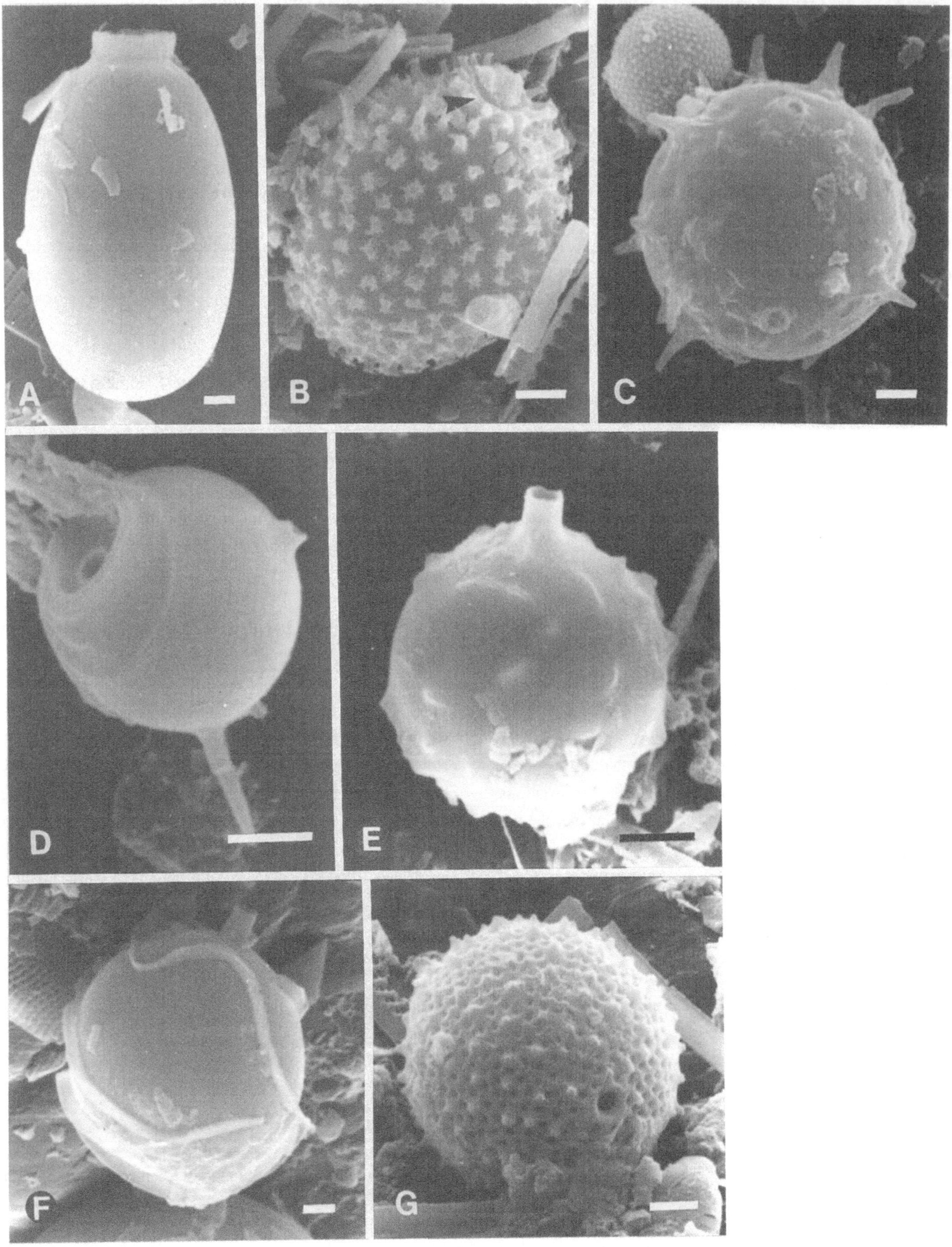
A
B
C
D
E
F
G

with a low, wide, cylindrical collar, and possibly several low conula.

Stomatocyst 25, Duff & Smol, 1988

This stomatocyst is ovate (12.0-14.4 μm) with a small conical collar, low ridges, and low, wide posterior verrucae.

Unidentified stomatocyst 3, Zeeb *et al.*, 1990

This is a spherical stomatocyst (diameter 13-15 μm) with a cylindrical or slightly obconical collar, and posterior conula.

Unidentified stomatocyst 5, Zeeb & Smol in Pienitz *et al.*, 1992

This is a small, spherical cyst (diameter 2.0-3.5 μm) with a pore and possibly low ridges; it may belong to a previously described cyst, but SEM confirmation is necessary.

Unidentified stomatocyst 6, Zeeb & Smol in Pienitz *et al.*, 1992

This morphotype is small and spherical with a low collar. It probably belongs to one of stomatocysts 50, 51, 52, or 110 (Figs. 12C, D), but SEM confirmation is necessary.

Unidentified stomatocyst 7, Zeeb & Smol, 1993a

This is a large obovate cyst (21.3 x 25.3 μm) with a low collar and a smooth surface.

Unidentified stomatocyst 9, Zeeb & Smol in Zeeb *et al.*, subm.

This cyst is spherical to widely oval (9.6-9.3 x 10.3-10.5 μm) with a complex collar. The primary collar is short and conical, and the secondary collar is wide, cylindrical to slightly obconical and asymmetric. The cyst is ornamented with a wide variable reticulum with roughly circular lacunae.

Unidentified stomatocyst 10, Zeeb & Smol in Zeeb *et al.*, subm.

This is a spherical stomatocyst (diameter 12.0-12.4 μm) with multiple cyst walls (at least four). The collar consists of three to four projections over the pore. Echinate spines with rounded apices cover the cyst surface.

Group 3: Morphotypes reassigned within our numerical scheme

Stomatocyst 7, Duff & Smol, 1988

This stomatocyst number included specimens of several different cyst morphs. The characteristic specimen is an immature cyst 101 (Fig. 131).

Stomatocyst 22, Duff & Smol, 1988

This morphotype was invalidly described based on LM specimens only. It has now been correctly identified as stomatocyst 35 (Fig. 138).

Stomatocyst 26, Duff & Smol, 1988

This cyst was invalidly described based on LM specimens only, and its specimens have been reassigned to stomatocyst 31 (Fig. 70).

Stomatocyst 43, Duff & Smol, 1991

These specimens have been reassigned to stomatocyst 1 (Figs. 6A, B).

Fig. 163. Stomatocyst morphotypes that lack sufficient SEM specimens for a full description. A: Stomatocyst 201. B: Stomatocyst 214. Collar indicated by arrow. C: Stomatocyst 216. Pore/collar not visible. D: Stomatocyst 220. E: Stomatocyst 225. F: Stomatocyst 226. G: Stomatocyst 230. Scale bars = 2 μm.

Stomatocyst 48, Duff & Smol, 1991

These specimens have been reassigned to either stomatocyst 49 (Fig. 9) or 120 (Figs. 7C, D, and H).

Stomatocyst 54, Duff & Smol, 1991

This number has been invalidated because the characteristic specimen has been reassigned to stomatocyst 53 (Fig. 29). Three other specimens have been reassigned elsewhere. The remaining two specimens are insufficient for a full description, and may not be the same morphotype.

Stomatocyst 63, Duff & Smol, 1991

These specimens have been reassigned to stomatocyst 62 (Fig. 92).

Stomatocyst 65, Duff & Smol, 1991

These specimens have been reassigned to stomatocyst 64 (Fig. 97).

Stomatocyst 66, Duff & Smol, 1991

These specimens have been reassigned to stomatocyst 4 (Fig. 84).

Stomatocyst 81, Duff & Smol, 1991

These specimens have been reassigned to stomatocyst 35 (Fig. 138).

Stomatocyst 82, Duff & Smol, 1991

These specimens have been reassigned to stomatocyst 35 (Fig. 138).

Stomatocyst 109, Zeeb *et al.*, 1990

These specimens have been reassigned to stomatocyst 1 (Figs. 6A, B).

Stomatocyst 124, Duff & Smol in Duff *et al.*, 1992

These specimens have been reassigned to either stomatocyst 50 (Fig. 24) or 51 (Fig. 13).

Unidentified stomatocyst 1, Zeeb *et al.*, 1990

These specimens, observed with LM only, have now been identified as belonging to stomatocyst 111 (*Spiniferomonas trioralis* (Takahashi) Preisig & Hibberd - Fig. 106).

Unidentified stomatocyst 2, Zeeb *et al.*, 1990

This morph was described based on LM only. Its specimens belong to stomatocyst 41 (*Dinobryon cylindricum* Imhof - Fig. 20).

Unidentified stomatocyst 4, Duff & Smol in Duff *et al.*, 1992

This morph, which was previously described based on LM only, has been reassigned to stomatocyst 197 (Fig. 162F).

Unidentified stomatocyst 8, Zeeb & Smol, 1993a

This morph was described based on LM only. Its specimens have been reassigned to stomatocyst 228 (Fig. 117).

References

Aaronson, S., 1980. Descriptive biochemistry and physiology of the Chrysophyceae. In M. Levandowsky & S.H. Hutner (eds), Biochemistry and physiology of protozoa, vol. 3. Academic Press, New York: pp. 117-169.

Adam, D.P., 1980a. Locality data for some chrysophyte cysts. U.S. Geol. Surv. Open-file Rep. 80-651.

Adam, D.P., 1980b. Scanning electron micrographs of modern chrysomonad cysts from Haypress Meadows, Eldorado County, California. U.S. Geol. Surv. Open-file Rep. 80-1235.

Adam, D.P., 1980c. Scanning electron micrographs of upper Pleistocene chrysomonad cysts from Flagpole Peak, Eldorado County, California. U.S. Geol. Surv. Open-file Rep. 80-1239.

Adam, D.P., 1981. Scanning electron micrographs of modern and late Holocene chrysomonad cysts from Harden Lake Meadow, Yosemite National Park, California. U.S. Geol. Surv. Open-file Rep. 81-46.

Adam, D.P. & A.D. Mahood, 1979a. A preliminary annotated bibliography on siliceous algal cysts and scales. U.S. Geol. Surv. Open-file Rep. 79-1215.

Adam, D.P. & A.D. Mahood, 1979b. Modern and Holocene chrysomonad cysts from Upper Echo Lake, Eldorado County, California. U.S. Geol. Surv. Open-file Rep. 79-1461.

Adam, D.P. & A.D. Mahood, 1980a. Modern chrysomonad cysts from Fallen Leaf Lake, Eldorado County, California. U.S. Geol. Surv. Open-file Rep. 80-798.

Adam, D.P. & A.D. Mahood, 1980b. Modern chrysomonad cysts from Alta Morris Lake, Eldorado County, California. U.S. Geol. Surv. Open-file Rep. 80-822.

Adam, D.P. & A.D. Mahood, 1980c. Scanning electron micrographs of chrysomonad cysts from Suzie Lake, Eldorado County, California. U.S. Geol. Surv. Open-file Rep. 80-1250.

Adam, D.P. & A.D. Mahood, 1981a. Chrysophyte cysts as potential environmental indicators. Geol. Soc. Am. Bull. 92: 839-844.

Adam, D.P. & A.D. Mahood, 1981b. Scanning electron micrographs of chrysomonad cysts from Lake Aloha, Eldorado County, California. U.S. Geol. Surv. Open-file Rep. 81-45.

Adam, D.P. & P.J. Mehringer, Jr., 1980a. Modern and Holocene chrysomonad cysts from Lost Trail Pass Bog, Montana. U.S. Geol. Surv. Open-file Rep. 80-797.

Adam, D.P. & P.J. Mehringer, Jr., 1980b. Scanning electron micrographs of modern chrysomonad cysts from Castor Pond, Jemez Mountains, New Mexico. U.S. Geol. Surv. Open-file Rep. 80-1231.

Adam, D.P. & P.J. Mehringer, Jr., 1980c. Scanning electron micrographs of modern and Holocene chrysomonad cysts from Fish Lake, Steens Mountains, Oregon. U.S. Geol. Surv. Open-file Rep. 80-1249.

Andersen, R.A., 1982. A light and electron microscopical investigation of *Ochromonas sphaerocystis* Matvienko (Chrysophyceae): the statospore, vegetative cell and its peripheral vesicles. Phycologia 21: 390-398.

Andrieu, B., 1936. Note sur les Chrysostomatacées d'une tourbe de l'Ile Kerguelen. Bull. Soc. Fr. Microscopie 5: 51-60.

Andrieu, B., 1937. Les Chrysostomatacées d'Auvergne. I. Dépôt de Verneuge (Puy-de-Dôme). Bull. Soc. Fr. Microscopie 6: 49-58.

Asmund, B., 1955a. Occurrence of *Dinobryon crenulatum* Wm. & G.S. West in some Danish ponds and remarks on its morphology, cyst formation, and ecology. Hydrobiologia 7: 75-87.

Asmund, B., 1955b. Electron microscope observations on *Mallomonas caudata* and some remarks on its occurrence in four Danish ponds. Bot. Tidsskr. 52: 163-168.

Asmund, B., 1956. Electron microscope observations on *Mallomonas* species and remarks on their occurrence in some Danish ponds II. Bot. Tidsskr. 53: 75-86.

Asmund, B., 1977. Two new species of *Mallomonas* (Chrysophyceae) in Swedish lakes. Bot. Tidsskr. 71: 253-258.

Asmund, B., G. Cronberg & M. Dürrschmidt, 1982. Revision of the *Mallomonas pumilio* group (Chrysophyceae). Nord. J. Bot. 2: 383-395.

Asmund, B. & J. Kristiansen, 1986. The genus *Mallomonas* (Chrysophyceae). Opera Bot. 85: 1-128.

Battarbee, R.W., 1973. A new method for estimating absolute microfossil numbers with special reference to diatoms. Limnol. Oceanogr. 18: 647-653.

Battarbee, R.W., 1981. Diatom and Chrysophyceae microstratigraphy of the annually laminated sediments of a small meromictic lake. Striae 14: 105-109.

Battarbee, R.W., 1986. Diatom analysis. In B.E. Berglund (ed), Handbook of Holocene palaeoecology and palaeohydrology. John Wiley & Sons, Chichester: 527-570.

Battarbee, R.W., G. Cronberg & S. Lowry, 1980. Observations on the occurrence of scales and bristles of *Mallomonas* spp. (Chrysophyceae) in the micro-laminated sediments of a small lake in Finnish North Karelia. Hydrobiologia 71: 225-232.

Belcher, J.H. & E.M.F. Swale, 1967. *Chromulina placentula* sp. nov. (Chrysophyceae), a freshwater nannoplankton flagellate. Br. Phycol. Bull. 3: 257-267.

Bird, D.F. & J. Kalff, 1986. Bacterial grazing by planktonic lake algae. Science 231: 493-495.

Bold, H.C. & M.J. Wynne, 1978. Introduction to the algae. Prentice-Hall, Englewood Cliffs, New Jersey: 720 pp.

Bourrelly, P., 1957. Recherches sur les Chrysophycées. Morphologie, phylogénie, systématique. Rev. Algol. Mém. Hors-Serie 1: 1-412.

Bourrelly, P., 1963. Loricae and cysts in the Chrysophyceae. Ann. N.Y. Acad. Sci. 108: 421-429.

Bourrelly, P., 1968. Les algues d'eau douce: initiation à la systématique. Tome II: Les algues jaunes et brunes. Chrysophycées, Phéophycées, Xanthophycées et Diatomées. Boubée, Paris: 109 pp.

Bradbury, J.P., 1988. A climatic-limnologic model of diatom succession for paleolimnological interpretation of varved sediments at Elk Lake, Minnesota. J. Paleolimnol. 1: 115-131.

Brown, K.M., 1994. Taxonomy and the ecological characterization of chrysophyte stomatocysts from the Yukon and Northwest Territories. M.Sc. Thesis, Queen's University.

Brown, K.M., M.S.V. Douglas & J.P. Smol, 1994. Siliceous microfossils in a Holocene, high arctic peat deposit (Nordvestø, northwestern Greenland). Can. J. Bot. 72: 208-216.

Carney, H.J., 1982. Algal dynamics and trophic interactions in the recent history of Frains Lake, Michigan. Ecology 63: 1814-1826.

Carney, H.J. & C.D. Sandgren, 1983. Chrysophycean cysts: indicators of eutrophication in the recent sediments of Frains Lake, Michigan, U.S.A. Hydrobiologia 101: 195-202.

Carney, H.J., M.C. Whiting, K.E. Duff & D.R. Whitehead, 1992. Chrysophycean cysts in Sierra Nevada (California) lake sediments: paleoecological potential. J. Paleolimnol. 7: 73-94.

Charles, D.F. & J.P. Smol, 1988. New methods for using diatoms and chrysophytes to infer past pH of low-alkalinity lakes. Limnol. Oceanogr. 33: 1451-1462.

Charles, D.F. & J.P. Smol, 1994. Long-term chemical changes in lakes: quantitative inferences using biotic remains in the sediment record. In L. Baker (ed), Environmental chemistry of lakes and reservoirs. Advances in Chemistry Series 237, American Chemical Society Books, Washington, D.C.: 3-31.

Christie, C.E., J.P. Smol, J. Meriläinen & P. Huttunen, 1988. Chrysophyte scales recorded in lake sediments from eastern Finland. Hydrobiologia 161: 237-243.

Conrad, W., 1933. Revision du genre *Mallomonas* Perty (1851) incl. *Pseudomallomonas* Chodat (1920). Mem. Mus. Hist. Nat. Belg. 56: 1-82.

Conrad, W., 1938. Notes Protistologiques. V. Observations sur *Uroglena soniaca* n. sp. et remarques sur le genre *Uroglena* Ehr. (incl. *Uroglenopsis* Lemm.). Bull. Mus. Roy. Hist. Nat. Belg. 14: 1-27.

Conrad, W., 1940. Notes Protistologiques. XVII. Chrysomonadées fossiles des collections du Musée royal d'Histoire naturelle de Belgique. Bull. Mus. Roy. Hist. Nat. Belg. 16: 1-15.

Cronberg, G., 1973. Development of cysts in *Mallomonas eoa* examined by scanning electron microscopy. Hydrobiologia 43: 29-38.

Cronberg, G., 1980. Cyst development in different species of *Mallomonas* (Chrysophyceae) studied by scanning electron microscopy. Arch. Hydrobiol. Suppl. 56: 421-434.

Cronberg, G., 1986a. Blue-green algae, green algae and chrysophyceae in sediments. In B.E. Berglund (ed), Handbook of Holocene palaeoecology and palaeohydrology. John Wiley & Sons, Chichester: 507-526.

Cronberg, G., 1986b. Chrysophycean cysts and scales in lake sediments: a review. In J. Kristiansen & R.A. Andersen (eds), Chrysophytes: aspects and problems. Cambridge University Press, Cambridge: 281-315.

Cronberg, G., 1988. Variability in size and ultrastructure of the statospore of *Mallomonas caudata*. Hydrobiologia 161: 31-39.

Cronberg, G., 1989a. Scaled chrysophytes from the tropics. Beih. Nova Hedwigia 95: 191-232.

Cronberg, G., 1989b. Stomatocysts of *Mallomonas hamata* and *M. heterospina* (Mallomonadaceae, Synurophyceae) from South Swedish lakes. Nord. J. Bot. 8: 683-692.

Cronberg, G., 1992. *Uroglena dendracantha* n. sp. (Chrysophyceae) from central Småland, Sweden. Nord. J. Bot. 12: 507-512.

Cronberg, G. & J. Kristiansen, 1980. Synuraceae and other Chrysophyceae from central Småland, Sweden. Bot. Notiser 133: 595-618.

Cronberg, G. & C.D. Sandgren, 1986. A proposal for the development of standardized nomenclature and terminology for chrysophycean statospores. In J. Kristiansen & R.A. Andersen (eds), Chrysophytes: aspects and problems. Cambridge University Press, Cambridge: 317-328.

Croome, R.L. & P.A. Tyler, 1985. Distribution of silica-scaled Chrysophyceae (Paraphysomonadaceae and Mallomonadaceae) in Australian inland waters. Aust. J. Mar. Freshw. Res. 36: 839-853.

Croome, R.L., H.U. Ling & P.A. Tyler, 1988. *Dinobryon unguentariforme* (Chrysophyceae), a new species from Australia. Br. Phycol. J. 23: 129-133.

Cumming, B.F. & J.P. Smol, 1993. Scaled chrysophytes and pH inference models: the effects of converting scale counts to cell counts and other species transformations. J. Paleolimnol. 9: 147-153.

Cumming, B.F., J.P. Smol & H.J.B. Birks, 1991. The relationship between sedimentary chrysophyte scales (Chrysophyceae and Synurophyceae) and limnological characteristics in 25 Norwegian lakes. Nord. J. Bot. 11: 231-242.

Cumming, B.F., J.P. Smol & H.J.B. Birks, 1992a. Scaled chrysophytes (Chrysophyceae and Synurophyceae) from Adirondack (N.Y., U.S.A.) drainage lakes and their relationship to measured environmental variables, with special reference to lakewater pH and labile monomeric aluminum. J. Phycol. 28: 162-178.

Cumming, B.F., J.P. Smol, J.C. Kingston, D.F. Charles, H.J.B. Birks, K.E. Camburn, S.S. Dixit, A.J. Uutala & A.R. Selle, 1992b. How much acidification has occurred in Adirondack region (New York, U.S.A.) lakes since pre-industrial times? Can. J. Fish. Aquat. Sci. 49: 128-141.

Cumming, B.F., S.E. Wilson & J.P. Smol, 1993. Paleolimnological potential of chrysophyte cysts and scales and of sponge spicules as indicators of lake salinity. Int. J. Salt Lake Res. 2: 87-92.

Davis, C.C., 1973. A seasonal quantitative study of the plankton of Bauline Long Pond, a Newfoundland lake. Nat. Can. 100: 85-105.

Deflandre, G., 1932. Archaeomonadaceae, une famille nouvelle de Protistes fossiles marins à loge siliceuse. C. R. Acad. Sci. [Paris] 194: 1859-1861.

Deflandre, G., 1934. Sur l'abus de l'emploi, en paléontologie, du nom de genre *Trachelomonas*. Ann. Protistol. 4: 151-165.

Deflandre, G., 1936. Les Flagellés fossiles. Apercu biologique et paleontologique. Role geologique. Actual. Sc. and Indust. Expos. Geol., Paris 355: 8-97.

Deflandre, G. 1952. Chrysomonadines fossiles. In P. Grassé (ed), Traité de Zoologie, vol. 1. Masson et cie, Paris: 560-570.

Dixit, S.S., A.S. Dixit & J.P. Smol, 1988. Scaled chrysophytes (Chrysophyceae) as indicators of pH in Sudbury, Ontario, lakes. Can. J. Fish. Aquat. Sci. 45: 1411-1421.

Dixit, S.S., A.S. Dixit & J.P. Smol, 1989a. Relationship between chrysophyte assemblages and environmental variables in 72 Sudbury lakes as examined by Canonical Correspondence Analysis (CCA). Can. J. Fish. Aquat. Sci. 46: 1667-1676.

Dixit, S.S., A.S. Dixit & R.D. Evans, 1989b. Paleolimnological evidence for trace metal sensitivity in scaled chrysophytes. Environ. Sci. Tech. 23: 110-115.

Dixit, S.S., A.S. Dixit & J.P. Smol, 1989c. Lake acidification recovery can be monitored using chrysophycean microfossils. Can. J. Fish. Aquat. Sci. 46: 1309-1312.

Dixit, S.S., J.P. Smol, D.S. Anderson & R.B. Davis, 1990a. Utility of scaled chrysophytes in predicting lakewater pH in northern New England lakes. J. Paleolimnol. 3: 269-286.

Dixit, S.S., A.S. Dixit & J.P. Smol, 1990b. Paleolimnological investigation of three manipulated lakes from Sudbury, Canada. Hydrobiologia 214: 245-252.

Donaldson, D.A. & J.R. Stein, 1984. Identification of planktonic Mallomonadaceae and other Chrysophyceae from selected lakes in the lower Fraser Valley, British Columbia (Canada). Can. J. Bot. 62: 525-539.

Dop, A.J., 1980. Benthic Chrysophyceae from The Netherlands. Ph.D. Thesis, Vrije University: 141 pp.

Douglas, M.S.V. & J.P. Smol, 1994. Limnology of high arctic ponds (Cape Herschel, Ellesmere Island, N.W.T.). Arch. Hydrobiol.: in press.

Duff, K.E., 1994. Relationships of sedimentary chrysophycean stomatocyst assemblages in lake sediments to environmental gradients. Ph.D. Thesis, Queen's University.

Duff, K.E. & J.P. Smol, 1988. Chrysophycean stomatocysts from the postglacial sediments of a High Arctic lake. Can. J. Bot. 66: 1117-1128.

Duff, K.E. & J.P. Smol, 1989. Chrysophycean stomatocysts from the postglacial sediments of Tasikutaaq Lake, Baffin Island, N.W.T. Can. J. Bot. 67: 1649-1656.

Duff, K.E. & J.P. Smol, 1991. Morphological descriptions and stratigraphic distributions of the chrysophycean stomatocysts from a recently acidified lake (Adirondack Park, N.Y.). J. Paleolimnol. 5: 73-113.

Duff, K.E. & J.P. Smol, 1994. Chrysophycean cyst flora from British Columbia (Canada) lakes. Nova Hedwigia 58: 353-389.

Duff, K.E., M.S.V. Douglas & J.P. Smol, 1992. Chrysophyte cysts in 36 Canadian high arctic ponds. Nord. J. Bot. 12: 471-499.

Dürrschmidt, M., 1980. Studies on the Chrysophyceae from Rio Cruces, Prov. Voldinia, South Chile by scanning and transmission microscopy. Nova Hedwigia 33: 353-388.

Dürrschmidt, M., 1982a. *Mallomonas parvula* sp. nov. and *Mallomonas retifera* sp. nov. (Chrysophyceae, Synuraceae) from South Chile. Can. J. Bot. 60: 651-656.

Dürrschmidt, M., 1982b. Studies on the Chrysophyceae from South Chilean inland waters by means of scanning and transmission electron microscopy II. Arch. Hydrobiol. Suppl. 63: 121-163.

Dürrschmidt, M., 1983. A taxonomic study of the *Mallomonas mangofera* group (Mallomonadaceae, Chrysophyceae), including the description of four new taxa. Plant Syst. Evol. 143: 175-196.

Dürrschmidt, M., 1984. Studies on scale-bearing Chrysophyceae from the Giessen area, Federal Republic of Germany. Nord. J. Bot. 4: 123-143.

Duthie, H.C. & M.L. Ostrofsky, 1974. Plankton, chemistry and physics of lakes in the Churchill Falls Region of Labrador. J. Fish. Res. Board Can. 31: 1105-1117.

Ehrenberg, C.G., 1854. Microgeologie. Leopold Voss, Leipzig.

Elner, J.K. & C.M. Happey-Wood, 1978. Diatom and chrysophycean cyst profiles in sediment cores from 2 linked but contrasting Welsh lakes. Br. Phycol. J. 13: 341-360.

Eloranta, P., 1986. Phytoplankton structure in different lake types in central Finland. Holarctic Ecology 9: 214-224.

Eloranta, P., 1989. Scaled chrysophytes (Chrysophyceae and Synurophyceae) from national park lakes in southern and central Finland. Nord. J. Bot. 8: 671-681.

Frenguelli, G., 1925. Sopra alcuni microrganismi a guscio siliceo. Roma Soc. Geol. Ital. Boll. 44: 1-8.

Frenguelli, G.(J.), 1929. *Trachelomonas* de los esteros de la región del Yberá en la provincia de Corrientes, Argentina. Revista Chilena de Historia Natural 33: 563-568.

Frenguelli, G.(J.), 1931. Analisis microscopico de una muestra de Tripoli de Angostura (Provincia de Colchagua, Chile). Revista Chilena de Historia Natural 35: 9-14.

Frenguelli, G., 1932. Trachelomonadi del Pliocene Argentino. Memorie Soc. Geol. Ital. 1: 1-44.

Frenguelli, G.(J.), 1935a. Einige Bemerkunden zu den Archaeomonadaceen. Arch. Protistenk. 84: 232-241.

Frenguelli, G.(J.), 1935b. Traquelomónadas del Platense de la Costa Atlántica de la Provincia de Buenos Aires. Notas del Museo de la Plata 1: 35-44.

Frenguelli, G.(J.), 1936. Crisostomatáceas del Neuquén. Notas del Museo de la Plata 1: 247-275.

Frenguelli, G.(J.), 1938a. *Deflandreia*, nuevo género de Crisostomatáceas. Notas del Museo de la Plata 3: 47-54.

Frenguelli, G.(J.), 1938b. Crisostomatáceas Platenses. Acta Geographica 14: 149-154.

Glew, J.R., 1991. Miniature gravity corer for recovering short sediment cores. J. Paleolimnol. 5: 285-287.

Gayral, P. & C. Billard, 1986. A survey of the marine Chrysophyceae with special reference to the Sarcinochrysidales. In J. Kristiansen & R.A. Andersen (eds), Chrysophytes: aspects and problems. Cambridge University Press, Cambridge: 37-48.

Gretz, M.R., M.R. Sommerfeld & D.E. Wujek, 1979. Scaled Chrysophyceae of Arizona: a preliminary survey. J. Ariz. -Nev. Acad. Sci. 14: 75-80.

Gritten, M.M., 1977. On the fine structure of some Chrysophycean cysts. Hydrobiologia 53: 239-252.

Grönlund, E., H. Simola & P. Huttunen, 1986. Paleolimnological reflections of fiber-plant retting in the sediment of a small clearwater lake. Hydrobiologia 143: 425-531.

Hajós, M., 1970. Kieselgurvorkommen im Tertiärbecken von Aflenz (Steiermark). Mitt. Geol. Gesellschaft in Wien 63: 149-159.

Hajós, M., 1971. Diatomées du Pannonien Inférieur provenant du bassin Néogène de Csákvár. Ie partie. Acta Botanica Academiae Scientiarum Hungaricae 17: 59-82.

Hajós, M., 1973. Diatomées du Pannonien Inférieur provenant du bassin Néogène de Csákvár. IIe partie. Acta Botanica Academiae Scientiarum Hungaricae 18: 95-118.

Hajós, M., 1974. A pulai Put-3. sz. fúrás felsöpannóniai képzödményeinek Diatoma flórája. Magyar Állami Földtani Intézet Évi Jelentése: 263-285.

Hajós, M. & G. Radócz, 1969. Diatomás rétegek a bükkalji alsópannonból. Magyar Állami Földtani Intézet Évi Jelentése: 271-297.

Hällfors, G. & S. Hällfors, 1988. Records of chrysophytes with siliceous scales (Mallomonadaceae and Paraphysomonadaceae) from Finnish inland waters. Hydrobiologia 161: 1-29.

Harris, K., 1953. A contribution to our knowledge of *Mallomonas*. Linn. Soc. Lond. J. 55: 88-102.

Harris, K., 1958. A study of *Mallomonas insignis* and *Mallomonas akrokomos*. J. Gen. Microbiol. 19: 55-64.

Harris, K., 1966. The genus *Mallomonopsis*. J. Gen. Microbiol. 42: 175-184.

Harris, K., 1967. Variability in *Mallomonas*. J. Gen. Microbiol. 46: 185-191.

Harris, K., 1970. Species of the *Torquata* group of *Mallomonas*. J. Gen. Microbiol. 61: 77-80.

Harris, K. & D.E. Bradley, 1957. III. An examination of the scales and bristles of *Mallomonas* in the electron microscope using carbon replicas. Roy. Micro. Soc. Lond. J. 76: 37-46.

Harris, K. & D.E. Bradley, 1958. Some unusual Chrysophyceae studied in the electron microscope. J. Gen. Microbiol. 18: 71-83.

Harris, K. & D.E. Bradley, 1960. A taxonomic study of *Mallomonas*. J. Gen. Microbiol. 22: 750-777.

Harwood, D.M., 1986. Do diatoms beneath the Greenland Ice Sheet indicate interglacials warmer than present? Arctic 39: 304-308.

Hibberd, D.J., 1977. Ultrastructure of cyst formation in *Ochromonas tuberculata* (Chrysophyceae). J. Phycol. 13: 309-320.

Hickel, B., 1975. The application of the scanning electron microscope to freshwater phytoplankton taxonomy and morphology. Arch. Hydrobiol. 76: 218-228.

Hickel, B. & I. Maaβ, 1989. Scaled chrysophytes, including heterotrophic nanoflagellates from the lake district in Holstein, northern Germany. Beih. Nova Hedwigia 95: 233-257.

Hilliard, D.K., 1959. Notes on the phytoplankton of Karluk Lake, Kodiak Island, Alaska. Can. Field Nat. 73: 135-143.

Hilliard, D.K., 1966. Studies on Chrysophyceae from some ponds and lakes in Alaska. V. Notes on the taxonomy and occurrence of phytoplankton in an Alaskan pond. Hydrobiologia 28: 553-576.

Hilliard, D.K. & B. Asmund, 1963. Studies on Chrysophyceae from some ponds and lakes in Alaska. II. Notes on the genera *Dinobryon*, *Hyalobryon* and *Epipyxis* with descriptions of new species. Hydrobiologia 22: 331-397.

Huber-Pestalozzi, G., 1941. Das Phytoplankton des Süsswassers. Systematik und Biologie. Teil 2, Hälfte 1. Chrysophyceen, Farblose Flagellaten, Heteroconten. Die Binnengewässer 16, Schweizerbartsche Verlagsbuchhandlung, Stuttgart: 366 pp.

Kalina, T., 1969. Submicroscopic structure of silica scales in some *Mallomonas* and *Mallomonopsis* species. Preslia 41: 227-228.

Krieger, W., 1932. Untersuchungen über Plankton-Chrysomonaden. Die Gattungen *Mallomonas* und *Dinobryon* in monographischer Bearbeitung. Bot. Arch. 29: 257-329.

Kristiansen, J., 1964. Flagellates from Finnish Lappland. Bot. Tidsskr. 59: 315-333.

Kristiansen, J., 1965. Occurrence and ecology of *Chrysolykos planctonicus*, a chrysomonad with sexual reproduction. Bot. Tidsskr. 61: 98-105.

Kristiansen, J., 1976. Studies on the Chrysophyceae of Bornholm. Bot. Tidsskr. 70: 126-142.

Kristiansen, J., 1980. Chrysophyceae from some Greek lakes. Nova Hedwigia 33: 167-194.

Kristiansen, J., 1986. Silica-scale bearing chrysophytes as environmental indicators. Br. Phycol. J. 21: 425-436.

Kristiansen, J., 1988. Seasonal occurrence of silica-scaled chrysophytes under eutrophic conditions. Hydrobiologia 161: 171-184.

Kristiansen, J., 1990. Phylum Chrysophyta. In L. Margulis, J.O. Corliss, M. Melkonian & D.J. Chapman (eds), Handbook of protoctista. Jones and Bartlett Publishers, Boston: 438-453.

Kristiansen, J., 1994. History of research on chrysophyte algae. In C.D. Sandgren, J.P. Smol & J. Kristiansen (eds), Chrysophyte algae: ecology, phylogeny and development. Cambridge University Press, Cambridge: in press.

Kristiansen, J. & R. A. Andersen (eds.), 1986. Chrysophytes: aspects and problems. Cambridge University Press, Cambridge: 337 pp.

Kristiansen, J. & E. Takahashi, 1982. Chrysophyceae: introduction and bibliography. In J.R. Rosowski & B.C. Parker (eds), Selected papers in phycology II. Phycological Society of America, Inc., Lawrence, Kansas: 698-704.

Kristiansen, J. & D. Tong, 1989. Studies on silica-scaled chrysophytes from Wuhan, Hangzhou and Beijing , P.R. China. Nova Hedwigia 49: 183-202.

Kristiansen, J., G. Cronberg & U. Geissler (eds.), 1989. Chrysophytes: developments and perspectives. J. Cramer, Berlin: 287 pp.

Leventhal, E.A., 1970. The Chrysomonadina. Transactions of the American Philosophical Society 60: 123-142.

Livingstone, D.A., 1955. A lightweight piston sampler for lake deposits. Ecology 36: 137-139.

Loeblich, A.R. & L.A. Loeblich, 1977. Division Chrysophyta. In A.I. Laskin & H.A. Lechevalier (eds), CRC handbook of microbiology, vol. II. CRC Press, West Palm Beach, Fla.: 411-423.

Lund, J.W.G., 1942. Contributions to our knowledge of British Chrysophyceae. New Phytologist 41: 274-292.

Lund, J.W.G., 1953. New or rare British Chrysophyceae. II. *Hyalobryon polymorphum* n. sp. and *Chrysonebula holmesii* n. gen., n. sp. New Phytologist 52: 114-123.

Mahood, A.D. & D.P. Adam, 1979. Late Pleistocene cysts from core 7, Clear Lake, Lake County, California. U.S. Geol. Surv. Open-file Rep. 79-971.

Marsicano, L.J. & P.A. Siver, 1993. A paleolimnological assessment of lake acidification in five Connecticut lakes. J. Paleolimnol. 9: 209-221.

McKenzie, C. & H. Kling, 1989. Scale-bearing Chrysophyceae (Mallomonadaceae and Paraphysomonadaceae) from Mackenzie Delta area lakes, Northwest Territories, Canada. Nord. J. Bot. 9: 103-112.

Momeu, L. & L.S. Péterfi, 1983. Electron microscopical study of some *Mallomonas* species from Romania. Nova Hedwigia 38: 369-400.

Moore, J.W., 1978. Distribution and abundance of phytoplankton in 153 lakes, rivers, and pools in the Northwest Territories. Can. J. Bot. 56: 1765-1773.

Moore, J.W., 1979. Factors influencing the diversity, species composition and abundance of phytoplankton in twenty-one Arctic and Subarctic lakes. Int. Rev. Gesamten Hydrobiol. 64: 485-497.

Moore, J.W., 1981. Patterns of distribution of phytoplankton in northern Canada. Nova Hedwigia 34: 599-621.

Moss, B., 1972. Studies on Gull Lake, Michigan. I. Seasonal and depth distribution of phytoplankton. Freshwater Biol. 2: 289-307.

Munch, C.S., 1980. Fossil diatoms and scales of Chrysophyceae in the recent history of Hall Lake, Washington, U.S.A. Nord. J. Bot. 5: 505-510.

Nicholls, K.H., 1981. *Chrysococcus furcatus* (Dolg.) comb. nov.: a new name for *Chrysastrella furcata* (Dolg.) Defl. based on the discovery of the vegetative stage. Phycologia 20: 16-21.

Nicholls, K.H., 1984. *Spiniferomonas septispina* and *S. enigmata*, two new algal species confusing the distinction between *Spiniferomonas* and *Chrysosphaerella* (Chrysophyceae). Pl. Syst. Evol. 148: 103-117.

Nicholls, K.H., 1994. Chrysophyte blooms in the plankton and neuston of marine and freshwater systems. In C.D. Sandgren, J.P. Smol & J. Kristiansen (eds), Chrysophyte algae: ecology, phylogeny and development. Cambridge University Press, Cambridge: in press.

Nygaard, G., 1956. Ancient and recent flora of diatoms and Chrysophyceae in Lake Gribsø. Folia Limnol. Scand. 8: 32-262.

Nygaard, G., 1977. New or interesting plankton algae with a contribution on their ecology. Kongl. Danske Vidensk. Selskab, Biol. Skr. 21: 1-107.

Nygaard, G., 1979. Freshwater phytoplankton from the Narssaq Area, South Greenland. Bot. Tidsskrift 73: 191-238.

Pantocsek, J., 1912. A fertö to kovamozat viranga: Bacillariae locus Peisonis. Pozxony, Wigand K.F. Könyvnyomdeja.

Pantocsek, J., 1913. Die im Andesittuffe von Kopacsee vorkommenden Bacillarien. Botanikai Közlemenyek 12: 126-137.

Peglar, S.M., S.C. Fritz, T. Alapieti, M. Saarnisto & H.J.B. Birks, 1984. Composition and formation of laminated sediments in Diss Mere, Norfolk, England. Boreas 13: 13-28.

Péterfi, L.S., 1965. Four *Mallomonas* species of Rumania studied in light and electron microscope. Advancing Frontiers of Plant Science 10: 135-140.

Péterfi, L.S., 1967. Studies on the Rumanian Chrysophyceae. I. Nova Hedwigia 13: 117-137.

Péterfi, L.S. & L. Momeu, 1976. Romanian *Mallomonas* species studied in light and electron microscopes. Nova Hedwigia 27: 353-392.

Peters, M.C. & R.A. Andersen, 1993. The fine structure and scale formation of *Chrysolepidomonas dendrolepidota* gen. et sp. nov. (Chrysolepidomonadaceae fam. nov., Chrysophyceae). J. Phycol. 29: 469-475.

Pienitz, R., I.R. Walker, B.A. Zeeb, J.P. Smol & P.R. Leavitt, 1992. Biomonitoring past salinity changes in an athalassic subarctic lake. Int. J. Salt Lake Res. 1: 91-123.

Pinter, I.J. & L. Provasoli, 1963. Nutritional characteristics of some chrysomonads. In C.H. Oppenheimer (ed), Symposium on marine microbiology. Thomas, Springfield: 114-121.

Preisig, H.R. & D.J. Hibberd, 1982a. Ultrastructure and taxonomy of *Paraphysomonas* (Chrysophyceae) and related genera 1. Nord. J. Bot. 2: 397-420.

Preisig, H.R. & D.J. Hibberd, 1982b. Ultrastructure and taxonomy of *Paraphysomonas* (Chrysophyceae) and related genera 2. Nord. J. Bot. 2: 601-638.

Preisig, H.R. & D.J. Hibberd, 1983. Ultrastructure and taxonomy of *Paraphysomonas* (Chrysophyceae) and related genera 3. Nord. J. Bot. 3: 695-723.

Preisig, H.R. & E. Takahashi, 1978. *Chrysosphaerella (Pseudochrysosphaerella) solitaria* spec. nov. (Chrysophyceae). Pl. Syst. Evol. 129: 135-142.

Pringsheim, E.G., 1946. On iron flagellates. Phil. Trans. Roy. Soc. (Lond.), Ser. B. 232: 311-342.

Ramberg, L., 1978. Some rare Chrysophyta from Swedish oligotrophic lakes. Br. Phycol. J. 13: 141-147.

Roijackers, R.M.M., 1986. Development and succession of scale-bearing Chrysophyceae in two shallow freshwater bodies near Nijmegen, The Netherlands. In J. Kristiansen & R.A. Andersen (eds), Chrysophytes: aspects and problems. Cambridge University Press, Cambridge: 241-258.

Roijackers, R.M.M. & H. Kessels, 1981. Chrysophyceae from freshwater localities near Nijmegen, The Netherlands II. Hydrobiologia 80: 231-239.

Roijackers, R.M.M. & H. Kessels, 1986. Ecological characteristics of scale-bearing Chrysophyceae from The Netherlands. Nord. J. Bot. 6: 373-385.

Rott, E., 1988. Some aspects of the seasonal distribution of flagellates in mountain lakes. Hydrobiologia 161: 159-170.

Rull, V., 1986. Diatomeas y crisofíceas en los sedimentos acuáticos de una depresión cárstica del Pirineo catalán. Oecologia aquatica 8: 11-24.

Rull, V., 1991. Palaeoecological significance of chrysophycean stomatocysts: a statistical approach. Hydrobiologia 220: 161-165.

Rybak, M., 1986. The chrysophycean paleocyst flora of the bottom sediments of Kortowskie Lake (Poland) and its ecological significance. Hydrobiologia 140: 67-84.

Rybak, M., 1987. Fossil chrysophycean cyst flora of Racze Lake, Wolin Island (Poland) in relation to paleoenvironmental conditions. Hydrobiologia 150: 257-272.

Rybak, M., I. Rybak & M. Dickman, 1987. Fossil chrysophycean cyst flora in a small meromictic lake in southern Ontario, and its paleoecological interpretation. Can. J. Bot. 65: 2425-2440.

Rybak, M., I. Rybak & K. Nicholls, 1991. Sedimentary chrysophycean cyst assemblages as paleoindicators in acid sensitive lakes. J. Paleolimnol. 5: 19-72.

Salonen, K. & S. Jokinen, 1988. Flagellate grazing on bacteria in a small dystrophic lake. Hydrobiologia 161: 203-209.

Sandgren, C.D., 1980a. Resting cyst formation in selected chrysophyte flagellates: an ultrastructural survey including a proposal for the phylogenetic significance of interspecific variations in the encystment process. Protistologica 16: 289-303.

Sandgren, C.D., 1980b. An ultrastructural investigation of resting cyst formation in *Dinobryon cylindricum* Imhof (Chrysophyceae, Chrysophycota). Protistologica 16: 259-276.

Sandgren, C.D., 1981. Characteristics of sexual and asexual resting cyst (statospore) formation in *Dinobryon cylindricum* Imhof. (Chrysophyta). J. Phycol. 17: 199-210.

Sandgren, C.D., 1983a. Morphological variability in populations of chrysophycean resting cysts. I. Genetic (interclonal) and encystment temperature effects on morphology. J. Phycol. 19: 64-70.

Sandgren, C.D., 1983b. Survival strategies of chrysophycean flagellates: reproduction and the formation of resistant resting cysts. In G.A. Fryxell (ed), Survival strategies of the algae. Cambridge University Press, Cambridge: 23-48.

Sandgren, C.D., 1988. The ecology of chrysophyte flagellates: their growth and perennation strategies as freshwater phytoplankton. In C.D. Sandgren (ed), Growth and reproductive strategies of freshwater phytoplankton. Cambridge University Press, Cambridge: 9-104.

Sandgren, C.D., 1989. SEM investigations of statospore (stomatocyst) development in diverse members of the Chrysophyceae and Synurophyceae. Beih. Nova Hedwigia 95: 45-69.

Sandgren, C.D., 1991. Chrysophyte reproduction and resting cysts: a paleolimnologist's primer. J. Paleolimnol. 5: 1-9.

Sandgren, C.D. & S.B. Barlow, 1989. Siliceous scale production in chrysophytes algae. II. SEM observations regarding the effects of metabolic inhibitors on scale regeneration in laboratory population of scale-free *Synura petersenii* cells. Beih. Nova Hedwigia 95: 27-44.

Sandgren, C.D. & H.J. Carney, 1983. A flora of fossil Chrysophycean cysts from the recent sediments of Frains Lake, Michigan, U.S.A. Nova Hedwigia 38: 129-163.

Sandgren, C.D. & J. Flanagin, 1986. Heterothallic sexuality and density dependent encystment in the Chrysophycean alga *Synura petersenii* Korsh. J. Phycol. 22: 206-216.

Sandgren, C.D. & J.P. Smol, 1991. Reply to Dr. Simola's comment on the taxonomy of 'snowflakes' (chrysophyte cysts). J. Paleolimnol. 6: 261-263.

Sandgren, C.D., J.P. Smol & J. Kristiansen (eds), 1994. Chrysophyte algae: ecology, phylogeny and development. Cambridge University Press, Cambridge: in press.

Sheath, R.G., J.A. Hellebust & T. Sawa, 1975. The statospore of *Dinobryon divergens* Imhof.: formation and germination in a subarctic lake. J. Phycol. 11: 131-138.

Simola, H., 1991. Chrysophyte stomatocyst taxonomy - classification of snowflakes? J. Paleolimnol. 6: 261-263.

Simola, H., 1993. Reply to Sandgren and Smol: further discussion on chrysophyte cyst taxonomy. J. Paleolimnol. 9: 63-68.

Siver, P.A., 1988a. Distribution of scaled chrysophytes in 17 Adirondack (New York) lakes with special reference to pH. Can. J. Bot. 66: 1391-1403.

Siver, P.A., 1988b. The distribution and ecology of *Spiniferomonas* (Chrysophyceae) in Connecticut, U.S.A. Nord. J. Bot. 8: 205-212.

Siver, P.A., 1989a. The distribution of scaled chrysophytes along a pH gradient. Can. J. Bot. 67: 2120-2130.

Siver, P.A., 1989b. The separation of *Mallomonas acaroides* v. *acaroides* and v. *muskokana* (Synurophyceae) along a pH gradient. Beih. Nova Hedwigia 95: 111-117.

Siver, P.A., 1991a. The biology of *Mallomonas*: morphology, taxonomy and ecology. Kluwer Academic Publishers, Dordrecht: 230 pp.

Siver, P.A., 1991b. Implications for improving paleolimnological inference models utilizing scale-bearing siliceous algae: transforming scale counts to cell counts. J. Paleolimnol. 5: 219-225.

Siver, P.A., 1991c. The stomatocyst of *Mallomonas acaroides* v. *muskokana* (Chrysophyceae). J. Paleolimnol. 5: 11-17.

Siver, P.A., 1993. Inferring lakewater specific conductivity with scaled chrysophytes. Limnol. Oceanogr. 38: 1480-1492.

Siver, P.A., 1994. The distribution of chrysophytes along environmental gradients: their use as biological indicators. In C.D. Sandgren, J.P. Smol & J. Kristiansen (eds), Chrysophyte algae: ecology, phylogeny and development. Cambridge University Press, Cambridge: in press.

Siver, P.A. & J.S. Chock, 1986. Phytoplankton dynamics in a chrysophycean lake. In J. Kristiansen & R.A. Andersen (eds), Chrysophytes: aspects and problems. Cambridge University Press, Cambridge: 165-183.

Siver, P.A. & J.S. Hamer, 1990. Use of extant populations of scaled chrysophytes for the inference of lakewater pH. Can. J. Fish. Aquat. Sci. 47: 1339-1347.

Siver, P.A. & J.S. Hamer, 1992. Seasonal periodicity of Chrysophyceae and Synurophyceae in a small New England lake: implications for paleolimnological research. J. Phycol. 28: 186-198.

Siver, P.A. & J.P. Smol, 1993. The use of scaled chrysophytes in long term monitoring programs for the detection of changes in lakewater acidity. Water, Air & Soil Pollution 71: 357-376.

Skogstad, A., 1984. Vegetative cells and cysts of *Mallomonas intermedia* (Mallomonadaceae, Chrysophyceae). Nord. J. Bot. 4: 275-278.

Skogstad, A., 1986. *Chromophysomonas* (Chrysophyceae) from twenty-seven localities in the Oslo area. In J. Kristiansen & R.A. Andersen (eds), Chrysophytes: aspects and problems. Cambridge University Press, Cambridge: 259-269.

Skogstad, A. & O.L. Reymond, 1989. An ultrastructural study of vegetative cells, encystment, and mature statospores in *Spiniferomonas bourrellyi* (Chrysophyceae). Beih. Nova Hedwigia 95: 71-79.

Smith, M.A. & M.J. White, 1985. Observations on lakes near Mount St. Helens: phytoplankton. Arch. Hydrobiol. 104: 345-362.

Smol, J.P., 1980. Fossil synuracean (Chrysophyceae) scales in lake sediments: a new group of paleoindicators. Can. J. Bot. 58: 458-465.

Smol, J.P., 1983. Paleophycology of a high arctic lake near Cape Herschel, Ellesmere Island. Can. J. Bot. 61: 2195-2204.

Smol, J.P., 1984. The statospore of *Mallomonas pseudocoronata* (Mallomonadaceae, Chrysophyceae). Nord. J. Bot. 4: 827-831.

Smol, J.P., 1985. The ratio of diatom frustules to chrysophycean statospores: a useful paleolimnological index. Hydrobiologia 123: 199-208.

Smol, J.P., 1986. Chrysophycean microfossils as indicators of lakewater pH. In J.P. Smol, R.W. Battarbee, R.B. Davis, J. Meriläinen (eds), Diatoms and lake acidity. Dr. W. Junk Publishers, Dordrecht: 275-287 pp.

Smol, J.P., 1987. Methods in Quaternary ecology - freshwater algae. Geoscience Canada 14: 208-217.

Smol, J.P., 1988a. Chrysophycean microfossils in paleolimnological studies. Palaeogeography, Palaeoclimatology, Palaeoecology 62: 287-297.

Smol, J.P., 1988b. The North American 'endemic' *Mallomonas pseudocoronata* (Mallomonadaceae, Chrysophyta) in an Austrian lake. Phycologia 27: 427-429.

Smol, J.P., 1988c. Paleoclimate proxy data from freshwater arctic diatoms. Int. Ver. Theor. Angew. Limnol. Verh. 23: 837-844.

Smol, J.P., 1990. Diatoms and chrysophytes - a useful combination in palaeolimnological studies: report of a workshop and a working bibliography. In H. Simola (ed), Proceedings of the Tenth International Diatom Symposium. Koeltz Scientific Books, Koenigstein: 585-592 pp.

Smol, J.P., 1994. Application of chrysophytes to problems in paleoecology. In C.D. Sandgren, J.P. Smol & J. Kristiansen (eds), Chrysophyte algae: ecology, phylogeny and development. Cambridge University Press, Cambridge: in press.

Smol, J.P. & J.R. Glew, 1992. Paleolimnology. Encyclopedia of Earth System Science 3: 551-564.

Srivastava, S.K. & P.L. Binda, 1984. Siliceous and silicified microfossils from the Maastrichtian Battle Formation of southern Alberta, Canada. Paleobiologie Continentale 14: 1-24.

Stoermer, E.F., J.A. Wolin, C.L. Schelske & D.J. Conley, 1985. Postsettlement diatom succession in the Bay of Quinte, Lake Ontario. Can. J. Fish. Aquat. Sci. 42: 754-767.

Swale, E.M.F. & J.H. Belcher, 1966. *Ochromonas ostreaformis* nov. sp., a large compressed chrysomonad. New Phytologist 65: 267-272.

Takahashi, E., 1978. Electron microscopical studies of the Synuraceae (Chrysophyceae) in Japan. Taxonomy and ecology. Tokai University Press, Tokyo: 194 pp.

Takahashi, E., 1981. Floristic study of ice algae in the sea ice of a lagoon, Lake Saroma, Hokkaido, Japan. Mem. Nat. Inst. Polar Res. Sec. E. Biol. Med. Sci. 34: 49-56.

Takahashi, E., 1987. Loricate and scale-bearing protists from Lützhow-Holm Bay, Antarctica II. Four marine species of *Paraphysomonas* (Chrysophyceae) including two new species from the fast-ice covered coastal area. Jpn. J. Phycol. 35: 155-166.

Takahashi, E., K. Watanabe & H. Satoh, 1986. Siliceous cysts from Kita-no-seto Strait, north of Syowa Station, Antarctica. Mem. Natl. Inst. Polar Res., Spec. Issue 40: 84-95.

Tippett, R., 1964. An investigation into the nature of the layering of deep-water sediments in two eastern Ontario lakes. Can. J. Bot. 42: 1693-1709.

VanLandingham, S.L., 1964. Chrysophyta cysts from the Yakima basalt (Miocene) in south-central Washington. J. Paleontol. 38: 729-739.

Vigna, M.S., 1984. Estudia de la estatospora de *Mallomonas areolata* Nygaard con microscopio electronico de barrido (Chrysophyceae). Physis, Secc. B. 42: 99-101.

Vigna, M.S., 1988. Contribution to the knowledge of Argentine Mallomonadaceae. Nova Hedwigia 47: 129-144.

Wallen, D.G. & R. Allen, 1982. Variations in phytoplankton communities in Canadian Arctic ponds. Nat. Can. 109: 213-221.

Wawrik, F., 1977. Phytoplankton aus new angelegten Streckteichen (Walkviertel, Neiderösterreich). Int. Revue Ges. Hydrobiol. 62: 295-313.

Wawrik, F., 1980. Algologische und ökologische Beobachtungen zur Zeit des Eisschließens und Eisbrechens 1978/79 in Teichen des niederösterreichischen Waldviertels. Nova Hedwigia 33: 789-810.

Wee, J.L., 1982. Studies on the Synuraceae (Chrysophyceae) of Iowa. Bibliotecha Phycologica 62: 1-183.

Willén, T., 1992. *Dinobryon faculiferum*, a new name for *Dinobryon petiolatum* (Chrysophyceae: Dinobryaceae). Taxon 41: 62-63.

Withers, N.W., A. Fiksdahl, R.C. Tuttle & S. Liaaen-Jensen, 1981. Carotenoids of the Chrysophyceae. Comp. Biochem. Physiol. B Comp. Biochem. 68: 345-349.

Wujek, D.E., 1969. Ultrastructure of flagellated chrysophytes. I. *Dinobryon*. Cytologia 34: 71-79.

Zeeb, B.A., 1994. Assessment of chrysophycean stomatocysts as paleolimnological markers of environmental change. Ph.D. Thesis, Queen's University.

Zeeb, B.A. & J.P. Smol, 1991. Paleolimnological investigation of the effects of road salt seepage on scaled chrysophytes in Fonda Lake, Michigan. J. Paleolimnol. 5: 263-266.

Zeeb, B.A. & J.P. Smol, 1993a. Chrysophycean stomatocyst flora from Elk Lake, Clearwater County, Minnesota. Can. J. Bot. 71: 737-756.

Zeeb, B.A. & J.P. Smol, 1993b. Postglacial chrysophycean cyst record from Elk Lake, Minnesota. In J.P. Bradbury & W.E. Dean (eds), Elk Lake, Minnesota: evidence for rapid climate change in the north-central United States. Geological Society of America Special Paper 276, Boulder: 239-249 pp.

Zeeb, B.A., K.E. Duff & J.P. Smol, 1990. Morphological descriptions and stratigraphic profiles of chrysophycean stomatocysts from the recent sediments of Little Round Lake, Ontario. Nova Hedwigia 51: 361-380.

Zeeb, B.A., C.E. Christie, J.P. Smol, D. Findlay, H. Kling & H.J.B. Birks, 1994. Responses of diatom and chrysophyte assemblages in Lake 227 to experimental eutrophication. Can. J. Fish. Aquat. Sci. : in press.

Zeeb, B.A., J.P. Smol & S.L. VanLandingham. Pliocene chrysophycean stomatocysts from the Sonoma Volcanics, Napa County, California. Submitted.

Stomatocyst Index

Pages with scanning electron micrographs shown in boldface.

Species Index

GPSR Compliance

The European Union's (EU) General Product Safety Regulation (GPSR) is a set of rules that requires consumer products to be safe and our obligations to ensure this.

If you have any concerns about our products, you can contact us on ProductSafety@springernature.com

In case Publisher is established outside the EU, the EU authorized representative is:

Springer Nature Customer Service Center GmbH
Europaplatz 3
69115 Heidelberg, Germany

Batch number: 10824518

Printed by Printforce, the Netherlands